Agricultural English

Agricultural English

Edited by

Georgeta Raţă, Florin Sala and Ionel Samfira

CAMBRIDGE
SCHOLARS
PUBLISHING

Agricultural English,
Edited by Georgeta Raţă, Florin Sala and Ionel Samfira

This book first published 2012

Cambridge Scholars Publishing

12 Back Chapman Street, Newcastle upon Tyne, NE6 2XX, UK

British Library Cataloguing in Publication Data
A catalogue record for this book is available from the British Library

ISBN (10): 1-4438-3890-X, ISBN (13): 978-1-4438-3890-0

TABLE OF CONTENTS

Chapter Two: Syntax

Chapter Three: Lexis

Chapter Four: Semantics

Chapter Five: Pragmatics

Chapter Six: Etymology

Chapter Seven: A Comparative Approach

LIST OF TABLES

LIST OF ILLUSTRATIONS

FOREWORD

Agricultural English is a collection of essays on the English of Agriculture. It would appeal to *agriculturists, animal breeders, lexicographers, professors, researchers, students, and translators from Croatian-, English-, French-, German-, and Romanian-speaking countries, active in their own countries or abroad.*

The approach is a linguistic one with focus on **stylistic features** and **technical lexis**. The different aspects of the English used in the ***field of agriculture*** (agricultural practices, agricultural systems) and in some ***fields related to agriculture*** (agricultural zoology, agri-tourism, biology, botany, ecology, entomology, gastronomy, land measurement, plant pathology, and zoology) are analysed from several points of view.

Any language for specific purposes relies on several **morphological** ways of building up its own inventory of terms – abbreviation, affixation (prefixation, suffixation, and multiple affixation), backformation, change of morphological accent, composition / compounding, conscious / deliberate coinage, contraction, conversion, corruption, deflection, derivation from proper or personal names, folk / popular etymology. Only two of these procedures seem to be extremely productive in Agricultural English: *affixation* – the vocabulary of pesticides analysed by Georgeta Raţă and Anica Perković; *combination / compounding* – some of the practices associated with sustainable agricultural systems analysed by Anica Perković, Georgeta Raţă and Florin Sala, the combining forms with *bi(o)-* analysed by Alina-Andreea Dragoescu and Diana-Andreea Boc-Sînmărghiţan, the combining forms with *eco-* analysed by Anica Perković and Georgeta Raţă, the combining forms with *culture* analysed by Georgeta Raţă, Ionel Samfira and Anica Perković, and the combining forms with *tree* analysed by Georgeta Raţă; or a *combination of the two* – the vocabulary of ecology analysed by Georgeta Raţă and the cereals names analysed by Georgeta Raţă and Laura-Constantina Micu. The English verbs of animal communication analysed by Georgeta Raţă and Elena-Mirela Samfira are, no matter the source, the imitative / onomatopoeic.

From a **syntactical point of view**, all three teams of authors focused on agricultural entomology. Thus, Georgeta Raţă focused on verbal nouns in -*ing*, Georgeta Raţă and Laura-Constantina Micu analysed 'verbal Noun +

Noun' compounds containing participial adjectives, verbal nouns, or having both values, while Anica Perković and Georgeta Raţă focused on the 'Noun + Verbal nouns in *-ing* + Noun' names of pests.

As far as the **lexicology** and **lexicography** are concerned, the authors focused on the terminology of precision agriculture (Georgeta Raţă, Florin Sala and Anica Perković), on the names of weeds (Anica Perković and Georgeta Raţă), on the vocabulary of irrigation (Georgeta Raţă, Cornelia Petroman and Ioan Petroman), on the names of fruits ending in *-berry* (Georgeta Raţă and Anica Perković), and on the various types of tourism practiced in the countryside (Georgeta Raţă, Anica Perković and Ioan Petroman).

Semantics is represented by papers in which the focus is on homonymy (Georgeta Raţă), homonymy and synonymy (Oana Boldea), lexical meaning (Oana Boldea), semantic fields (Anica Perković and Georgeta Raţă), specific meaning (Georgeta Raţă and Anica Perković; Cornelia Petroman, Ioan Petroman and Snježana Tolić).

Three papers deal with **pragmatics** issues. Thus, Georgeta Raţă, Ioan Petroman and Scott Hollifield focus on the academic discourse and on professional rhetoric in the field of botanical nomenclature, Oana Boldea analyses English idioms containing names of plants, and Alina-Andreea Dragoescu and Petru Dragoescu focus on creative metaphors in the naming of plants.

Etymology is very much under scrutiny since it is an important tool in understanding terminology. The Latin heritage is studied by Georgeta Raţă and Scott Hollifield in botanical English, by Georgeta Raţă and Camelia Giuchici in the field of plant pathology, and by Georgeta Raţă in zoological English, while Georgeta Raţă, Ionel Samfira and Anica Perković analyse the vocabulary of seeds and seedling, Astrid-Simone Groszler analyses the names of spices, and Georgeta Raţă and Florin Sala analyse the vocabulary of land measurement units – all from an etymological perspective.

The **contrastive approach** is illustrated by a large number of essays. Andreea Varga writes about agricultural terminology in the context of multicultural communication. English, Croatian and Romanian are compared by Anica Perković, Georgeta Raţă and Martina Perković (who analyse names of dog breeds); English and French are compared by Georgeta Raţă, Ionel Samfira and Camelia Giuchici (who write about names of plant diseases); English, French and Romanian are compared by Georgeta Raţă and Iasmina Iosim (who analyse the verbs of animal communication); English and German are compared by Astrid-Simone Groszler and Biljana Ivanovska (who write about plant names); English and Romanian are compared by Georgeta Raţă and Anica Perković (who focus on the land improvement vocabulary), by Georgeta Raţă and Elena-Mirela Samfira (who analyse agricultural entomology), by Oana Boldea (who studies names of fodder plants, names of wild flowers, names of plants in food additive guides, names of pests), by Astrid-Simone Groszler and Biljana Ivanovska (who focus on plant names), by Georgeta Raţă (who deals with true and false "berries"), by Andreea Varga and Astrid-Simone Groszler (who deal with animal idioms), by Astrid-Simone Groszler (who focuses on animal idioms), and by Georgeta Raţă, Cornelia Petroman and Ioan Petroman (who study the Romanian of agri-tourism Internet sites); Romanian – English are compared by Oana Boldea (who writes about agricultural terms), by Alina-Andreea Dragoescu (who studies metaphorical plant names), and by Georgeta Raţă, Maria-Adriana Proca and Camelia Giuchici (who write about names of plant diseases).

The book would appeal to academic teaching staff, researchers and students in the field of **agriculture** and of some related fields – ***agricultural zoology***, ***agri-tourism***, ***biology***, ***botany***, ***ecology***, ***entomology***, ***gastronomy***, ***land measurement***, ***plant pathology***, and ***zoology*** – as well as in the field of **English for Specific Purposes (ESP)**.

The Editors

Chapter One

Morphology

PRACTICES ASSOCIATED WITH SUSTAINABLE AGRICULTURAL SYSTEMS

ANICA PERKOVIĆ, GEORGETA RAŢĂ AND FLORIN SALA

Introduction

Sustainable agriculture has addressed, ever since the concept appeared, ecological, economic, social, and philosophical issues in its battle against the prevailing agricultural system, variously called "conventional farming", "modern agriculture", or "industrial farming" that, it is true, has delivered tremendous gains in productivity and efficiency, but also a series of **concerns**. If agriculture profoundly affects many ecological systems through the negative effects of current practices, if economic and social problems associated with agriculture cannot be separated from external economic and social pressures because of the barriers to a sustainable and equitable food supply system, if there are potential hazards tied to sub-therapeutic use of antibiotics in animal production, and pesticide and nitrate contamination of water and food in humans, if the challenge of defining and dealing with the problems associated with today's food production system is inherently laden with controversy and emotion, things do not get simpler with the blooming of new agricultural concepts and practices whose definitions are inevitably compromises among differing world views, sets of values, etc. One thing is sure: we can analyse the terms (words or phrases) defining them to make them easier to understand.

Material and Methods

We have inventoried 79 such concepts and practices (some of which are synonyms); we then selected only the words formed with **combining forms**. The terms thus selected were analysed from the point of view of their structure (**combining form** and **basic word**) to see if we can draw any conclusion at all about the trends in modern agricultural nomenclature.

Results and Discussion

Of the 79 *concepts and practices associated with sustainable agricultural systems*, 9 are designated by words formed with **combining forms** (11%), while the rest are **compounds** with 2, 3 or more elements (89%).

A **combining form** is, in grammar, 'a linguistic form that occurs only in combination with other forms' (WEUDEL) (the definition does not mention if it is placed before or after the basic word), or 'a bound form (or bound morpheme) used in conjunction with another linguistic element in the formation of a word' (Chalker & Weiner 1994).

In **word formation**, a **combining form** may conjoin with:

- an **independent word** (*mini-* + ***skirt***);
- another **combining form** (*photo-* + ***-graphy***), or
- an **affix** (*cephal-* + ***-ic***).

This distinguishes it from an **affix** (prefix or suffix) that can be added to either a free word or a **combining form** but not solely to another affix (WEUDEL), adjusting the sense of a base (e.g. *ex-*, *un-*) or changing the word-class of the base (e.g. *-ation*, *-ise*) (Chalker & Weiner 1994).

There are three types of **combining forms**:

- forms borrowed from Greek or Latin that are **derivatives** of independent nouns, adjectives, or verbs in those languages: these **combining forms**, used in the formation of learned coinages, often semantically parallel independent words in English (e.g. *cardio-* in relation to *heart*, *-phile* in relation to *lover*) and usually appear only in combination with other **combining forms** of Greek or Latin origin (*bibliophile* and not**bookophile*);
- forms of free-standing English words: such **combining forms** usually have only single, restricted senses of the free words, and may differ from the words phonetically (*-land*, *-man*, *-proof*, *-wide*, *-worthy*);
- forms extracted from existing free words and used as bound forms, typically maintaining the meaning of the free words, or some facet of them (*-aholic*, *-gate*, *heli-*, *mini-*, *-orama*, *para-*).

The newly-formed words in our corpus are as follows:

- ***agrobiodiversity*** [?1997] (< *agro-* < Gk *agrós* 'tilled land', 'a combining form meaning 'field', 'soil', 'crop production' used in the formation of compound words' + *bio-* 'a combining form meaning 'life' occurring in **loanwords** from Greek and used in the formation of compound words' + *diversity* 'the state or fact of being diverse') is defined as 'a fundamental feature of farming systems around the world' encompassing many types of biological resources tied to agriculture, such as: genetic resources (the essential living materials of plants and animals); edible plants and crops (including traditional varieties, cultivars, hybrids, and other genetic material developed breeders); livestock (small and large, lineal breeds or thoroughbreds) and freshwater fish; soil organisms vital to soil fertility, structure, quality, and soil health; naturally occurring insects, bacteria, and fungi that control insect pests and diseases of domesticated plants and animals; agroecosystem components and types (poly-cultural / monocultural, small / large scale, rain fed / irrigated, etc.) indispensable for nutrient cycling, stability, and productivity; 'wild' resources (species and elements) of natural habitats and landscapes that can provide services (e.g. pest control and ecosystem stability) to agriculture (Thrupp in Gold 1999);

- ***agroecology*** [?1987] (< *agro-* < Gk *agrós* 'tilled land', 'a combining form meaning 'field', 'soil', 'crop production' used in the formation of compound words' + ecology 'the branch of biology dealing with the relations and interactions between organisms and their environment, including other organisms') has a broad definition implying a number of features about society and production that go well beyond the limits of the agricultural field 'a more environmentally and socially sensitive approach to agriculture, one that focuses not only on production, but also on the ecological sustainability of the productive system', and a narrow one 'the study of purely ecological phenomena within the crop field, such as predator/prey relations, or crop / weed competition' (Hecht, in Gold 1999);
- ***bio-control*** [1920-1925] (< *bio-* 'a combining form meaning 'life' occurring in loanwords from Greek and used in the formation of compound words' + control 'prevention of the flourishing or spread of something undesirable') is defined by language dictionaries as 'the control of pests by interference with their ecological status, as by introducing a natural enemy or a pathogen into the environment' (WEUDEL) or as 'man's use of a specially chosen living organism (predator, parasite, or disease) to control a particular pest (weeds, plant pathogens, vertebrates and insects)' (Orr, in Gold 1999);
- ***biodiversity*** [?] (< *bio-* 'a combining form meaning 'life' occurring in loanwords from Greek and used in the formation of compound words' + *diversity* 'the state or fact of being diverse') is, at its simplest level,

'the sum total of all the plants, animals, fungi and micro-organisms in the world, or in a particular area; all of their individual variation; and all the interactions between them (Raven, in Gold 1999);

- ***biodynamics*** [?](< *bio-* 'a combining form meaning 'life' occurring in loanwords from Greek and used in the formation of compound words' + *dynamics* 'the branch of mechanics that deals with the motion and equilibrium of systems under the action of forces, usually from outside the system') is defined by language dictionaries as 'the branch of biology dealing with energy or the activity of living organisms' (WEUDEL) or as 'a biodynamic method in which certain herbal preparations that guide the decomposition processes in manures and compost are central' (1985-1986 Year End Report, in Gold 1999);
- ***biotechnology*** [1940-1945] (< *bio-* 'a combining form meaning 'life' occurring in loanwords from Greek and used in the formation of compound words' + *technology* 'the branch of knowledge that deals with the creation and use of technical means and their interrelation with life, society, and the environment, drawing upon such subjects as industrial art, engineering, applied science, and pure science') is defined by language dictionaries as 'the use of living organisms or other biological systems in the manufacture of drugs or other products or fro environmental management, as in waste recycling (micro-organisms to degrade oil slicks or organic waste, genetically engineered bacteria to produce human hormones, and monoclonal antibodies to identify antigens)' (WEUDEL). More recently, products such as plants engineered for herbicide tolerance or insect resistance, and bacteria engineered to produce drugs for livestock may point to reduced chemical use and other sustainable applications in agriculture (Shaping an Agriculture for the Twenty-First Century: Biotechnology, in Gold, 1999);
- ***mini-farming*** [1995] (< *mini-* 'a combining form obtained by shortening of miniature, minimal, or minimum, with the meanings of 'small or reduced size in comparison with others of its kind; limited ins cope, intensity, or duration' + *farming* 'the business of operating a farm') is defined as 'a production system (including double-dug, raised beds, intensive planting, composting, companion planting, and whole system synergy) that makes it possible for one person to grow all of his or her family's food using truly sustainable methods that maintain the fertility of the soil without relying on non-renewable resources like petrochemicals or imported organic matter' (Jeavons in Gold 1999).

Two other coinages have been formed with **combining forms** that have not yet acquired this status, but that behave as such:

- ***no-till(age)*** [1965-1970] (< *no* 'used before a noun to convey the opposite of the noun's meaning' + *tillage* 'the operation, practice, or art of tilling land') is defined by language dictionaries as 'the planting of crops by direct seeding without ploughing, using herbicides as necessary to control weeds' (WEUDEL) or as 'a specific type of conservation tillage (a broad range of soil tillage systems that leave residue cover on the soil surface, substantially reducing the effects of soil erosion from wind and water, minimising nutrient loss, decreased water storage capacity, crop damage, and decreased farmability, leaving the soil undisturbed from harvest to planting except for nutrient amendment, and accomplishing weed control primarily with herbicides, limited cultivation, and, in more sustainable systems, with cover crops') (Conservation Technology Information Centre, in Gold 1999);
- ***permaculture*** [late 1970s] (< *perm(a)-* 'permanent' + *culture* 'the art or practice of cultivating the soil; tillage') is defined as 'an alternative sustainable agriculture system emphasising the location of each element in a landscape, and the evolution of landscape over time, and aiming at producing an efficient, low-maintenance integration of plants, animals, people and structure, etc., applied at the scale of a home garden, all the way through to a large farm' (Quinney, Jeeves, and Mollison in Gold 1999).

The most productive **combining form** is *bio-* (4 occurrences), followed by *agro-* (2 occurrences) and *mini-* (1 occurrence). Thus, there are 2 **combining forms** borrowed from Greek (*bio-* and *agro-*) with 6 occurrences (67%), and 1 **combining form** extracted from an existing free word, *mini-* (1 occurrence). Though they have not yet acquired the status of **combining form**, *no-* (1 occurrence) and *perm(a)-* (1 occurrence) behave as such (22%). These **combining forms** have conjoined only with independent words – *control*, *culture*, *diversity* (2 occurrences), *dynamics*, *ecology*, *farming*, *technology*, and *tillage* – closely related to agricultural practices, and never with other **combining forms** or with affixes.

Conclusions

All these concepts and practices have very literal meanings that have been coloured by their historic use and practitioners' experiences. Such as they are, these terms designate concepts and practices that simply defy definition, but that have provided 'talking points' not only for agriculturists, but also for linguists. Most of the newly-formed words are, naturally, learned coinages (67%) that continue the long-lasting tradition of renewing scientific vocabulary "to supply new needs for technical vocabulary that arose partly from the revival of learning in western Europe in the fifteenth and sixteenth centuries known as the Renaissance, and partly from the industrial revolution of the eighteenth century and its scientific spin-offs" (Carstairs-McCarthy 2002). But the large share of words formed with **combining forms** extracted from existing **free words** or with developing **combining forms** (33%) shows that agricultural English can also appeal to unorthodox means of enriching its vocabulary in its seek for new words to designate new realities in agriculture, i.e. new agricultural concepts and practices associated with sustainable agricultural systems, proving, once again, "the versatility and vigour of English word-formation processes" (Carstairs-McCarthy 2002).

References

Adams, Valerie. (1987). *An Introduction to Modern English Word-formation*. London – New York: Longman.

Carstairs-McCarthy, A. (2002). *An Introduction to English Morphology. Words and Their Structure*. Edinburgh: Edinburgh University Press.

Chalker, Sylvia & Weiner, E. (1994). *The Oxford Dictionary of English Grammar*. London - New York - Sydney - Toronto: BCA.

Gold, Mary. (1999). *Sustainable Agriculture: Definitions and Terms*. Online: http://www.nal.usda.gov/afsic.

Leviţchi, L. & Bantaş, A. (1995). *Dicţionar englez-român*. [English-Romanian Dictionary]. Bucureşti: Ed. Teora.

Perković, Anica & Raţă, Georgeta. (2006). On Word Formation in the English of Agriculture: Practices Associated with Sustainable Agricultural Systems. *Lucrări ştiinţifice. Facultatea de Agricultură XXXVIII*: 475-480.

Plag, I. (2002). *Word-formation in English*. Cambridge: Cambridge University Press.

Sheehan, M. J. (2000). *Word Parts Dictionary. Standard and Reverse Listings of Prefixes, Suffixes, Roots and Combining Forms*. Jefferson,

NC – London: McFarland & Company, Inc., Publishers.
Štekauer, P. & Lieber, Rochelle. (2005). *Handbook of Word-Formation*. Dordrecht: Springer.
Webster Encyclopaedic Unabridged Dictionary of the English Language. (1996). New York: Gramercy Books. (WEUDEL)

THE VOCABULARY OF ECOLOGY

GEORGETA RAȚĂ

Material and Methods

Words can take a new form and have a new grammatical function through one of the following transformations: **abbreviation**, **affixation (prefixation, suffixation, and multiple affixation)**, **backformation**, **change of morphological accent**, **composition**, **conscious / deliberate coinage**, **contraction**, **conversion**, **corruption**, **deflection**, **derivation from proper or personal names**, **folk-etymology** (Levițchi 1970). The focus here is on transformations of these types that considerably enriched *contemporary English*. Our analysis of the ***English of ecology*** has been made on a corpus of English words and phrases used in ecology. To do so, we have inventoried terms specific to the ***English of ecology*** in some of the most comprehensive encyclopaedias and dictionaries of the world, and in some basic works of ecology. We then have tried to see how these terms formed.

Results and Discussion

1. Affixation. An ***affix*** is, in grammar, 'a bound element, as a prefix or suffix, added to a base or stem to form a fresh stem or a word, as *-ed* added to *want* to form *wanted*, or *im-* added to *possible* to form *impossible*' (RHDEL).

1.1. Prefixation. A ***prefix*** is, in grammar, 'an affix placed before a base or another prefix, as *un-* in *unkind*, *un-* and *re-* in *unrewarding*' (RHDEL). Prefixes productive in the ***English of ecology*** are of different origins (Latin, Greek, Middle English), and form both nouns and adjectives. The only ***Noun-forming prefix*** is of Latin origin: ***trans***- 'a prefix occurring in loan words from Latin (*transcend*, *transfix*); on this model used with the meaning 'across', 'beyond', 'through', 'changing thoroughly', 'transverse', in combination with elements of any origin: *transempirical*, *transvalue*': ***transpiration*** (1545-1555, *trans-* + L *spiration-*) '*Bot.* the passage of water through a plant from the roots through the vascular system to the atmosphere' (RHDEL), and 'loss of water vapour from a plant to the outside atmosphere' (Smith & Smith 1998).

1.2. Suffixation. A ***suffix*** is, in grammar, 'an affix that follows the element to which it is added, as *-ly* in *kindly*' (RHDEL). Suffixes productive in the ***English of ecology*** are of different origins (English, French, Germanic, Greek, Latin, Middle English, and Romance) and form both nouns and adjectives.

1.2.1. ***Noun-Forming Suffixes*** are of Latin, English, French, Germanic, Greek, and Romance origins: ***-ance*** 'a [Romance] suffix used to form nouns either from adjectives in - *ant* or from verbs': *abund**ance*** (< ME < MF < L *abundantia*) 'an extremely plentiful or over sufficient quantity or supply; affluence, wealth' (RHDEL), 'an extremely plentiful or over sufficient quantity or supply' (Smith & Smith 1998), 'the number of individuals of a species in a given area' (WCD); ***-(a)tion*** 'a [Latin] combination of *-ate* and *-ion*, used to form nouns from stems in *-ate*; on this model, used independently to form nouns from stems of other origin': *adapt**ation*** (< ML *adaptation-*) '*Biol.* any alteration in the structure or function of an organism or any of its parts by which the organism becomes better fitted to survive in its environment; a form of structure modified to fit changed environment' (RHDEL), '*Biol.* any alteration in the structure or function of an organism or any of its parts that results from natural selection and by which the organism becomes better fitted to survive and multiply in its environment; a form of structure modified to fit a changed environment; the ability of a species to survive in a particular ecological niche, especially because of alterations of form or behaviour brought about through natural selection' (Smith & Smith 1998), and 'a genetically determined characteristic [behavioural, morphological, or physiological] that improves an organism's ability to survive and reproduce under prevailing environmental conditions' (WCD); *assimil**ation*** (< L *assimilation-*) '*Bot.* the total process of plant nutrition, including absorption of external foods and photosynthesis' (RHDEL), '*Bot.* The total process of plant nutrition, including photosynthesis and the absorption of raw materials' (Smith & Smith 1998), and 'transformation or incorporation of a substance by organisms; absorption and conversion of energy and nutrients into constituents of an organism' (WCD); *(biological / ecological* or *biodiversity / nature / species / wildlife) conserv**ation*** *(activity / biology / practice* or *of ecosystems / individuals / populations)*, *(environment / nature / wildlife) protec**tion***; ***-cy*** 'a [Latin] suffix used to form abstract nouns from adjectives with stems in *-t*, *-te*, *-tic*, and especially *-nt*, also forming nouns from other adjectives and from nouns':

(nature) conservancy; ***-ist*** 'a suffix of nouns, often accompanying verbs ending in *-ise* or nouns ending in *-ism*, denoting a person who practices or is concerned with something, or holds certain principles, doctrines, etc.: *conservation**ist**, ecolog**ist**, protection**ist***; ***-ity*** 'a [Latin] suffix used to form abstract nouns expressing state or condition': *commun**ity*** (< L *comunitat-*) '*Ecol.* the plant and animal populations occupying a given area' (RHDEL), '*Ecol.* an assembling if interacting populations occupying a given area' (Smith & Smith 1998), and 'a group of interacting plants and animals inhabiting a given area' (WCD); ***-ment*** 'a [Romance] suffix of nouns, often concrete, denoting an action or resulting state, a product, or means': ***environment*** (1595-1605, *environ + -ment*) 'the aggregate of surrounding things, conditions, or influences' (RHDEL), 'the aggregate of all external and internal conditions affecting the existence, growth, and welfare of organisms' (Smith & Smith 1998), and 'the total surroundings of an organism, including other plants and animals and those of its own kind' (WCD).

1.2.2. The only ***Adjective-forming suffix*** is of Latin origin: ***-al*** 'a [Latin] adjectival suffix occurring in loan words from Latin; on this model, used in the formation of adjectives from other sources': *ecologic**al*** [as in ***ecological*** *action / commandment / complexity / consciousness / conservation / criterion / culture / education / ethics / ground / knowledge / law / management / monitoring / movement / needs / perspectives / phenomenon / philosophy / policy / process / pyramid / restoration / retrogression / salvation / science / significance / space / study*].

2. Composition is, in grammar, 'the formation of ***compounds***: *the composition of 'bootblack' consists of 'boot' and 'black'*', i.e., of words 'consisting of two or more parts that are bases, or that include a **combining form** and a base (e.g., *biochemistry*), two **combining forms** (e.g., *ethnography*), or a **combining form** and a **non-inflectional affix** (e.g., *aviary*, *dentoid*)'. In everything that follows, only the first part of the definition will be taken into account – the one referring to the semantic-grammatical combination of two or more words (also called roots or stems). Here again, only **compound nouns** and **compound adjectives** will be taken into account, as there are no compound pronouns, verbs, adverbs or prepositions in our list of the ***English of ecology*** words.

2.1. Compound Nouns are extremely numerous in contemporary ***English of ecology***. They are sometimes written in a single word, sometimes in words written separately, or separated by means of a hyphen - and no definite rules can be given in this respect.

The elements **compound nouns** are formed of are as follows:

- COMBINING FORMS + NOUN: ***agro***- 'a learned borrowing from Greek meaning 'soil', 'crop production' used in the formation of compound words': ***agro****ecology* 'the branch of biology dealing with the relations between organisms and their environment' (RHDEL); ***eco***- from the Greek *oīko(s)* 'house': ***eco****development* < *eco*- [short for ***ecology****?*] a learned borrowing from Greek meaning 'house', 'habitat' + *development* 'the act or process of developing; progress' and 'a developed state, form, or product' (RHDEL), ***eco****system* (*eco*- + *system*) '*Ecol.* a system formed by the interaction of a community of organisms with their environment' (RHDEL), '*Ecol.* a system formed by the interaction of a community of organisms with their environment' (Smith & Smith 1998), and 'the biotic community and its abiotic environment functioning as a system' (WCD); ***eco****sophy*, a word developed in the second half of the 20th century (in the 1970's) ***eco****sophy* < *eco*- [short for ***ecology***?] a learned borrowing from Greek meaning 'house', 'habitat' + *-sophy* an element occurring in loan words from Greek [*philo****sophy***, *theo****sophy***], on this model used with the meaning 'science of' in the formation of compound words [*anthropo****sophy***]' (RHDEL); ***homeo***- 'a learned borrowing from Gk *homeo*- 'similar' used in the formation of compound words: ***homeo****static*: ***homeo****stasis* (1925-1930, *homeo*- + *stasis*) 'the tendency of a system, especially the physiological system of higher animals, to maintain internal stability, owing to the coordinated response of its parts to any disruptive situation or stimulus' (RHDEL), 'the tendency of a system, especially the physiological system of higher animals, to maintain internal stability, owing to the coordinated response of its parts to any situation or stimulus tending to disturb its normal condition or function' (Smith & Smith 1998), and 'maintenance of nearly constant conditions in function of an organism or in interaction among individuals in a population' (WCD); ***meta***- 'a learned borrowing from Greek meaning 'after', 'along' 'with', 'beyond', 'among', 'behind', and often denoting change, used in the formation of compound words': *meta****ecology*** < *meta*- a learned borrowing from Greek meaning 'after', 'along with', 'beyond', 'among', 'behind' and often denoting change, used in the formation of compound words + ***ecology*** 'the branch of biology dealing with the relations between organisms and their environment' (RHDEL); ***micro***- 'a learned borrowing from Greek, where it meant 'small': *microcosm*': ***micro****climate* (1820-1825, *micro*- + *climate*) 'the climate of a small area, as of confined spaces such as caves or houses, of plant communities, or of urban communities' (RHDEL), 'the climate of a small area, as of confined spaces such as caves or houses [cryptoclimate], of plant communities, wooded areas, etc. [phytoclimate] or of urban communities, which may be different from that in the general region' (Smith & Smith 1998), and 'climate on a very local scale, which differs from the general climate of the area; influences the presence and distribution of organisms' (WCD); ***non-*** 'a combining [Latin] form meaning 'not', freely used as an English formative, usually with a simple negative force as implying mere negation or absence of something (rather than the opposite or reverse of it, as often expressed by *un-*)': *non**ecologist*** < *non-* 'a combining form meaning 'not', freely used as an English formative, usually with a simple negative force as implying mere negation or absence of something [rather than the opposite or reverse of it, as often expressed by *un-*]' + ***ecologist*** [it does not appear in language dictionaries]; ***photo-*** 'a learned borrowing from Greek meaning 'light': *photoelectric*: ***photo**synthesis* (1895-1900, *photo-* + *synthesis*) '*Biol.* the synthesis of complex organic materials, especially carbohydrates, from carbon dioxide, water, and inorganic salts, using sunlight as the source of energy and with the aid of a catalyst, such as chlorophyll' (RHDEL), '*Biol.* the synthesis of complex organic materials, especially carbohydrates, from carbon dioxide, water, and inorganic salts, using sunlight as the source of energy and with the aid of chlorophyll and associated pigments' (Smith & Smith 1998), and 'use of light energy by plants to convert carbon dioxide and water into simple sugars' (WCD); ***-vore*** 'a learned borrowing from Latin meaning 'eating', used in the formation of compound words: *carnivore*': ***detritivore*** (1975-1980, *detrit(us)* + *i* + *-vore*) '*Ecol.* an organism that uses organic waste as a food source, as certain insects' (Smith & Smith 1998), 'organism that feeds on dead organic matter; usually applies to detritus-feeding organisms other than bacteria and fungi' (WCD);

- COMBINING FORM + COMBINING FORM: ***eco-*** from the Greek *oĩko(s)* 'house'+ ***-logy*** 'a combining form used in the names of sciences or bodies of knowledge: *palaeontology*, *theology*': ***ecology*** 'the branch of biology dealing with the relations between organisms and their environment' (RHDEL), 'the branch of biology dealing with the relations and interactions between organisms and their environment, including other organisms' (Smith & Smith 1998), and 'the study of relations between organisms and their natural environment functioning as a system' (WCD); ***meso-*** 'a learned borrowing from Greek meaning 'middle' used in the formation of compound words: *mesocephalic'* + ***-phyll*** 'final element of compound words: *chlorophyll*': ***mesophyll*** (1830-1840, *meso-* + *-phyll*) '*Bot*. the parenchyma, usually containing chlorophyll, that forms the interior parts of a leaf' (RHDEL), '*Bot*. the parenchyma, usually containing chlorophyll, that forms the interior parts of a leaf' (Smith & Smith 1998), and 'specialised tissue located between the epidermal layers of a leaf' (WCD).

2.2. Compound Adjectives. A **compound part of speech** (e.g., *free-market*) or a **free combination** (e.g., *just-in-time*) is often used attributively with a noun, characterising it and becoming adjectives.

The elements **compound adjectives** are formed of are: COMBINING FORMS + ADJECTIVE: *anti**ecological*** [as in *anti**ecological** activity*] < *anti-* a prefix from Greek meaning 'against', 'opposite of', freely combining with elements of any origin and used here with the meaning 'not', 'un-' + ***ecological*** [it does not appear in language dictionaries]; ***eco-*** from the Greek *oīko(s)* 'house': ***eco**logical (action / bad / commandment / complexity / consciousness / conservation / criterion / culture / education / ethics / ground / knowledge / law / management / monitoring / movement / needs / perspective / phenomenon / philosophy / policy / process / pyramid / restoration / retrogression / salvation / science / significance / space / study).*

Conclusions

Of the transformations resulting in new forms, only the following two are represented in the ***English of ecology***: **affixation** (*prefixation* and *suffixation*) and **composition**. In **affixation**, there is only one Noun-forming prefix (*trans-*), and no Adjective-forming prefix. There are also a lot of Noun-forming suffixes *(-ance*, *-ation*, *-cy*, *-ist*, *-ity*, *-ment*), but only one Adjective-forming suffix (*-al*). **Composition** is illustrated only by the COMBINING FORM (*agro-*, *eco-*, *homeo-*, *meta-*, *micro-*, *non-*, *photo-*, and *-vore*) + NOUN, and the COMBINING FORM (*eco-*, *meso-*) + COMBINING FORM (*-sophy*, *-phyll*) types to produce nouns, and by the COMBINING FORMS (*anti-*, *eco-*) + ADJECTIVE (*ecological*, *logical*) type to produce adjectives. They are all technical terms of scientific vocabulary, coined self-consciously out of non-English elements, mostly from Latin and Greek (Carstairs-McCarthy 2002). All this accounts for *English for special purpose information*, which is not surprising, if we bear in mind that ECOLOGY as a science has only emerged recently.

References

Adams, Valerie. (1987). *An Introduction to Modern English Word-formation*. London – New York: Longman.

Benson, M., Benson, Evelyn & Ilson, R. (1990). *The Combinatory Dictionary of English*. Amsterdam – Philadelphia: John Benjamins Publishing Company.

Carstairs-McCarthy, A. (2002). *An Introduction to English Morphology. Words and Their Structure*. Edinburgh: Edinburgh University Press.

Chalker, Sylvia & Weiner, E. (1994). *The Oxford Dictionary of English Grammar*. London - New York - Sydney - Toronto: BCA.

Grzega, J. & Schöner, Marion. (2007). *English and General Historical Lexicology. Materials for Onomasiology Seminars*. Katholische Universität Eichstätt-Ingolstadt.

Guidelines for Protected Area Management Categories. Interpretation and Application of the Protected Area Management Categories in Europe. IUCN World Commission on Protected Areas, and EUROPARC Federation with the assistance of the World Conservation Monitoring Centre. (1999).

Johnson, B. R. & Hill, K. (Eds.). (2001). *Ecology and Design. Frameworks for Learning*. Washington, D.C.: Island Press.

Leviţchi, L. & Bantaş, A. (1995). *Dicţionar englez-român*. [English-Romanian Dictionary]. Bucureşti: Ed. Teora.

Leviţchi, L. (Ed.). (2004). *Dicţionar englez-român*. [English-Romanian Dictionary].Bucureşti: Ed. Univers Enciclopedic.

Leviţchi, L. D. (1970). *Limba engleză contemporană. Lexicologie*. [Contemporary English. Lexicology]. Bucureşti: E.D.P.

Liu, J. & Taylor, W. W. (Eds.). (2002). *Integrating Landscape Ecology into Natural Resource Management*. Cambridge: Cambridge University Press.

Longman Dictionary of English Language and Culture (2003). London: Longman.

Maarel, E. van der. (Ed.). (2004). *Vegetation Ecology*. Hoboken, NJ: Wiley-Blackwell.

Perković, Anica & Rață, Georgeta. (2005). On the Vocabulary of Ecology: Ageing and Renewing. *Lucrări ştiinţifice. Facultatea de Agricultură, XXXVII*: 633-634.

Plag, I. (2002). *Word-formation in English*. Cambridge: Cambridge University Press.

Rață, Georgeta & Perković, Anica. (2006). Environmental Language and/or Medical Language? *Proceedings. Linguistics and Theory of*

Literature. Pedagogy and Psychology 45, Rousse, Bulgaria: 55-60.

Rață, Georgeta & Perković, Anica. (2006). Environmental Management Categories in Europe: A Lexicological Approach. *Proceedings of The 3rd Conference Energy efficiency and agricultural engineering, June 7-9, 2006, Rousse, Bulgaria*: 99-103

Rață, Georgeta. (2005). On the Vocabulary of Ecology: Renewing through New Forms. *Scientific Works L (6), Plovdiv, Bulgaria*: 137-142.

Schulze, E.-D., Beck, E. & Muller-Hohenstein, K. (2005). *Plant Ecology*. Berlin - Heidelberg: Springer.

Slobodkin, L. B. (2003). *A Citizen's Guide to Ecology*. Oxford: Oxford University Press.

Smith, R. L. & Smith, T. M. (1998). *Elements of Ecology*. San Francisco: Benjamin/Cummings Science Publishing.

The Random House Dictionary of the English Language. (1968). New York: Random House. (RHDEL)

Webster Comprehensive Dictionary. (1995). Chicago: J. G. Ferguson Publishing Company. (WCD)

Webster Encyclopaedic Unabridged Dictionary of the English Language. (1996). New York: Gramercy Books.

Webster's Universal Dictionary & Thesaurus (2003). New Lanark: Geddes & Grosset.

COMPOUNDS WITH AND DERIVATIVES OF CEREAL NAMES

GEORGETA RAŢĂ
AND LAURA-CONSTANTINA MICU

Introduction

A quick look at the dictionary definition of the word ***corn*** will be enough to justify our motivation in carrying out the present research. Thus, according to English language dictionaries, ***corn*** is defined as '**1**. Also called ***Indian corn***; especially technical and British, ***maize***, U.S., Canadian, Australian. **a**. a tall, annual cereal plant, *Zea mays*, having a jointed, solid stem and bearing the kernels on large ears. **b**. the kernels, used as food. **c**. the ears. **2**. The edible seed of certain other cereal plants, especially wheat in England and oats in Scotland. **3**. The plants themselves. **4**. U.S. (loosely) sweet corn. **5**. Corn whiskey. **6**. *Slang*. Old-fashioned, trite, or mawkishly sentimental entertainment material' (*Webster Comprehensive Dictionary*, 1995). This dictionary entry alone creates problems because of the ambiguity of ***corn*** 'maize' and ***corn*** 'cereal'. We focused on ***cereal*** **compounds** and **derivatives** to see if there is any pattern at all that could be used in approaching the vocabulary specific to cereals.

Material and Method

We have inventoried 170 **compound** and **derivative forms** (i.e. entries) of cereal nouns in some of the most outstanding language dictionaries. We then compared the different forms and draw a few conclusions. The method we used was quantitative. We started from the hypothesis that the different **compound** and **derivative forms** of cereal names must share certain patterns which are the result of a long process of creating words and/or phrases and assimilating them in English.

Results and Discussion

Cereals or grains appear in the following collocations:

- ***barley*** (17): ***barley*** *bird*, ***barley*** *bread*, ***barley***-*bree* / ***barley****bree*, ***barley*** *broth*, ***barley*** *coal*, ***barley*** *corn* / ***barley****corn*, ***barley*** *candy*, ***barley****hood*, ***barley*** *meal*, ***barley*** *mill*, ***barley*** *sack*, ***barley*** *sorting*, ***barley*** *stripe*, ***barley*** *sugar*, ***barley*** *water*, ***barley*** *wine*, *French / peeled* ***barley***;
- ***corn*** (92): ***corn****ball*, ***corn*** *beef*, ***Corn*** *Belt*, ***corn*** *binder*, ***corn*** *border*, *(European)* ***corn*** *borer*, ***corn*** *bread*, ***corn*** *broom*, ***corn*** *cake*, ***corn*** *chandler*, ***corn*** *chip*, ***corn*** *cleaner*, ***corn*** *cob* / ***corn****cob*, ***corn****cob (pipe)*, ***corn*** *cockle*, ***corn*** *colour*, ***corn*** *cracker*, ***corn*** *crake*, ***corn****crib*, ***corn*** *cutter*, ***corn*** *dealer*, ***corn*** *dodger* / ***corn****dodger*, ***corn*** *dog*, ***corn*** *earworm*, ***corned***, ***corned*** *beef*, ***corned*** *leather*, ***corn*** *exchange*, ***corn*** *factor*, ***corn*** *failure*, ***corn*** *fed* / ***corn****fed*, ***corn***-*field* / ***corn****field*, ***corn****field ant*, ***corn*** *flag* / ***corn****flag*, ***corn*** *flakes* / ***corn****flakes*, ***corn*** *floor*, ***corn*** *flour*, ***corn*** *flower* / ***corn****flower*, ***corn*** *gluten*, ***corn*** *grass*, ***corn*** *grit(s)*, ***corn*** *harvester*, ***corn*** *house* / ***corn****house*, ***corn****husk*, ***corn*** *husker* / ***corn***-*husker* / ***corn****husker*, ***Corn****husker State*, ***corn****husking*, ***corning***, ***Corn*** *Law* / ***corn*** *laws*, ***corn*** *lift*, ***corn*** *lily*, ***corn*** *loft*, ***corn*** *liquor*, ***corn*** *marigold*, ***corn*** *market*, ***corn*** *meal* / ***corn****meal*, ***corn*** *middlings*, ***corn*** *mill*, ***corn*** *mint*, ***corn*** *muffin*, ***corn*** *oil*, ***corn*** *picker*, ***corn*** *picking*, ***corn*** *pimpernel*, ***corn***-*pipe*, ***corn*** *plant*, ***corn*** *planter*, ***corn*** *pone* / ***corn***-*pone*, ***corn*** *poppy*, ***corn*** *reaper*, ***corn*** *rent*, ***corn***-*root aphid*, ***corn*** *rootworm*, ***corn****row*, ***corn*** *salad*, ***corn*** *shock*, ***corn*** *silk*, ***corn*** *sieve*, ***corn*** *smut*, ***corn*** *snake*, ***corn*** *snow*, ***corn*** *stack*, ***corn*** *stalk* / ***corn****stalk*, ***corn****starch*, ***corn*** *stick*, ***corn*** *sugar*, ***corn*** *syrup*, ***corn*** *van*, ***corn*** *weevil*, ***corn*** *whiskey*, ***corny***;
- ***maize*** (5): ***maize*** *flour*, ***maize*** *husk*, ***maiz****ena*, ***maize*** *oil*, ***maize*** *starch*;
- ***millet*** (3): ***millet*** *ale*, ***millet*** *beer*, ***millet*** *grass*;
- ***oat(s)*** (10): ***oat***-*cake* / ***oat****cake*, ***oaten***, ***oater***, ***oat*** *grass*, ***oat*** *like*, ***oat*** *malt*, ***oat****meal*, *bearded / Hungarian / oriental / Tartarian* ***oat***, *false* ***oat***, *wild* ***oat***;
- ***rice*** (18): ***rice*** *bean*, ***rice****bird*, ***rice*** *blast*, ***rice*** *coal*, *rice corn*, ***rice*** *paddy*, ***rice*** *paper*, ***rice***-*paper tree*, *rice plantation*, ***rice*** *pudding*, ***ricer***, ***rice*** *rat*, ***rice***-*water*, ***rice*** *weevil*, ***ricey***, *ground* ***rice***, *husked* ***rice***, *rough* ***rice***;
- ***rye*** (6): ***rye*** *bread*, ***rye*** *flour*, ***rye****grass*, ***rye****peck*, ***rye*** *whiskey*, *perennial* ***rye****grass*;
- ***sorg(h)o / sorghum*** (2): *sugar* ***sorghum***, *sweet* ***sorghum***;
- ***wheat*** (17): ***wheat*** *berry*, ***wheat*** *bran*, ***wheat*** *bread*, ***wheat*** *cake*, ***wheat***

dampener, ***wheat**ear*, ***wheaten***, *(soft coated)* ***wheaten*** *terrier*, ***wheat*** *flour*, ***wheat*** *germ*, ***wheat*** *grass*, ***wheat*** *meal*, ***wheat*** *paddock*, ***wheat*** *rust*, ***wheat*** *separator*, ***wheat*** *worm*, *Turkey / Turkish* ***wheat***.

It is interesting to see that a series of 15 nouns seem to be more productive than other nouns (they have produced 2 and more than 2 **compound forms** with cereals names in English) in **compounding** (Table 1-1).

Table 1-1. Compounds with cereal names

Nouns forming the compound	**Name of the cereal**								
	barley	**corn**	**maize**	**millet**	**oat**	**rice**	**rye**	**sorghum**	**wheat**
bird	+					+			
bread	+	+					+		+
cake		+			+				+
coal	+					+			
corn	+					+			
flour		+	+				+		+
grass		+		+	+		+		+
husk		+	+						
meal	+	+			+				+
oil		+	+						
starch		+	+						
sugar	+	+						+	
water	+								
weevil		+							
whiskey		+					+		

The **compounds** can be grouped depending on the common element in a series of *cereal* **compounds** as follows:

- cereal + *bird* (2): ***barley bird*** (*Jynx torquilla*), ***ricebird*** 'Southern U.S. the bobolink (a common North American passerine songbird, *Dolichonyx oryzivorus*);
- cereal + *bread* (4): ***barley bread***, ***corn bread*** 'a bread made of corn meal', ***rye (bread)*** 'bread made entirely or partly from rye flour, often with caraway seeds; rye', ***wheat bread***;
- cereal + *cake* (3): ***corn cake*** 'a cake made of corn meal', ***oatcake*** 'a

cake, usually thin and brittle, made of oatmeal', ***wheat cake*** 'a pancake made of wheat flour';

- cereal + *coal* (2): ***barley coal*** 'anthracite in sizes ranging from 1.2 mm to 4.8 mm', ***rice coal*** 'anthracite in sizes ranging from 0.79 cm to 0.48 cm';
- cereal + *corn* (2): ***barleycorn*** '1. Barley or a grain of barley; 2. A measure equal to 1/3 of an inch', ***rice corn*** (*Sorgum vulgare*);
- cereal + *flou*r (4): ***corn flour*** '1. Flour prepared from corn. 2. Chiefly Brit. cornstarch', ***maize flour***, ***rye flour***, ***wheat flour***;
- cereal + *grass* (5): ***corn grass*** (*Aira caespitosa*), ***millet grass*** (*Panicum milliaceum*), ***oat grass*** '1. Any of certain oat like grasses. 2. Any wild species of oat' (*Avena fatua*)', ***rye grass*** 'any of several European grasses of the genus *Lolium*, as *L. perenne* (***perennial ryegrass***) grown for forage in the U.S.', ***wheat grass*** ' any of several wheat-like grasses of the genus *Agropyron*, grown for forage in the western U.S. (*Triticum repens*)';
- cereal + *husk* (2): cornhusk 'the husk of an ear of corn', maize husk;
- cereal + *meal* (4): ***barley meal***, ***corn meal / cornmeal*** '1. Meal made of corn. 2. Scot. Oatmeal', ***oatmeal*** '1. Meal made from oats. 2 a cooked breakfast food made from this. 3. Made with or containing oatmeal: oatmeal cookies', ***wheat meal***;
- cereal + *oil* (2): ***corn oil*** 'the oil obtained by expressing the germs of corn kernels, used in the preparation of foods, especially salad dressing, lubricants, soaps, and hairdressings', ***maize oil*** 'corn oil';
- cereal + *starch* (2): ***cornstarch*** 'a starch or a starchy flour made from corn and used for thickening gravies and sauces, making puddings, etc.', ***maize starch***;
- cereal + *sugar* (3): ***barley sugar*** (there is also ***barley candy***) 'a brittle, amber-coloured, citrus-flavoured candy, usually twisted into strips or moulded into a variety of shapes', ***corn sugar*** 'dextrose', ***sugar sorghum*** 'sorgo';
- cereal + *water* (2): ***barley water*** 'a decoction of barley, used especially as a medicament in the treatment of diarrhoea in infants', ***rice-water***;
- cereal + *weevil* (2): ***corn weevil*** (*Colandra granaries*), ***rice weevil*** 'a brown weevil, *Sitophitus oryzae*, that infests stored grains, especially rice';
- cereal + *whiskey* (2): ***corn whiskey*** 'whiskey made from a mash having at least 80 percent corn', ***rye whiskey*** 'a straight whiskey distilled from a mash containing 51 percent or more rye grain; a blended whiskey'.

In all these **compounds**, the word ***cereal*** ranks first, except for the **compounds** with ***sugar***, in which they rank twice first ad once second.

Conclusions

The number of occurrences of "cereal" names in noun phrases shows the importance of particular cereals in the life of the English people over the centuries. Thus, ***corn*** (28% of the **compounds** and **derivatives**) must have always been the most important of all, because of its rarity. Second comes ***barley*** (17%), followed by ***rice*** and ***wheat*** (12% of the **compounds** and **derivatives** each), ***maize*** and ***rye*** (10% of the **compounds** and **derivatives** each), ***oat*** (7% of the **compounds** and **derivatives**), ***millet*** and ***sorghum*** (2% of the **compounds** and **derivatives** each) (Figure 1-1).

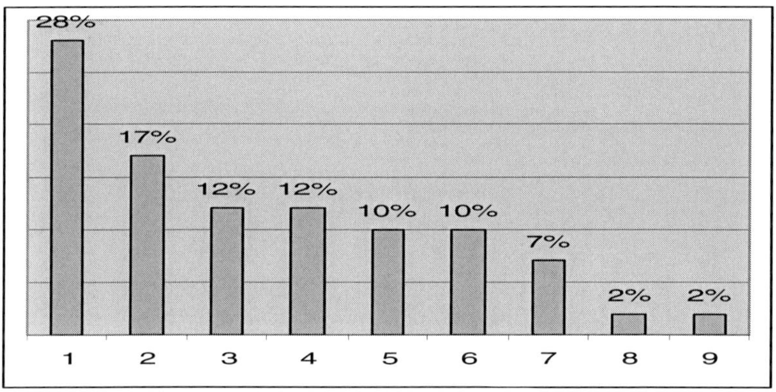

Figure 1-1. Number of occurrences of cereals names in noun phrases: 28% *corn*, 17% *barley*, 12% *rice* and *wheat*, 10% *maize* and *rye*, 7% *oat*, 2% *millet* and *sorghum*

The number of occurrences of other nouns in the **compounds** with cereals names shows the importance of the "objects" in the everyday life of the English people. Thus, ***grass*** comes first (11% of the **compounds**), followed by ***bread***, ***flour***, and ***meal*** (10% of the **compounds** each), ***cake*** and ***sugar*** (7% of the **compounds** each), and by ***bird***, ***coal***, ***corn***, ***husk***, ***oil***, ***starch***, ***water***, ***weevil***, and ***whiskey*** (5% of the **compounds** each) (Figure 1-2). To note that ***flour*** and ***meal*** are synonyms, which ranks them first (8 occurrences).

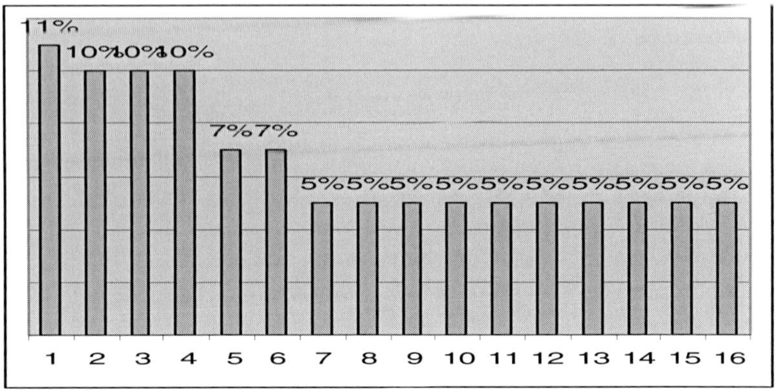

Figure 1-2. Number of occurrences of other nouns in cereals names: 11% *grass*, 10% *bread*, *flour* and *meal*, 7% *cake* and *sugar*, 5% *bird*, *coal*, *corn*, *husk*, *oil*, *starch*, *water*, *weevil*, and *whiskey*

References

Benson, M., Benson, Evelyn & Ilson, R. (1990). *The Combinatory Dictionary of English*. Amsterdam – Philadelphia: John Benjamins Publishing Company.

Chalker, Sylvia & Weiner, E. (1994). *The Oxford Dictionary of English Grammar*. London - New York - Sydney - Toronto: BCA.

Chiţoran, D. (1973). *Elements of English Structural Semantics*. Bucureşti: E.D.P.

Dicţionarul explicativ al limbii române. [The Explanatory Dictionary of the Romanian Language]. (1998). Bucureşti: Editura U.E.

Raţă, Georgeta & Micu, Laura-Constantina. (2004). Notes on *Cereal* Compounds and Derivatives: A Lexicological Approach. *Tradiţie şi modernitate în ştiinţele umaniste şi sociale*. Timişoara: Editura Mirton. 127-132.

Raţă, Georgeta, Samfira, I., Boc-Sânmărghiţan, Diana-Andreea & Butnariu, Monica. (2012). *Compendium of Forage Technical Terms in English, French and Romanian*. Newcastle upon Tyne: Cambridge Scholars Publishing.

Webster Comprehensive Dictionary. (1995). Chicago: J. G. Ferguson Publishing Company.

Webster Encyclopaedic Unabridged Dictionary of the English Language. (1996). New York: Gramercy Books.

DERIVATIVES WITH –*CIDE* IN THE VOCABULARY OF PESTICIDES

GEORGETA RAȚĂ AND ANICA PERKOVIĆ

Introduction

The study of any subject should start with the study of its specific terminology. Thus, in the study of pesticides, "Understanding the proper use of pesticides is imperative to their effectiveness and to your safety." (*Arizona Master Gardener Manual: An Essential Reference for Gardening in the Desert Southwest*). Mentions such as "The wording 'insecticides and pesticides' is incorrect because insecticides are pesticide." (*Idem*) or "herbicides kill plants, not just weeds" (*Idem*) are also important for a better understanding of the terms.

Therefore, any introduction to the terminology specific to the field of pesticides should include the following:

- the types of pesticides according to their effect: ***acaricides*** (that control mites, ticks, and spiders), ***avicides*** (that control birds), ***bactericides*** (that control bacteria), ***fungicides*** (that control fungi), ***herbicides*** (that control plants), ***insecticides*** (that control insects), ***miticides*** (that control mites), ***molluscicides*** (that control molluscs, such as slugs and snails), ***nematicides*** (that control nematodes), ***piscicides*** (that control fish), ***predacides*** (that control pest animals), ***rodenticides*** (that control rodents), on the one hand, and ***antitranspirants*** and ***antidesiccants*** (that reduce water loss from plants), ***attractants*** (that lure pests), ***desiccants*** and ***defoliants*** (that remove or kill leaves and stems), ***growth regulators*** (that stop, speed up, or otherwise change normal plant processes), and ***repellents*** (that keep pests away), on the other hand;
- the types of pesticides according to the way they work: ***contact poisons*** (that kill pests simply by touching them), ***fumigants*** (that kill when they are inhaled or otherwise absorbed by pests), ***nonselective pesticides*** (that kill most plants or animals), ***selective pesticides*** (that

kill only certain kinds of plants or animals, e.g. 2, 4-D used for lawn weed control, kills broadleaved plants but does not harm grass), ***stomach poisons*** (that kill when swallowed), ***systemics*** (that kill best by being taken into the blood of the animal or sap of the plant upon which the pest is feeding), and ***translocated herbicides*** (that move from the point of initial application to circulate throughout the plant, the circulation of toxin ensures the kill of the entire plant);

- the types of pesticides according to the time they are applied: ***post-emergent*** (used after the crop or weeds have germinated), ***pre-emergent*** (used before plants emerge from soil), and ***pre-planting*** (used before crop is planted by applying to the soil);
- the terms describing how to use pesticides: ***band*** (application to a strip over or along each crop row), ***broadcast*** (uniform application to an entire, specific area by scattering), ***dip*** (immersion of a plant in a pesticide), ***directed*** (aiming the pesticide at a portion of a plant, animal or structure), ***drench*** (saturating the soil with a pesticide), ***foliar*** (application to the leaves of plants), ***in-furrow*** (application to or in the furrow in which a plant is growing), ***side-dress*** (application along the side of a crop row), and ***spot treatment*** (application of a pesticide to a small section or area of a crop).

Material and Method

We have analysed the terms ending in *-cide* 'to kill', 'killer', 'to murder, to cause death, to slayer, to cut down', a Latin suffix that should not be mistaken for *-cide* 'to cut', although the two suffixes are related.

We have left aside 71 of these **derivatives** that have nothing to do with the killing of any organisms dangerous for humans, crops, or domestic animals (***algiocide*** 'pain killer', ***cytocide*** 'the killing of cells', ***episcopacide*** 'the killing of a bishop, etc.).

We have also left aside the **derivatives** ***avenicide***, ***botrycide***, ***corvicide***, ***graminicide***, ***limacide***, ***ovicide***, ***tileticide***, and ***zoocide***, as well as any other possible **derivatives** in *-cide* that, though used in the field of agriculture, are not mentioned in our corpus (*Arizona Master Gardener Manual: An Essential Reference for Gardening in the Desert Southwest*).

Results and Discussion

The 91 **derivatives** ending in *-cide* and concerning the field of "pests" (we understand by "pests" any living organism damaging animals and/or humans, in any way possible) have the following meanings:

- 51 **derivatives** (58%) designate 'the act of killing pests': ***adulticide*** 'the killing of adult insects (as opposed to larvae)', ***anophelicide*** 'the killing of anopheles mosquitoes', ***antharcocide*** 'the killing of anthrax bacteria', ***aphicide / aphidicide*** 'the killing of aphids', ***apicide*** 'the killing of bees', ***apricide*** 'the killing of boars', ***arachnicide*** 'the killing of spiders, scorpions, etc.', ***avicide*** 'the killing of birds', ***cervicide*** 'the killing of deer', ***ceticide*** 'the killing of whales or other cetaceans', ***cimicide*** 'the killing of bedbugs', ***elephanticide*** 'the killing of elephants', ***felicide*** 'the killing of cats', ***formicicide / formicide*** 'the killing of ants', ***gallinicide*** 'the killing of hens, turkeys, or poultry, in general', ***gametocide*** 'the killing of malarial parasites', ***gonoccocide / gonococcide*** 'the killing of gonococci', ***helminthicide*** 'the killing of intestinal worms', ***herpecide / herpicide*** 'the killing of reptiles', ***hiricide*** 'the killing of goats', ***hirudicide*** 'the killing of leeches', ***imagicide / imagocide*** 'the killing of adult insects, especially mosquitoes', ***leporicide*** 'the killing of hares / rabbits', ***lumbricide*** 'the killing of roundworms', ***lupicide*** 'the killing of wolves', ***microbicide*** 'the killing of microbes', ***muricide*** 'the killing of mice', ***muscacide / muscicide*** 'the killing of flies', ***myrmecide*** 'the killing of ants', ***perdricide*** 'the killing of partridges', ***plasmodicide*** 'the killing of malarial parasites', ***poultrycide*** 'the killing of poultry', ***protozoacide*** 'the killing of protozoans', ***pulicide / pilicicide*** 'the killing of fleas', ***scabicide / scabieticide*** 'the killing of organisms that cause scabies', ***schistosomacide / schistosomicide*** 'the killing of schistosomes or blood flukes', ***schizonticide*** 'the killing of malarial parasites', ***serpenticide*** 'the killing of serpents', ***sporicide*** 'the killing of spores', ***staphylocide / staphylococcicide / staphylococcide*** 'the killing of organisms that cause staphylococcus injections', ***streprococcicide*** 'the killing of streptococci', ***talpicide*** 'the killing of moles', ***tauricide*** 'the killing of bulls', ***treponimicide*** 'the killing of parasitical bacteria', ***trichomonacide*** 'the killing of parasites that cause trichomoniasis / diarrhoea', ***trypanocide / trypanosomacide*** 'the killing of organisms that cause sleeping sickness', ***tuberculicide / tuberculocide*** 'the killing of tuberculosis bacilli', ***ursicide*** 'the killing of bears', ***vaccicide*** 'the killing of cows', ***vespacide*** 'the killing of wasps', and ***vulpicide*** 'the killing of foxes';
- 24 **derivatives** (23%) designate 'the agent (substance) that kills pests': ***acaricide*** 'a substance that kills mites or ticks', ***amebicide / amoebicide*** 'a substance that kills amoebae', ***bacillicide*** 'a substance that kills bacilli', ***bacteriacide / bactericide / bacteriocide*** 'a substance that kills bacteria', ***biocide*** 'a substance that kills micro-organisms',

biopesticide 'a natural substance that kills insects', ***culicide*** 'an agent that kills mosquitoes, especially culicide mosquitoes', ***floricide*** 'a substance that kills flowers', ***gallicide*** 'a substance that kills fowls', ***germicide*** 'a substance that kills germs', ***herbicide*** 'a substance that kills plants', ***larvicide*** 'a substance that kills larvae', ***miticide*** 'a substance that kills mites', ***molluscicide*** 'a substance that kills molluscs', ***mosquitocide*** 'an agent that is destructive to mosquitoes', ***mycocide*** 'an agent that destroys fungi, ***nemacide / nematicide / nematocide*** 'a substance that kills nematodes', ***ovicide*** 'a substance that kills insects' eggs', ***parasiticide*** 'an agent or preparation that destroys parasites', ***pesticide*** 'a substance that kills pests, especially insects', ***phytocide*** 'a substance that kills plants', ***predacide*** 'a substance that kills epsts', ***silvicide*** 'a substance that kills trees or forests', and ***spirillicide*** 'an agent that is destructive to spirilla';

- 16 **derivatives** (20%) designate both 'the act of killing pests' and 'the agent (substance) that kills pests': ***algaecide / algicide*** 'the killing of algae' + 'a substance that kills algae', ***arboricide*** 'the killing of trees' + 'a substance that kills trees', ***ascaricide*** 'the killing of roundworms' + 'a substance that kills worms of the genus *Ascaris*', ***canicide*** 'the killing of dogs' + 'a substance that kills dogs', ***epozoicide*** 'the killing of external-animal parasites' + 'a substance that kills epizoa', ***fungicide*** 'the killing of fungi' + a substance that kills fungi', ***insecticide*** 'the killing of insects' + 'a substance that kills insects', ***microbicide*** 'the killing of microbes' + 'a substance that kills microbes', ***oxyuricide*** 'the killing of worms of the genus *Oxyurus*' + 'a substance that kills pinworms', ***pediculicide*** 'the killing of lice' + 'a substance that kills lice', ***piscicide*** 'the killing of fish' + 'a substance that kills fish', ***rodenticide*** 'the killing of rodents' + 'a substance that kills rodents, especially rats and mice', ***spirochaeticide / spirocheticide*** 'the killing of spirochetes' + 'a substance that kills spirochetes', ***taeniacide / teniacide / tenicide*** 'the killing of tapeworms' + 'a substance that kills tapeworms', ***vermicide / verminicide*** 'the killing of vermin or intestinal worms' + 'a substance that kills worms', and ***viricide / virudice / viruscide*** 'the killing of viruses' + 'a substance that kills viruses.'

Conclusions

The following conclusions can be drawn from the analysis above:

- the number of **derivatives** in *-cide* that have nothing to do with the use of pesticides (i.e., 71) is almost equal to that of the **derivatives** in *-cide* designating pesticides (i.e., 91), pointing to an astonishing prolificacy of this suffix in modern English (particularly in the second half of the 20th century, as mentioned by English language dictionaries);
- the number of **derivatives** in *-cide* designating 'the act of killing a pest' is almost twice larger (51 + 16 = 67) than that of the **derivatives** in *-cide* designating 'a substance that kills a pest' (24 + 16 = 40) (Figure 1-3);
- the scarce presence of the **derivatives** in *-cide* in English language dictionaries points to the recentness of these terms, which have not acquired yet the status of English words, despite their frequent use in ***Agricultural English***.

Here again, the main influence of Greek has been in its use in the coinage of scientific and technical words (Carstairs-McCarthy 2002).

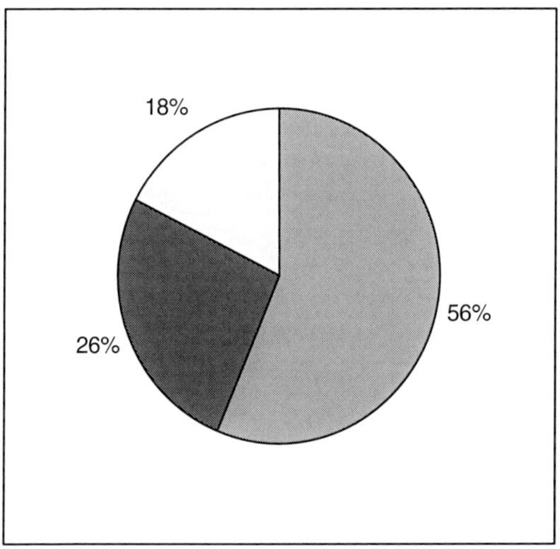

Figure 1-3. Share of derivatives in *-cide* in the English of pesticides: 56% 'the act of killing a pest', 26% 'a substance that kills a pest', 18% 'the act of killing a pest' + 'a substance that kills a pest'

References

Adams, Valerie. (1987). *An Introduction to Modern English Word-formation*. London – New York: Longman.

Arizona Master Gardener Manual: An Essential Reference for Gardening in the Desert Southwest. Cooperative Extension, College of Agriculture and Life Sciences, The University of Arizona. Online: http://ag.arozona.edu/pubs/garden/mg/index/html.

Carstairs-McCarthy, A. (2002). *An Introduction to English Morphology. Words and Their Structure*. Edinburgh: Edinburgh University Press.

Grzega, J. & Schöner, Marion. (2007). *English and General Historical Lexicology. Materials for Onomasiology Seminars*. Katholische Universität Eichstätt-Ingolstadt.

Merriam-Webster Dictionary. (2011). Online: www.merriam-webster.com/dictionary.

Plag, I. (2002). *Word-formation in English*. Cambridge: Cambridge University Press.

Rață, Georgeta & Perković, Anica. (2008). The Vocabulary of Pesticides: A Terminological Approach. *Proceedings of the 43rd Croatian and 3rd International Symposium on Agriculture, 18-21 February 2008, Opatija, Croatia*: 725-729.

Sheehan, M. J. (2000). *Word Parts Dictionary. Standard and Reverse Listings of Prefixes, Suffixes, Roots and Combining Forms*. Jefferson, NC – London: McFarland & Company, Inc., Publishers.

Štekauer, P. & Lieber, Rochelle. (2005). *Handbook of Word-Formation*. Dordrecht: Springer.

The American Heritage Dictionary of the English Language. (2008). Boston: Houghton Mifflin.

The Random House Dictionary of the English Language. (1968). New York: Random House.

Webster Comprehensive Dictionary. (1995). Chicago: J. G. Ferguson Publishing Company.

*Words ending in **-cide***. Online: http://www.wordinfo.info/words/index/info/view_unit.

COMBINING FORMS: *BI(O)-*

ALINA-ANDREEA DRAGOESCU
AND DIANA-ANDREEA BOC-SÎNMĂRGHIȚAN

Introduction

In the general context of globalization, when English has become the language of reference in science and in numerous other areas of communication and technology, it is interesting to note that most words, even newly coined ones, are based on the previous *lingua franca*, i.e. Latin. Unsurprisingly, the Neo Latin element becomes the cornerstone on which both English and other European languages are based, thus generating an easy understanding of different linguistic corpora.

Moreover, organic and biodynamic methods of farming, as well as ecological issues have become ever more imperative exigencies in the world we live in. All of the latest trends in biology, ecology, and agriculture underline the importance of life and protecting biotic systems. Therefore, notions such as ***bioconversion***, ***biodegradation***, ***biodiversity***, etc., all based on the **combining form** *bi(o)-*, are an indispensable input for all the students in the fields mentioned above.

From a linguistic point of view, this paper deals with **word formation**, especially by means of **compounding**, based on the **combining form** *bi(o)-* 'life', which is fundamental to several fields of science. According to Carstairs-McCarthy (2002), a **compound** is 'a word containing more than one root or **combining form** producing a new meaning'. A **combining form** is a bound base or 'a modified form of an independent word in English or in a language such as Greek or Latin from which English has borrowed that occurs only in combination with other forms' (*The American Heritage Dictionary of the English Language*, 2008). It combines with other **combining forms**, **affixes**, or **free words**, to form **compounds**. As a result, the classical **compounds** they help form may be paraphrased (e.g. *biology* 'the science of life', or *bioprocess* 'a biological process'). Elements of **compounds** can be considered **combining forms** if they cannot stand alone as free words. However, there are several exceptions, especially in the late 20th century, when short forms are very recurrent. Our goal here is to prove that *bi(o)-* is, actually, a **combining form**.

Material and Method

The issue of language enrichment has been the target of our analysis, as language is enhanced by means of **compounding** and **derivation**. We have inventoried 80 words with *bi(o)-* in *The American Heritage Dictionary of the English Language* (2008), leaving aside very recent words with *bio-* such as ***biodiesel*** [< *bio-* + *diesel* (< Rudolf Diesel, German mechanical engineer and inventor)], ***biofuel*** 'fuel such as methane produced from renewable biological resources such as plant biomass and treated municipal and industrial waste' [< *bio-* + *fuel*], or ***bioterrorism*** [< *bio-* + *terrorism*]. The words in our corpus are of utmost importance to students, researchers, or anyone working in the fields of biology, agriculture, and ecology. Their meanings, as well as their etymology are useful tools in understanding notions that specialists in the field must certainly be aware of. The main sources that have been used for this purpose are English language dictionaries and encyclopaedias, as well as literature in the field of linguistics. The methods used in this research pertain to applied linguistics, namely etymological and morphological analysis.

Results

Bi(o)- (< Gk *bios* 'life'), a first element in the Modern Latin words of our corpus of 80 terms, generally indicates or involves life or living organisms. It is considered by linguists:

- a **combining form** 'a modified form of an independent word in English or in a language such as Greek or Latin from which English has borrowed that occurs only in combination with other forms: it combines with words, affixes, or other combining forms to form compounds or derivatives, as *electro-* (from *electric*) in *electromagnet* or *geo-* in *geochemistry*' (*The American Heritage Dictionary of the English Language*, 2008) by *Collins English Dictionary* (2003);
- a **prefix** 'an affix, such as *dis-* in *disbelieve,* put before a word to produce a derivative word or an inflected form' (*The American

Heritage Dictionary of the English Language, 2008) by *The American Heritage Dictionary of the English Language* (2008).

In our corpus, there are also, beside **compounds** (words that consist either of two or more elements that are independent words, such as *loudspeaker, baby-sit,* or *high school,* or of specially modified **combining forms** of words, such as Gk *philosophia,* from *philo-* 'loving' and *sophia,* 'wisdom' – *The American Heritage Dictionary of the English Language*, 2008), **blends** or **portmanteaus** (words produced by combining parts of other words, as *smog* from *smoke* and *fog* – *The American Heritage Dictionary of the English Language*, 2008), **borrowings** or **loanwords** (words adopted from another language and completely or partially naturalized, as *very* and *hors d'oeuvre,* both from French – *The American Heritage Dictionary of the English Language*, 2008), and **calques** or **loan translations** (borrowings from one language to another whereby the semantic components of a given term are literally translated into their equivalents in the borrowing language: English *superman,* for example, is a loan translation from German *Übermensch* – *The American Heritage Dictionary of the English Language*, 2008). Thus:

- 59 of the words in our corpus (72.50%) are **compounds** of the 'combining form *bio-* + Noun' type: ***bioaccumulation*** 'the accumulation of a substance, such as a toxic chemical, in various tissues of a living organism' [< *bio-* + *accumulation*]; ***bioacoustics*** 'the study of sounds produced by or affecting living organisms, especially those sounds involved in communication' [< *bio-* + *acoustics*]; ***bioactivity*** 'the effect of a given agent, such as a vaccine, upon a living organism or on living tissue' [< *bio-* + *-activity*]; ***bioassay*** 'determination of the strength or biological activity of a substance, such as a drug or hormone, by comparing its effects with those of a standard preparation on a test organism' [< *bio-* + *assay*]; ***bioastronautics*** 'determination of the strength or biological activity of a substance, such as a drug or hormone, by comparing its effects with those of a standard preparation on a test organism' [< *bio-* + *astronautics*]; ***bioavailability*** 'the degree to which a drug or other substance becomes available at the physiological site of activity after administration' [< *bio-* + *availability*]; ***biobibliography*** 'a book or article combining an account of a person's life with a discussion of works written by or about that person' [< *bio-* + *bibliography*]; ***biocatalyst*** 'a substance, especially an enzyme, that initiates or modifies the rate of a chemical reaction in a living body; a biochemical catalyst' [< *bio-* + *catalyst*]; ***biocenose*** / ***biocenosis*** 'a community of biologically integrated and interdependent plants and animals' [< *bio-* + Gk *koinōsis* 'a mingling']; ***biochemistry*** 'the study of the chemical substances and vital processes occurring in living organisms; biological chemistry; physiological chemistry' [< *bio-* + *chemistry*]; ***biochip*** 'a computer chip made from organic molecules rather than silicon or germanium' [< *bio-* + *chip*]; ***bioclimatology*** 'the study of the effects of climatic conditions on living organisms' [< *bio-* + *climatology*]; ***biocompatibility*** 'the property of being biologically compatible by not producing a toxic, injurious, or immunological response in living tissue' [< *bio-* + *compatibility*]; ***bioconversion*** 'the conversion of organic materials, such as plant or animal waste, into usable products or energy sources by biological processes or agents, such as certain microorganisms' [< *bio-* + *conversion*]; ***biodiversity*** 'the variety of organisms found within a specified geographic region' [< *bio-* + *diversity*]; ***biodynamics*** '1. The study of the effects of dynamic processes, such as motion or acceleration, on living organisms. 2. The science of the force or energy of living matter and physiological processes. 3. A method of organic gardening and crop cultivation in which certain factors, such as planetary and seasonal cycles, are considered' [< *bio-* + *dynamics*]; ***bioelectricity*** 'an electric current that is generated by living tissue, such as nerve and muscle' [< *bio-* + *electricity*]; ***bioelectronics*** '1. The application of the principles of electronics to biology and medicine. 2. The study of the role of intermolecular electron transfer in physiological processes' [< *bio-* + *electronics*]; ***bioenergetics*** 'the study of the flow and transformation of energy in and between living organisms and between living organisms and their environment' [< *bio-* + *energetics*]; ***bioengineering*** '1. The application of engineering principles to the fields of biology and medicine, as in the development of aids or replacements for defective or missing body organs. Also called biomedical engineering 2. Genetic engineering' [< *bio-* + *engineering*]; ***bioethics*** 'the study of the ethical and moral implications of new biological discoveries and biomedical advances, as in the fields of genetic engineering and drug research' [< *bio-* + *ethics*]; ***biofeedback*** 'the technique of using monitoring devices to furnish information regarding an autonomic bodily function, such as heart rate or blood pressure, in an attempt to gain some voluntary control over that function' [< *bio-* + *feedback*]; ***bioflavonoid*** 'any of a group of biologically active substances found in plants and functioning in the maintenance of the walls of small blood vessels in mammals' [< *bio-* + *flavonoid*]; ***biogas*** 'a mixture of methane and carbon dioxide produced by bacterial degradation of organic matter and used as a fuel'

[< *bio-* + *gas*]; ***biogenesis*** '1. The principle that living organisms develop only from other living organisms and not from nonliving matter. 2. Generation of living organisms from other living organisms. 3. Biosynthesis. 4. The supposed recurrence of the evolutionary stages of a species during the embryonic development and differentiation of a member of that species' [< *bio-* + *genesis*]; ***biogeochemistry*** 'the study of the relationship between the geochemistry of a region and the animal and plant life in that region' [< *bio-* + *geochemistry*]; ***biogeography*** 'the study of the geographic distribution of organisms' [< *bio-* + *geography*]; ***biohazard*** '1. A biological agent, such as an infectious microorganism, or a condition that constitutes a threat to human beings, especially in biological research or experimentation. 2. The potential danger, risk, or harm from exposure to such an agent or condition' [< *bio-* + *hazard*]; ***bioinstrumentation*** '1. Use of instruments for the recording or transmission of physiological information, such as breathing rate or heart rate. 2. The instruments so used' [< *bio-* + *instrumentation*]; ***bioluminescence*** 'emission of visible light by living organisms such as the firefly and various fish, fungi, and bacteria' [< *bio-* + *luminescence*]; ***biomarker*** 'a specific physical trait used to measure or indicate the effects or progress of a disease or condition' [< *bio-* + *marker*]; ***biomass*** '1. The total mass of living matter within a given unit of environmental area. 2. Plant material, vegetation, or agricultural waste used as a fuel or energy source' [< *bio-* + *mass*]; ***biomathematics*** 'the application of mathematical principles to biological processes' [< *bio-* + *mathematics*]; ***biomechanics*** '1. The study of the mechanics of a living body, especially of the forces exerted by muscles and gravity on the skeletal structure. 2. The mechanics of a part or function of a living body, such as of the heart or of locomotion' [< *bio-* + *mechanics*]; ***biomedicine*** '1. The branch of medical science that deals with the ability of human beings to tolerate environmental stresses and variations, as in space travel. 2. The application of the principles of the natural sciences, especially biology and physiology, to clinical medicine' [< *bio-* + *medicine*]; ***biometeorology*** 'The study of the relationship between atmospheric conditions, such as temperature and humidity, and living organisms' [< *bio-* + *meteorology*]; ***biophysics*** 'the science that deals with the application of physics to biological processes and phenomena' [< *bio-* + *physics*]; ***biopic*** 'a film or television biography, often with fictionalized episodes' [*bio-* + *pic* 'motion picture']; ***biopolymer*** 'a macromolecule, such as a protein or nucleic acid, that is formed in a living organism' [< *bio-* + *polymer*]; ***bioprocess*** 'a. A technique that produces a biological material, such as a genetically engineered microbial strain, for commercial use. b. Production of a commercially useful chemical or fuel by a biological process, such as microbial fermentation or degradation' [< *bio-* + *process*]; ***biopsychology*** 'psychobiology' [< *bio-* + *psychology*]; ***bioreactor*** '1. A container, such as a large fermentation chamber, for growing living organisms that are used in the industrial production of substances such as pharmaceuticals, antibodies, or vaccines. 2. A living organism, such as a bacterium or yeast, that is used in the biotechnological production of substances such as pharmaceuticals, antibodies, or vaccines' [< *bio-* + *reactor*]; ***bioregion*** 'An area constituting a natural ecological community with characteristic flora, fauna, and environmental conditions and bounded by natural rather than artificial borders' [< *bio-* + *region*]; ***bioregionalism*** 'The belief that social organization and environmental policies should be based on the bioregion rather than on a region determined by political or economic boundaries' [< *bio-* + *regionalism*]; ***bioresearch*** 'research in the biological sciences' [< *bio-* + *research*]; ***biorhythm*** 'an innate, cyclical biological process or function' [< *bio-* + *rhythm*]; ***biosatellite*** 'an artificial, recoverable satellite that is designed to carry and support humans, animals, or other living organisms' [< *bio-* + *satellite*]; ***bioscience*** 'life science' [< *bio-* + *science*]; ***biosensor*** '1. A device that detects, records, and transmits information regarding a physiological change or process. 2. A device that uses biological materials to monitor the presence of various chemicals in a substance' [< *bio-* + *sensor*]; ***biosphere*** '**1.** The part of the earth and its atmosphere in which living organisms exist or that is capable of supporting life. **2.** The living organisms and their environment composing the biosphere' [< *bio-* + *sphere*]; ***biostatistics*** 'application of statistics to the analysis of biological and medical data' [< *bio-* + *statistics*]; ***biosynthesis*** 'formation of a chemical compound by a living organism' [< *bio-* + *synthesis*]; ***biosystematics*** 'the use of data obtained from cytogenetic, biochemical, and other experimental studies to assess the taxonomic relationships of organisms or populations, especially within an evolutionary framework'[< *bio-* + *systematics*]; ***biotechnology*** '1. The use of microorganisms, such as bacteria or yeasts, or biological substances, such as enzymes, to perform specific industrial or manufacturing processes. 2. a. The application of the principles of engineering and technology to the life sciences; bioengineering. b. Ergonomics' [< *bio-* + *technology*]; ***biotelemetry*** 'The monitoring, recording, and measuring of a living organism's basic physiological functions, such as heart rate, muscle activity, and body temperature, by the use of telemetry techniques' [< *bio-* + *telemetry*]; ***biotherapy*** 'Treatment of disease with biologicals, such as certain drugs, vaccines, or antitoxins' [< *bio-* + *therapy*]; ***biotope*** 'an area that is uniform in environmental conditions and in its distribution of animal and plant life' [< *bio-* + Gk *topos* 'place']; ***biotransformation*** 'chemical alteration of a substance within the body, as by the action of enzymes' [< *bio-* + *transformation*]; and ***biotype*** 'a group of organisms having the same genotype' [< *bio-* + *type*];

- 11 of the words in our corpus (13.75%) are **compounds** of the 'combining form *bio-* + affix' type: ***biocide*** 'a chemical agent, such as a pesticide, that is capable of destroying living organisms' [< *bio-* + *-cide* 'killer; act of killing']; ***biogeny*** 'biogenesis' [< *bio-* + *-geny*]; ***biographee*** 'the subject of a biography' [< *biograph(y)* + *-ee* 'one that performs a specified action']; ***biography*** '1. An account of a person's life written, composed, or produced by another. 2. Biographies considered as a group, especially when regarded as a genre. 3. The writing, composition, or production of biographies' [< *bio-* + *-graphy*]; ***biolysis*** '1. Death of a living organism or tissue caused or accompanied by lysis. 2. The decomposition of organic material by living organisms, such as microorganisms' [< *bio-* + *-lysis*]; ***biome*** 'a major regional or global biotic community, such as a grassland or desert, characterized chiefly by the dominant forms of plant life and the prevailing climate' [< *bi-* + *-ome*]; ***biometrics*** 'the statistical study of biological phenomena' [< *bio-* + *-metrics*]; ***biopsy*** 'the removal and examination of a sample of tissue from a living body for diagnostic purposes' [< *bi-* + *-opsy*]; ***bioscope*** 'an early movie projector' [< *bio-* + *-scope*]; ***bioscopy*** 'medical examination of a body to determine the presence or absence of life' [< *bio-* + *-scopy*]; ***biotron*** 'a climate-control chamber used for studying a living organism's response to specific environmental conditions' [< *bio-* + *-tron*];
- 5 of the words in our corpus (6.25%) are **compounds** of the **blend** or **portmanteau** type: ***bioc(o)enology*** 'the branch of ecology concerned with the relationships and interactions between the members of a natural community ' [< *biocen(osis)* + *-o-* + *-logy*, a blend word (*The Random House Dictionary of the English Language*, 1968); or < *bio-* + *ceno-* < Gk *koinos* 'common' + *-logy* (*English Collins Dictionary*, 2003)]; ***biographer*** 'one who writes, composes, or produces biography' [< *biograph(y)* + *-er*]; ***biomaterial*** 'a biocompatible material that is used to construct artificial organs, rehabilitation devices, or prostheses and replace natural body tissues' [< *bio(compatible)* + *material*]; ***biometry*** 'biometrics' [< *biometr(ics)* + -

y]; ***bionics*** 'application of biological principles to the study and design of engineering systems, especially electronic systems' [< *bi-* + *(electr)onics*]; some sources suggest that bioethics also belongs to this type of **word formation**: ***bioethics*** 'the study of the ethical and moral implications of new biological discoveries and biomedical advances, as in the fields of genetic engineering and drug research' [< *bio(logy)* + *ethics*];

- 4 of the words in our corpus (5.00%) are **loanwords** from French, German and Greek: ***biology*** '1. The science of life and of living organisms, including their structure, function, growth, origin, evolution, and distribution. 2. The life processes or characteristic phenomena of a group or category of living organism. 3. The plant and animal life of a specific area or region' [< G *Biologie* < *bi-* + *-logie* '-logy']; ***bionomics*** 'ecology' [< *bionomic* probably < F *bionomique* < F *bionomie* 'ecology' (*Merriam-Webster Dictionary*, 2011)]; ***biota*** 'the combined flora and fauna of a region' [< NL < Gk *biotē* 'life']; ***biotite*** 'a dark-brown to black mica, found in igneous and metamorphic rocks' [< F *biotite* < Jean Baptiste *Biot* (1774-1862), French mineralogist and physicist];
- 1 word of our corpus (1.25%) is a **calque** (or **loantranslation**): ***biotin*** 'a colourless crystalline vitamin of the vitamin B complex' [< *vitamin*].

Discussion

After analyzing the structure of the words with *bi(o)-* in our corpus, we have come to the conclusion that *bi(o)-* is a **combining form** and not a **prefix** because it can combine not only with other words, but also with **affixes** (there are 11 such **compounds** in our corpus).

Several of the **compounds** we have analyzed may be considered **portmanteau** or **blend words** because they combine other words or parts of words with the meaning not merely of *bio-* "life", but also of other words composed with *bi(o)-*, which the latter replaces as a clipped form. Such is the case with ***bioassay***, which may be considered a **blend** of *biology* + *assay*, rather than a **compound** from the two **combining forms** *bi(o)-* + *-assay*. It is more accurate to consider the meaning of *bioassay* 'conducting a test to determine the biological activity of a substance', *i.e.* a biological trial, more closely connected to the applied adjective *biological* than to the abstract form *bi(o)-*. Likewise, ***biodegradation*** might be considered a clipping of "biological degradation". Also, though debatable, *The Random House Dictionary of the English Language* (1968) considers that ***biocenology*** is made up from a condensed variant of *biocenosis* + *-o-* + *-logy*, which makes it a portmanteau word.

We have come to the conclusion that *bi(o)-* is a significant and meaningful **combining form**. It has both a traditional straightforward meaning and a modern telescopic one. For instance, in biology, *bio-* means 'life', but in ***biodegradable***, it telescopes "biologically". This is also the case with present-day shortened forms like *bio-*, used as a free morpheme, which is, however, a telescopic abbreviation of another word. This may also be considered an instance of lexicalization, as in this particular case, *bio* (especially referring to food and agriculture) acquires independent lexical status and sometimes functions as a free form. However, this is merely an exception to the rule, as in all the other cases *bi(o)-* is a bound morpheme. This discussion proves that it is indeed a **combining form**, which can combine with bound roots and with words to make up new words. Thus, the question whether *bi(o)-* should be considered a **combining form** or an affixal formation has hopefully been settled. A final remark is that language is surely enriched by the huge productivity of this **combining form**. Various **compounds** based on such **combining forms** as *bio-* may be more easily understood if etymological analysis is considered.

Conclusions

The major effect of **compounding** is undoubtedly the enrichment of language. Studying **combing forms** and **compounds** which make up the vocabulary of biology and of related areas of research highlights various origins of modern English language, especially Latin and ancient Greek (Carstairs-McCarthy 2002), as well as French and German. Obviously, it may be remarked that the classical **compounds** have been acquired into English by three means: through French from Latin and Greek, directly from Latin and Greek, or by coinage in English, but still essentially based upon Greek or Latin patterns. Therefore, the **combining form** *bi(o)-* and the **compounds** built with it are not only part of English as well as Latin and Greek, but also part of French, Italian, Romanian and other Indo-European languages which employ them. In conclusion, it is important to be aware of this international linguistic resource, to study it accordingly and to make an informed use of such words.

References

Adams, Valerie. (1987). *An Introduction to Modern English Word-formation*. London – New York: Longman.

Carstairs-McCarthy, A. (2002). *An Introduction to English Morphology. Words and Their Structure*. Edinburgh: Edinburgh University Press.

Collins English Dictionary. (2003). Online: http://www.thefreedictionary.com.

Columbia Electronic Encyclopedia. (2004). Online: http://www.reference.com.

Dragoescu, Alina-Andreea & Boc-Sînmărghiţan, Diana-Andreea. (2011). The Combining Form *Bio-*. A Linguistic Approach. *Agrobuletin Agir 1(8)*: 113-117.

Harper, D. (2009). *Online Etymology Dictionary*. Online: http://www.etymonline.com.

Merriam-Webster Dictionary. (2011). Online: www.merriam-webster.com/dictionary.

Perković, Anica & Raţă, Georgeta. (2008). Notes on the Combining Form *eco-*. *Journal of Linguistic Studies 1*: 9-14.

Plag, I. (2002). *Word-formation in English*. Cambridge: Cambridge University Press.

Sheehan, M. J. (2000). *Word Parts Dictionary. Standard and Reverse Listings of Prefixes, Suffixes, Roots and Combining Forms*. Jefferson, NC – London: McFarland & Company, Inc., Publishers.

Štekauer, P. & Lieber, Rochelle. (2005). *Handbook of Word-Formation*. Dordrecht: Springer.

The American Heritage Dictionary of the English Language. (2008). Boston: Houghton Mifflin.

The Random House Dictionary of the English Language. (1968). New York: Random House.

Webster Comprehensive Dictionary. (1995). Chicago: J. G. Ferguson Publishing Company.

COMBINING FORMS: *ECO-*

ANICA PERKOVIĆ AND GEORGETA RAŢĂ

Introduction

Strengthening training in agribusiness, rural development, and agricultural public administration can no longer be conceived without strengthening training in environmental protection. From this perspective, it is imperative to get the necessary knowledge and know-how in the field of ecology. It is no longer possible for any E.U. Member State to improve competitiveness for farming and forestry, quality of life and diversification of rural economy, and the environment and countryside in the frame of strategic approaches and options without properly understanding "the language of environment and ecology" (both as English for special purpose).

Materials and Methods

We have inventoried all the terms containing the **combining form** ('a bound form or bound morpheme used in conjunction with another linguistic element in the formation of a word' – Chalker & Wiener 1994) ***eco-*** and its variants ***oec(o)-*** and ***oik(i)(o)-*** (from the Greek for 'house', 'household affairs' [environment, habitat], 'home', 'dwelling'; used in one extensive sense as environment) in one of the best English language dictionaries ever (*Webster Comprehensive Dictionary* 1995) and on the Internet (*Eco-words*. Online: http://www.wordinfo.info/words/index.E). We have analysed them from a *semantic* (i.e. from the point of view of their meaning) and *lexicological* (i.e. from the point of view of the stock of words being built on the same basis) perspective (Chalker & Wiener 1994).

Results

We have identified 114 terms containing the **combining form** ***eco-*** and its variants ***oec(o)-*** and ***oik(i)(o)-***.

A. Among them, 25 (22%) have nothing to do with ***ecology***, though they contain ***eco-*** and its variants ***oec(o)-*** and ***oik(i)(o)-***: ***ecofugic / oikofugic*** 'in psychiatry, a reference to or swayed by the impulse to wander or travel away from home'; ***ecomania / oecomania / oikomania*** '1. a morbid attitude toward the members of one's family [domineering behaviour at home and humility toward other persons in authority]; 2. a pathological dislike of the members of one's family often resulting in a feeling that one must get away from them'; ***econometric(al)*** '1. the branch of economics concerned with the application of mathematical economics to economic data by the use of statistical methods; 2. of, or relating to, or characterized by, the application of mathematics to economic data or theories'; ***econometrician*** 'a student of, or specialist in, econometrics'; ***econometrics*** 'the branch of economics concerned with the application of mathematical economics to economic data by the use of statistical methods'; ***econometrist*** 'econometrician'; ***economic(al)*** '1. pertaining to the management of a household, or to the ordering of private affairs; 2. relating to the science of economics; relating to the development and regulation of the material resources of a community or nation; 3. the science relating to the production and distribution of material wealth; sometimes used as equivalent to political economy, but more frequently with reference to practical and specific applications'; ***economically*** 'with economy'; ***economics*** 'the study of the production, distribution, and consumption of goods and services'; ***economise / economize*** 'to practice economy'; ***economiser*** 'a person who economises'; ***economism*** 'the theory or practice of assigning primary importance to the economy or to economic achievement'; ***economist*** '1. one who manages a household; a housekeeper; 2. someone who studies, works, or is an expert in the field of economics'; ***economization*** 'the action or process of economizing (force, material, etc.)'; ***economy / oeconomy*** '1. the production and consumption of goods and services of a community regarded as a whole; 2. the prudent managing of resources to avoid extravagant expenditure or waste; 3. a saving or attempt to reduce expenditure; 4. originally, the management of a household; 5. current usage is sometimes a reference to that which is intended to be less expensive or to give better value'; ***ecophobia / oik(i)ophobia*** '1. a morbid dislike of home or an abnormal fear of being home or in one's house; 2. a fear of home life or surroundings, including house-hold appliances, equipment, electricity, bathtubs, household chemicals, and many other common objects in the home'; ***ecotropic / oikotropic*** '1. homesick; a strong desire to return to one's home; 2. in virology, a retrovirus that can replicate only in the host of the species in which it originated'; ***macroeconomic*** 'related to macroeconomics'; ***macroeconomics*** '1. a branch of economics that focuses on the general features and processes that make up a national economy and the ways in which different segments of the economy are connected; 2. a branch of economics dealing with the broad and general aspects of an economy; such as, the relationship between the income and investments of a country as a whole'; ***macroeconomist*** 'a specialist in macroeconomics'; ***macroeconomy*** 'the economy viewed as a whole and in terms of all those factors that control its overall performance'; ***microeconomic*** 'related to microeconomy'; ***microeconomics*** 'the study of specific or localized aspects of an economy'; ***microeconomist*** 'a specialist in microeconomics'; ***oecophobia / oikophobia*** 'a fear of home surroundings and certain items in the home (electrical, etc.)'.

B. Other 88 terms (77%) containing ***eco-*** and its variants ***oec(o)-*** and ***oik(i)(o)-*** are closely related to ***ecology***. They are as follows: ***agroecological*** 'related to agroecology'; ***agroecologist*** 'an expert in or a student of agroecology'; ***agroecology*** '1. the study of the relationship between the environment and agricultural crops; 2. ecology as applied to agriculture'; ***aut(o)ecological*** 'related to aut(o)ecology'; ***aut(o)ecology*** '1. the ecology of an individual organism or species; 2. the study of the ecology of an individual plant or species; the opposite of *synecology*'; ***bioecologic(al)*** 'related to bioecology'; ***bioecologically*** 'from the point of view of bioecology'; ***bioecologist*** '1. a specialist who studies the relationships of organisms to their natural environments; 2. one who favours or specializes in bioecology; an *ecologist*'; ***bioecology*** 'the science of organisms as affected by the factors of their environments; study of the environment and life history of organisms; also, *ecology*'; ***di(o)ecious*** 'in biology, having the male and female reproductive organs in separate individuals; most animal species are dioecious, as are some plants, such as asparagus'; ***dioeciously*** 'in a dioecious way'; ***dioeciousness*** 'the property of being dioecious'; ***dioecism*** 'dioeciousness'; ***dioecy*** 'dioeciousness'; ***ecoactivist*** 'one who actively opposes the pollution, or destruction by other means, of the environment'; ***ecobabble*** 'using the technical language of ecology to make the user seem to be ecologically aware; ***ecobiology*** 'the study of the relationships of organisms to their natural environments; ***ecobiosis*** 'the conditions pertaining to a mode of life within a specific habitat'; ***ecobiotic*** 'related to ecobiology'; ***ecocatastrophe / ecocatastrophy*** 'major damage to the environment, especially when caused by human activity'; ***ecocentric*** '1. centring on the environment; emphasizing the importance of protecting the environment rather than the needs or rights of human beings; of or relating to ecocentrism; 2. an adherent or advocate of ecocentrism'; ***ecocentrism*** 'the view or belief that environmental concerns should take precedence over the needs and rights of human beings considered in isolation'; ***ecocidal*** 'designed or tending to destroy the environment'; ***ecocide*** 'destruction or damage of the environment; especially intentionally, e.g., by herbicides in war'; ***ecoclimate*** 'the climate as an ecological factor; the climate of a habitat'; ***ecoclimatology*** 'the science of ecoclimate'; ***ecocline*** 1. reflecting ecological conditions in general; 2. a more or less continuous character variation in a sequence of populations distributed along an ecological gradient, with each population exhibiting local adaptation to its particular segment of the gradient; 3. the differences in community structure resulting from changes in slope aspect around a mountain or ridge'; ***ecodeme*** 'a sub-specific group capable of interbreeding within a population'; ***ecofact*** 'in archaeology, a natural object or substance that has not been technologically altered but that has cultural significance; such as, a shell carried from the ocean to an inland settlement'; ***ecofallow*** 'a method of farming that diminishes weeds and conserves water by rotating crops and reducing or eliminating tillage'; ***ecofeminism*** 'a socio-political theory and movement that associates ecological (especially, environmental) concerns with feminist ones; especially, while regarding both as resulting from male dominance and exploitation'; ***ecofeminist*** 'of, relating to, or characteristic of *ecofeminism*; advocating or adhering to ecofeminism'; ***ecofreak*** 'a fanatical conservationist or environmentalist'; ***eco-friendliness*** 'the property of being eco-friendly'; ***eco-friendly*** 'ecologically "friendly"; not harmful to the environment; also applied to products manufactured with explicit regard to the environment'; ***ecogeographer*** 'a specialist in the geographical aspects of the ecology'; ***ecogeographic(al)*** 'related to the geographical aspects of ecology'; ***ecogeographically*** 'from the point of view of the geographical aspects of the ecology'; ***ecohazard*** 'any activity or substance that may constitute a threat to a habitat or environment'; ***ecolaw*** 'legislation dealing with the environment'; ***ecoline*** 'the rate of genetic change that occurs in an environment due to the merging of different varieties of a plant species'; ***ecologic(al)*** '1. of or relating to the environment or to the science of ecology; 2. relating to the wise use or beneficial management of natural resources and of the natural environment'; ***ecologist*** 'one who specializes in biological sciences that deal with the relationship between organisms and their environment'; ***ecology / oecology*** '1. the branch of the biological sciences that deals with the relationship between organisms and their environment, including their relationship with other organisms; 2. the science concerned with interactions between organisms and the environment on spatial scales ranging from parts of individuals to the biosphere as a whole'; ***ecomorphology*** 'the study of the relationship between the ecological relations of an individual and its morphology'; ***econiche*** 'niche'; ***ecoparasite*** 'a microparasite to which the host is normally immune or well adapted'; ***ecophysiological*** 'related to ecophysiology'; ***ecophysiologist*** 'a specialist in ecophysiology'; ***ecophysiology*** 'the study of the interrelationship between an organism's physical functioning and its environment'; ***ecopo(i)etist*** 'an aesthete in the introduction and selective elimination of species within a new or fractured ecology with the end goal of producing a self-sustaining dynamically balanced ecosystem that provides beauty and usefulness to humans'; ***ecopoiescience*** 'the scientific study of the breakdown of ecosystems and the processes involved in the re-diversification of species'; ***ecopoiesis*** '1. origin of ecosystems; 2. a type of planetary engineering that can be a major stage of terra formation': the primary stage of ecosystem creation is usually restricted to the initial seeding of microbial life; 3. the human creation of a self-sustaining ecosystem, or biosphere, on a lifeless planet'; ***ecosite / oecosite / oikosite*** 'an ecoparasite'; ***ecospecies*** '1. a taxonomic species considered in terms of its ecological characteristics and usually including several interbreeding ecotypes; 2. an index species that is characteristic of a particular biome or ecosystem'; ***ecospecific*** 'related to ecospecies'; ***ecospecifically*** 'from the point of view of ecospecies'; ***ecosphere*** '1. in ecology, the earth and the living organisms that inhabit it, along with all the environmental factors that operate on these organisms; biosphere; 2. in astronomy, the region of space around a star that is considered to be capable of supporting life; 3. the region of space, including planets, whose conditions are not incompatible with the existence of living things'; ***ecospheric*** 'related to the ecosphere'; ***ecosystem*** 'a system formed by the interaction of community of organisms with their environment'; ***ecotage*** 'sabotage aimed at polluters or destroyers of the natural environment'; ***ecotelemetry*** 'measurement and transmission of vital information; biotelemetry'; ***ecoterrorist*** 'a person involved in ecoterrorism'; ***ecoterrorism*** '1. the threat to use violent acts that would harm the quality of the environment in order to blackmail a group or society. It also includes the actual carrying out of the threats; 2. the sabotage of the activities of individuals or corporations, e.g., industrial companies, considered to be polluting or destroying the natural environment'; ***ecotonal*** 'related to ecotone'; ***ecotone*** '1. in ecology, a transition zone between two distinct habitats that contains species from each area, as well as organisms unique to it; 2. in anthropology, such an area of transition in which certain game or vegetation overlap; a region of primary importance for human subsistence'; ***ecotourism*** 'a form of tourism that strives to minimize ecological or other damage to areas visited for their natural or cultural interest'; ***ecotoxicologist*** 'a specialist in the harmful effects of chemicals to the natural environment'; ***ecotoxicology*** 'the scientific study of harmful effects caused by manmade chemicals to the natural environment, especially effects on populations, communities, and ecosystems; an essential part of ecotoxicology is the study of the movement of potentially toxic substances through food webs and through the water cycle, etc.'; ***ecotype*** '1. an organism that has adapted to its local environment through minor, genetically induced changes in its physiology; yet can still reproduce with other members of its species from other areas that have not undergone these changes; 2. a locally adapted population of a species with limited tolerance to changes in environmental factors'; ***ecotypic*** 'related to ecotype'; ***ecotypically*** 'from the point of view of an ecotype'; ***ecowarrior*** 'an activist who takes direct, often unlawful, action on an environmental issue'; ***ecozoiatry*** 'a branch of veterinary medicine dealing with domestic animals'; ***genecology*** 'in biology, the study of intraspecific variations and genetic compositions in relation to the environment'; ***heteroecious*** '1. a parasite occupying two or more different hosts at different stages of a life cycle; 2. a non hostspecific parasite; 3. a reference to a unisexual organism in which male and female gametes are produced by different individuals'; ***heteroeciously*** 'from the point of view of heteroecism'; ***heteroecism*** 'the development of different stages of a parasitic species on different host plants'; ***macroecology*** 'the ecology of a macrohabitat or larger generalized area'; ***microecological*** 'related to microecology'; ***microecology*** 'the ecology of a microhabitat (very small area)'; ***paleoecology*** 'ecology that deals with fossil organisms'; ***synecologic(al)*** 'related to synecology'; ***synecologically*** 'from the point of view of synecology'; ***synecologist*** 'a student of synecology'; ***synecology*** '1. the structure, development, and distribution of communities in relation to their environments; 2. the study of plant or animal communities'; ***zooecological*** 'related to zooecology'; ***zooecology*** 'the study of the relationships between animals and their environments; animal ecology'.

C. A single term (1%), ***ecopolitics*** '1. the study of politics as influenced by economy; 2. the study of the interrelation between politics and ecological issues and problems', can be ranged in either of the two groups above.

Discussion

As far as the terms containing the **combining form** ***eco-*** and its variants ***oec(o)-*** and ***oik(i)(o)-*** are concerned, 28 (32%) of them are either **compound words**, **derivatives** or **backformations** (Chalker & Wiener 1994).

The **compound words** in our corpus are words formed with the help of a **combining form** ('a linguistic form that occurs only in combination with other forms') (Chalker & Wiener 1994) and of an **independent word**. There are 8 such **combining forms** and 18 (64%) **compound words** in our corpus: ***bio-*** 'a combining form meaning *life*, used in the formation of compound words' (4): ***bioecologic(al)***, ***bioecologically***, ***bioecologist***, and ***bioecology***; ***agro-*** 'a combining form meaning *field*, *soil*, *crop production*, used in the formation of compound words' (3): ***agroecological***, ***agroecologist***, and ***agroecology***; ***hetero-*** 'a combining form meaning *different*, *other*, used in the formation of compound words; also, before a vowel, ***heter-***' (3): ***heteroecious***, ***heteroeciously***, and ***heteroecism***; ***aut(o)-*** 'a combining form meaning *self*, *same*, *spontaneous*, used in the formation of compound words; also, before a vowel, ***aut-***' (2): ***aut(o)ecological*** and ***aut(o)ecology***; ***micro-*** 'a combining form with the meanings *small*, *very small in comparison with others of the kind*, *too small to be seen by the unaided eye*, *dealing with extremely minute organisms, organic structures, or quantities of substance*, *localised, restricted in scope or area*, *containing or dealing with texts that require enlargement to be read*, *one millionth*' (2): ***microecological*** and ***microecology***; ***zoo-*** 'a combining form meaning *living being* or *animal*, used in the formation of compound words' (2): ***zooecological*** and ***zooecology***; ***macro-*** 'a combining form meaning large, long, great, excessive, used in the formation of compound words' (1): ***macroecology***; ***paleo-*** 'a combining form meaning *old* or *ancient*, especially in reference to former geologic time periods, used in the formation of compound words' (1): ***paleoecology***.

Other 9 words in our corpus are **derivatives**, i.e. words formed with the help of an **affix** (in our case, a **prefix** 'an affix placed before a base or another prefix') (Chalker & Wiener 1994). There are 2 such **prefixes** and 9 (32%) **derivatives** in our corpus: ***di-*** 'a prefix meaning *two*, *twice*, *double*, used in the formation of compound words and in chemical terms' (5): ***di(o)ecious***, ***dioeciously***, ***dioeciousness***, ***dioecism***, and ***dioecy***; ***syn-*** 'a prefix meaning *with* or *together*, used in the formation of compound words' (4): ***synecologic(al)***, ***synecologically***, ***synecologist***, and ***synecology***.

A single word (4%) in our corpus is a **backformation** ('a word formed from another word that appears to be a derived form of the first by modification') (Chalker & Wiener 1994): ***genecology*** (that must have had initially the form ***geneecology*** and that must have dropped one ***e*** maybe for spelling reasons).

The word ***ecopolitics***, though the only one to share both large groups of words containing ***eco-*** and its variants ***oec(o)-*** and ***oik(i)(o)-***, points by itself to the fact that ***ecology*** and ***politics*** are so intertwined that no policy of rural development could ever ignore environmental matters.

Conclusions

Though representing only 22% of the total number of terms containing the **combining form *eco-*** and its variants ***oec(o)-*** and ***oik(i)(o)-***, the terms having nothing to do with ***ecology*** and, therefore, nothing to do with the ***environment***, are numerous enough to be a nuisance for any non-native English-language speaking undergraduate or specialist in environmental matters. Therefore, this corpus of words should be taught as such to undergraduates in environmental ecology, underlying, at the same time, the huge richness of the English vocabulary and the wide range of means the English language can rely on in enriching its vocabulary.

References

Adams, Valerie. (1987). *An Introduction to Modern English Word-formation*. London – New York: Longman.

Chalker, Sylvia & Weiner, E. (1994). *The Oxford Dictionary of English Grammar*. London - New York - Sydney - Toronto: BCA.

Eco-words. Online: http://www.wordinfo.info/words/index.E.

Rață, Georgeta. (2005). On the Vocabulary of Ecology: Ageing and Renewing. *Lucrări ştiinţifice. Facultatea de Agricultură XXXVII*: 633-634.

—. (2005). On the Vocabulary of Ecology: Renewing Through New Forms. *Scientific Works L (6)*: 137-142.

Rață, Georgeta, Petroman, Cornelia, Petroman, I. & Perković, Anica. (2008). Eco-tourism Terms: A Hybrid Vocabulary. *Lucrări ştiinţifice. Facultatea de Agricultură 40 (3)*: 313-316.

Webster Comprehensive Dictionary. (1995). Chicago: J. G. Ferguson Publishing Company.

COMPOUNDS WITH *CULTURE*

GEORGETA RAŢĂ, IONEL SAMFIRA
AND ANICA PERKOVIĆ

Introduction

One of the first questions arising when reading a text in a foreign language is whether there is complete correspondence between a certain unit in one's mother tongue and the unit in the foreign language. From this point of view, the English language, which is a Germanic language, has different patterns of building up words, the relatively small number of words with a Latin etymology emerging in the late 18th and the 19th centuries, when scientific language was submerged by Latin and Greek terms called to fill linguistic voids caused by the emergence of new realities.

Material and Method

Teaching English for Special Purposes reveals the existence of a great number of nouns designating agricultural practices. Our feeling is that nouns coined with *-culture* (most of which have a Latin origin) are less numerous than **derivative lexemes** or than **compound lexemes**, two very common procedures of enriching vocabulary in English. We have used the quantitative and the comparative methods in our analysis of the terms in *-culture*. We have inventoried all the terms we could from both language dictionaries and agriculture-related literature or to its practice (be it type of product or farming method). We have used some of the most prominent English language dictionaries and encyclopaedias to better illustrate the field of nomenclature in the different agricultural branches.

Results and Discussion

There are a lot of nouns related to the practice of "agriculture" in English. Such nouns designating either "product types" or "farming methods" are: ***agronomy***, ***beekeeping***, ***farming***, ***forestry***, ***gardening***,

husbandry, ***hydroponics***, ***plantation***, ***pomology***, ***ranching***, etc. It is easy to notice that, with a few exceptions (***agronomy***), all these nouns are English **derivatives**.

Beside these nouns, there is also a series of noun phrases related to the practice of "agriculture". Such phrases designating either the "type of agriculture" or the "type of products" are **compounds** with *agriculture*.

As for the word **culture** '1. The act or practice of cultivating the soil; tillage. 2. The raising of plants or animals, especially with a view to their improvement', it has produced a considerable number of **compound lexemes**. The **compound lexemes** (**compound words** whose meaning cannot be deduced from their parts or similar combinations of words) related to agricultural practices are of two types: of the form ***fruit culture*** 'all fruit production from trees, shrubs, and vines' and ***greenhouse culture*** 'the growing of plants throughout the year', but their number must be larger if we take into account the fact that in theory, at least, one can associate the word "culture" with no matter what other word belonging or not to the field of agriculture and of the form ***agriculture*** (in which we identify a **combining form**, *agri-*, and a **free word**, *culture*).

The 17 **compounds** of the latter type we inventoried are as follows: ***agriculture*** [< L *agricultūra*] '1. The science or art of cultivating land in the raising of crops; husbandry; farming. 2. The production of crops, livestock, or poultry'; ***citriculture*** [< ?] 'the cultivation of citrus fruits'; ***apiculture*** [< L *api(s)* 'bee' + culture] 'beekeeping, especially on a commercial scale for the sale of honey'; ***aquaculture*** [< L *aqua* 'water' + (agri)culture] 'farming of aquatic organisms (fishes, crustaceans, molluscs, algae, and aquatic plants) in fresh, brackish or salt water', 'hydroponics', ***aquiculture*** [< L *aqui-* (combining form of *aqua* 'water') + (agri)culture] 'farming of aquatic organisms (fish such as mullet)', 'hydroponics'; ***arboriculture*** [< L *arbori-* (stem of *arbor* 'tree') + culture] 'the cultivation of trees and shrubs'; ***aviculture*** [< ?] 'The rearing or keeping of birds'; ***floriculture*** [< ?] 'cultivation of ornamental flowering plants for aesthetic purposes, whether grown in window boxes, greenhouses, or gardens', 'the cultivation of flowers or flowering plants, especially under glass'; ***horticulture*** [< L *hort(us)* 'garden' + (agri)culture] 'science and art of growing fruits, vegetables, flowers, shrubs, and trees', '1. The cultivation of flowers, fruits, vegetables, or ornamental plants. 2. The science and art of cultivating such plants'; ***monoculture*** [< ?] 'the use of land for growing only one type of crop'; ***olericulture*** [< L *oleri-* (stem of *olus* or *holus*) 'vegetable, kitchen herb' + culture] 'production of vegetable crops', 'the cultivation of vegetables'; ***pisciculture*** [< ?] 'fish farming, the raising of fish in ponds', 'the breeding, rearing, and transplantation of fish by artificial means'; ***sericulture*** [< Gk *sḗr* 'silkworm' + i + culture] 'the raising of silkworms', 'the raising of silkworms for the production of raw silk'; ***silviculture*** [L *sylvi* (combining form of *silva* 'woodland') + culture] 'study of the relationship of a forest to its environment, involving the development, care, and reproduction of stands of timber', 'the cultivation of forest trees; *forestry*' or ***sylviculture*** [L *sylv-* (stem of L *sylva, silva* 'forest') + i + culture] 'silviculture'; ***viniculture*** [< ?] 'the science or study of making wines'; ***viticulture*** [< L *vīti(s)* 'vine' + culture] 'wine growing', 'the culture or cultivation of grapevines; grape growing'.

Conclusions

The 17 **compounds** with *-culture* in English is rather small compared to languages such as French (Rață 2005a and 2005b), where there are 22 **compounds** designating different types of "culture", be it *type of product* (*algae*, *aquatic plants and animals*, *bulbs*, *carps*, *crustaceans*, *flax*, *fruit trees*, *geese*, *maize*, *molluscs*, *mushrooms*, *mussels*, *oysters*, *pearl oysters*, *pigeons*, *rabbits*, *salmon and trout*, *snails*, *trees for wood*, and *truffles*) or *farming method* (*practice of controlling sanitary status of future potato crops*, *practice of cultivating in greenhouses*, *practice of cultivating several crops on a single farm*, *use of engines in agriculture*, and *use of plastic in agriculture*), or even with Romanian. It is interesting to see that language dictionaries indicate the etymology for only 11 **compounds** with *-culture* (***agriculture***, ***apiculture***, ***aquaculture***, ***aquiculture***, ***arboriculture***, ***horticulture***, ***olericulture***, ***sericulture***, ***silviculture***, ***sylviculture***, and ***viticulture***), and not for other 6 **compounds**, perceived probably as easy to understand and to explain (***citriculture***, ***aviculture***, ***floriculture***, ***monoculture***, ***pisciculture***, and ***viniculture***). Again, it is interesting to note that only 10 of the 11 **compounds** for which the language dictionaries provide the etymology are Latin (***agriculture***, ***apiculture***, ***aquaculture***, ***aquiculture***, ***arboriculture***, ***horticulture***, ***olericulture***, ***silviculture***, ***sylviculture***, and ***viticulture***), while only 1 (***sericulture***) has a Greek etymology. Of the 17 English **compounds**, 14 (82%) indicate the *type of product* (***agriculture***, ***citriculture***, ***apiculture***, ***arboriculture***, ***aviculture***, ***floriculture***, ***horticulture***, ***olericulture***, ***pisciculture***, ***sericulture***, ***silviculture***, ***sylviculture***, ***viniculture***, and ***viticulture***) while only 3 (18%) indicate the *farming method* (***aquaculture***, ***aquiculture***, and ***monoculture***). The 17 English **compounds** are, in fact, 15, as 2 of the **compounds** have two different spelling forms (***aquaculture*** and ***aquiculture***, ***silviculture*** and ***sylviculture***). The **compounds** ***aquaculture*** and ***aquiculture*** have both common and different meanings.

References

Adams, Valerie. (1987). *An Introduction to Modern English Word-formation*. London – New York: Longman.

Benson, M., Benson, Evelyn & Ilson, R. (1990). *The Combinatory Dictionary of English*. Amsterdam – Philadelphia: John Benjamins Publishing Company.

Carstairs-McCarthy, A. (2002). *An Introduction to English Morphology. Words and Their Structure*. Edinburgh: Edinburgh University Press.

Chalker, Sylvia & Weiner, E. (1994). *The Oxford Dictionary of English Grammar*. London - New York - Sydney - Toronto: BCA.

New Standard Encyclopedia. (1995). Chicago: Standard Educational Corporation.

Perković, Anica, Rață, Georgeta & Petroman, I. (2008). The English of Agriculture: Compounds and Derivatives with *Culture*. *Scientific Papers. Faculty of Agriculture 40 (2)*: 289-292.

Plag, I. (2002). *Word-formation in English*. Cambridge: Cambridge University Press.

Rață, Georgeta & Perković, Anica. (2006). On Combining Forms with *Culture* in the English of Agriculture. *Proceedings. Linguistics and Theory of Literature. Pedagogy and Psychology 45, Rousse, Bulgaria*: 52-54.

Rață, Georgeta & Perković, Anica. (2008). Compounds with ‚Culture': A Semantic Approach. *Proceedings of the 43rd Croatian and 3rd International Symposium on Agriculture, 18-21 February 2008, Opatija, Croatia*: 715-719.

Rață, Georgeta. (2005a). Notes sur le français de l'agriculture: les composés avec *culture*. [Notes on Agricultural French : Compounds with *culture*]. *Lucrări științifice. Seria Horticultură 1 (48)*: 1061-1064.

Rață, Georgeta. (2010). Compuse cu termenul *-cultură* în lexicul românesc al agriculturii. [Compounds with -culture in Romanian Agricultural Vocabulary]. *Agrobuletin AGIR 6*: 15-19.

The Random House Dictionary of the English Language. (1968). New York: Random House.

Webster Encyclopedic Unabridged Dictionary of the English Language. (1996). New York: Gramercy Books.

COMPOUNDS WITH *TREE*

GEORGETA RAȚĂ

Introduction

The word ***tree*** 'a perennial plant having a permanent, woody, self-supporting main stem or trunk, ordinarily growing to a considerable height, and usually developing branches at some distance from the ground' may appear as a **free word** in **hyphemes** (**hyphened compound words** such as ***tree****-bordered*) or **solidemes** (**solid compound words** such as ***tree****top*), or as first or second element in two-word phrases (such as ***tree*** *guard*).

Material and Method

We have inventoried a number of 70 such **compounds** (i.e. entries) of ***tree*** in some very outstanding language dictionaries. We then grouped these combinations according to their type of **composition**.

Results and Discussion

Among the 70 entries inventoried, 32 are **hyphemes**, 1 is a **solideme**, 25 are **two-word phrases**, and 12 are different other **combinations** of ***tree***.

Hyphemes. There is considerable variation and inconsistency in the use of hyphens – the sign (-) used to join semantically or syntactically – in **compounds** (e.g. *coal field*, *coal-field*, *and coalfield*). Basically, hyphens are meant to aid comprehension. One useful convention is to separate with a hyphen vowels that would otherwise be run together in a word (e.g. *co-occur*). Another useful convention is to hyphenate words that would not normally be hyphenated in order to avoid ambiguity (e.g. *a spare room-heater* is not the same as *a spare-room heater*). Hyphens are also useful for showing a close connection between words that might otherwise be understood as separate and equal (e.g. *a black-bearded pilot*) and are normal when a sequence of words is, unusually, used in attributive position before a noun (e.g. *a **tree-bordered** alley*). Are hyphenated 32 **compounds** of ***tree***. In 21 of them, ***tree*** is used attributively together with a present/past participle and the result is an attribute: ***tree-bordered*** ‘bordered with trees’, ***tree-boring*** ‘(about insects) boring trees’, ***tree-clad*** ‘protected by trees’, ***tree-covered*** ‘covered with trees’, ***tree-crowned*** ‘bearing a tree in front’, ***tree-dotted*** ‘dotted with trees’, ***tree-dwelling*** ‘dwelling in trees’, ***tree-feeding*** ‘feeding on trees’, ***tree-fringed*** ‘fringed with trees’, ***tree-garnished*** ‘garnished with trees’, ***tree-girt*** ‘girt by trees’, ***tree-haunting*** ‘haunting in trees’, ***tree-inhabiting*** ‘inhabiting trees’, ***tree-lined*** ‘having a line or lines of trees’, ***tree-locked*** ‘locked by trees’, ***tree-loving*** ‘loving trees’, ***tree-marked*** ‘(of a road, etc.) marked with trees’, ***tree-planted*** ‘having trees planted on it’, ***tree-ripened*** ‘ripened while still on the tree’, ***tree-shaded*** ‘shaded by trees’, and ***tree-skirted*** ‘bordered with trees’; in 11 **compounds**, ***tree*** is still used attributively, but the result is a noun designating an ‘object’, an ‘operation’, or an ‘actor’: ***tree-climbing*** ‘getting to the top of a tree’, ***tree-hewing*** ‘the hewing of trees’, ***tree-holder*** ‘a device for holding trees’, ***tree-hopping*** ‘hopping over a wooden obstacle’, ***tree-nail*** (also, *trenail*, *trunnel*) ‘a wooden peg or nail of dry, hard wood which swells when wet, used for fastening timbers, especially in shipbuilding’, ***tree-planting*** ‘the planting of trees’, ***tree-protector*** ‘a small fence protecting a tree’, ***tree-pruning*** ‘the cutting or lopping superfluous undesired twigs, branches, or roots from’, ***tree-sawing*** ‘the sawing of trees’, ***tree-spraying*** ‘spraying trees to treat or prevent disease’, and ***tree-trimmer*** ‘one who trims trees’.

Solidemes. There is a single solideme in our corpus of **compounds** of ***tree***: ***treetop*** ‘the top of a tree; the uppermost branches of a tree’.

Two-word phrases. There are 25 two-word phrases in our corpus of **compounds** of ***tree***, and 24 of them are nouns denoting ‘objects’ and ‘operations’: ***tree agate*** ‘an agate with ramified traces’, ***tree brier*** ‘an evergreen shrub of southern Europe (*Erica arborea*) about 4 feet high, with white flowers’, ***tree calf*** ‘a calf skin bounding’, ***tree creeper*** ‘any animal creeping trees’, ***tree culture*** ‘the culture of trees’, ***tree dweller*** ‘any animal dwelling in trees’, ***tree farm*** ‘a tree-covered area managed as a business enterprise under a plan of reforestation that makes continuous production of timber possible’, ***tree fern*** ‘any of various ferns (families Cyathacaee and Dicksoniaceae) with large fronds and woody trunks that often attain a treelike size’, ***tree frog*** ‘an arboreal amphibian (Family Hylidae), having the toes dilated with viscous, adhesive disks’, ***tree goose*** (also barnacle) ‘any marine crustacean of the group Cirripedia, usually having a calcareous shell, being either stalked and found attached to ship bottoms and floating timber, or stalk-less and found attached to rocks, especially in the intertidal zone; a thing or person that clings tenaciously', ***tree guard*** 'a small fence for protecting a tree', ***tree heath*** 'tree brier', and ***tree house*** 'a small house, especially one for children to play in, built or placed in the branches of a tree', ***tree insulator*** 'a fence insulating a tree', ***tree ivy*** 'a species of ivy', ***tree kangaroo*** 'any of various kangaroos (genus *Dendrolagus*) of Australia and New Guinea adapted for tree-dwelling', ***tree planter*** 'a person who plants trees', ***tree resin*** 'a resin obtained by distilling wood', ***tree ring*** 'a growth ring', ***tree sparrow*** (also *Canada sparrow*) 'a North American sparrow (*Spizella arborea*) which nests in Canada and migrates southward in winter', ***tree surgeon*** 'one skilled in tree surgery', ***tree surgery*** 'the treatment of disease conditions and decay in trees by operative methods', ***tree tag*** 'a piece of strip of strong paper, etc., attached by one end to a tree as a mark or label', ***tree toad*** 'tree frog', and ***tree trunk*** 'the main stem of a tree; in 1 two-word phrase, ***tree*** is used attributively: ***tree like*** 'having the aspect of a tree'.

Other cases. ***Tree*** occurs in other 12 cases, as either first or second element of a two-word phrase. As first element, ***tree*** is no longer an attribute. It is particularised by the elements following it as in: ***tree of heaven*** (also *ailanthus*) 'a large ornamental tree (*Ailanthus altissima*) of eastern Asia, with large green flowers, those on the male tree being very ill-scented', ***tree of Knowledge*** (of good and evil) '(in the Bible) a tree in Eden whose fruit Adam and Eve were forbidden to eat', ***tree of life*** (also *arborvitae*) (in the Bible) the tree in the garden of Eden whose fruits conferred immortality; a similar tree in heaven'. In ***Tree-Planter-State*** 'nickname of Nebraska', it is part of a two-word phrase used attributively. In the phrases in which ***tree*** appears as second element – be they hyphenated words or two-word phrases – it is a notional noun whose meaning is particularised by the first term used attributively: ***Christmas tree*** 'an evergreen tree decorated at Christmas with ornaments and lights', ***clothes tree*** 'an upright pole with hooks near the top for hanging coats, hats, etc.', ***family tree*** 'a genealogical chart showing the ancestry, descent, and relationship of all members of a family', ***genealogical tree*** 'a family tree', ***rose-tree*** 'a rose bush', ***shade tree*** 'a tree providing shade', ***shoe tree*** 'one of a pair of foot-shaped devices, usually of metal or wood, for placing in a shoe to maintain its shape when it is not being worn'. In ***Christian tree antenna***, it is part of a two-word phrase used attributively.

Conclusions

No matter his/her knowledge in English, a non-native user of English could understand, by only having a short look at a **compound**, and with high precision, that in a hypheme, the basic form (e.g., *tree*) is used attributively and that the result is an attribute or a noun (46%); in a two-word phrase (no hyphen) (36%) and in more-than-two-word phrases (17%), the result is a noun. Therefore, the presence of a hyphen should warn the user on the presence of an attribute or a noun – the distinction is up to the user's skill of identifying present and past participles, in the first case, or nouns, in the second case, while its absence should guarantee the presence of a noun.

References

Adams, Valerie. (1987). *An Introduction to Modern English Word-formation*. London – New York: Longman.

Carstairs-McCarthy, A. (2002). *An Introduction to English Morphology. Words and Their Structure*. Edinburgh: Edinburgh University Press.

Grzega, J. & Schöner, Marion. (2007). *English and General Historical Lexicology. Materials for Onomasiology Seminars*. Katholische Universität Eichstätt-Ingolstadt.

Plag, I. (2002). *Word-formation in English*. Cambridge: Cambridge University Press.

Raţă, Georgeta. (2005). Notes on the Compound Forms of *Tree*. *Proceedings of the 40th Croatian Symposium on Agriculture, February 15-18, 2006, Opatija, Croatia*: 731-733.

—. (2006). *Tree* Compounds and Their Romanian Equivalents. *Romanian Journal of English Studies 3*: 171-176.

Sheehan, M. J. (2000). *Word Parts Dictionary. Standard and Reverse Listings of Prefixes, Suffixes, Roots and Combining Forms*. Jefferson, NC – London: McFarland & Company, Inc., Publishers.

Štekauer, P. & Lieber, Rochelle. (2005). *Handbook of Word-Formation*. Dordrecht: Springer.

The Random House Dictionary of the English Language. (1968). New York: Random House.

Webster Comprehensive Dictionary. (1995). Chicago: J. G. Ferguson Publishing Company.

ENGLISH VERBS OF ANIMAL COMMUNICATION: A POSSIBLE CLASSIFICATION

GEORGETA RAŢĂ
AND ELENA-MIRELA SAMFIRA

Introduction

Verbs of animal communication are, in any language spoken on Earth, verbs denoting the way in which humans perceive sound communication in animals (Raţă 2001). An analysis of their origin points to an evolution (heritage, **word borrowing**, **derivation**, mimicry) specific to each language apart.

Material and Method

We have used, in our corpus and classification of the **English verbs of animal communication** both works in the field of animal communication (Alcock 1089; Raţă 2001) and English language dictionaries.

The method used in the classification of **English verbs of animal communication** was the quantitative one based on the semantic analysis of the terms.

Results and Discussion

We have inventoried in our English language dictionaries 57 **English verbs of animal communication** (we have not taken into account phonetic variants for one and the same verb – e.g., *pur(r)*, *t(h)rill*, etc.) and we have grouped them into two categories – ***verbs of animal communication proper*** and ***unconventional verbs of animal communication***.

1. English verbs of animal communication proper are either *inherited from different periods of formation of the English language or* *borrowed from other languages*, *imitative*, or both.

1.1. ***English verbs of animal communication proper that have been inherited or borrowed*** come from the following periods in the development of European languages (for some of the verbs – such as ***bellow***, ***call***, or ***growl*** – we have considered the last period attested as certain):

- **Vulgar Latin**: ***cry*** 'to utter a characteristic sound or call (used of an animal)';
- **Medieval Latin** or **Old Norse**: ***bawl*** 'to cry or sob loudly; wail';
- **Old English** (before 1150): ***bark*** 'to utter the harsh, abrupt sound of a dog', ***bell*** 'to utter long, deep, resonant sounds; bellow', ***bleat*** 'to utter the characteristic cry of a goat or sheep', ***cluck*** 'to utter the characteristic sound of a hen', ***crow*** 'to utter the shrill cry characteristic of a cock or rooster', ***grunt*** 'to utter a deep, guttural sound, as a hog does', ***low*** 'to utter the sound made by cattle; moo', ***peep*** 'to utter short, soft, high-pitched sounds, like those of a baby bird; cheep', ***roar*** 'to breathe with a rasping sound (used of a horse)', ***sing*** 'to make melodious sounds (*birds singing outside the window*)', ***whistle*** 'to emit a shrill, sharp, high-pitched cry, as some birds and other animals', ***yelp*** 'to utter a short, sharp bark or cry (*excited dogs yelping*)';
- **Middle English** (1150-1500): ***bellow*** 'to make the deep roaring sound characteristic of a bull', ***boom*** 'to make a deep, resonant sound' [mention: imitative of a loud noise], ***burble*** 'to bubble; gurgle', ***call*** 'to utter a characteristic cry (*geese calling in early morning*)', ***chatter*** 'to utter a rapid series of short, inarticulate, speechlike sound (birds chattering in the trees) [mention: of imitative origin'], ***chirp*** 'to make a short, high-pitched sound' [mention: of imitative origin], ***chuck*** 'to make a clucking sound' [mention: of imitative origin], ***croak*** 'to utter a low, hoarse sound' [mention: probably of imitative origin], ***growl*** 'to emit a low, guttural sound or utterance', ***hiss*** 'to make a sharp, sibilant sound' [mention: of imitative origin], ***hoot*** 'to utter the characteristic cry of an owl' [mention: perhaps of imitative origin], ***howl*** 'to utter or emit a long, mournful, plaintive sound (used of dogs)', ***murmur*** 'to make a low, continuous, indistinct sound or succession of sounds', ***neigh*** 'to utter the characteristic sound of a horse; whinny' [mention: probably of imitative origin], ***quack*** 'to utter the characteristic sound of a duck' [mention: of imitative origin], ***rattle*** 'to make or emit a quick succession of short percussive sounds' [mention: probably of imitative origin], ***scream*** 'to utter a long, loud, piercing cry, as from pain or fear; to make a loud, piercing sound', ***snort*** 'to breathe noisily

and forcefully through the nostrils', ***squeal*** 'to utter or produce with a squeal' [mention: probably of imitative origin], ***tinkle*** 'to signal or call by tinkling' [mention: perhaps of imitative origin], ***twitter*** 'to utter a succession of light chirping or tremulous sounds; chirrup' [mention: ultimately of imitative origin], ***utter*** 'to send forth with the voice';

- **Old French** (before 1400): ***bay*** 'to utter a deep, prolonged bark', ***bray*** 'to utter the loud, harsh cry of a donkey';
- **Old North French**: ***warble*** 'to sing with trills, runs, or other melodic embellishments';
- **Germanic**: ***gaggle*** 'to cackle (of geese) [Germanic, of imitative origin]';
- **German**: ***yodel / yodle*** 'to sing so that the voice fluctuates rapidly between the normal chest voice and a falsetto';
- **Italian**: ***thrill / trill*** 'to produce or give forth a trill' [mention: ultimately probably of imitative origin];
- **Scandinavian**: ***wail*** 'to make a prolonged, high-pitched sound suggestive of a cry' [mention: probably of Scandinavian origin].

1.2. ***English verbs of animal communication that are imitative*** or ***onomatopoeic*** are verbs that approximate, by imitation, natural animal or non-animal sounds: ***baa*** 'to make a bleating sound, as a sheep or goat' [mention: imitative], ***cackle*** 'to make the shrill cry characteristic of a hen after laying an egg' [mention: of imitative origin], ***caw*** 'to utter the hoarse, raucous sound characteristic of a crow or similar bird' [mention: imitative], ***cheep*** 'to make a faint, shrill sound or sounds; chirp' [mention: imitative], ***chuckle*** 'to cluck or chuck, as a hen' [mention: probably frequentative of ***chuck***], ***coo*** 'to utter the murmuring sound of a dove or pigeon or a sound resembling it' [mention: imitative], ***gabble*** 'to make rapid, low muttering or quacking sounds, as a goose or duck' [mention: probably frequentative of ***gab***], ***honk*** 'to emit a honk (the sound characteristic of a wild goose)' [mention: imitative], ***meow / mew / miaou / miaow / miaul*** 'to make the crying sound of a cat' [mention: imitative], ***moo*** 'to emit the deep, bellowing sound made by a cow; low' [mention: imitative], ***pur / purr*** 'to make or utter a soft, vibrant sound (*the cat purred*)' [mention: imitative], ***whinny*** 'to neigh, as a horse, especially in a gentle tone', ***yap*** 'to bark sharply or shrill; yelp' [mention: probably imitative].

Two of these verbs are considered 'probably frequentative' (i.e., denoting a repeated action): ***chuckle*** 'to cluck or chuck, as a hen' [mention: probably frequentative of ***chuck***] and ***gabble*** 'to make rapid, low muttering or quacking sounds, as a goose or duck' [mention: probably frequentative of ***gab***].

1.3. Some ***English verbs of animal communication*** could be ranged in either of the two categories above:

- **Middle English** (1150-1500): ***boom*** 'to make a deep, resonant sound' [mention: Middle English, imitative of a loud noise], ***burble*** 'to bubble; gurgle', ***chatter*** 'to utter a rapid series of short, inarticulate, speechlike sound (birds chattering in the trees) [mention: Middle English, of imitative origin'], ***chirp*** 'to make a short, high-pitched sound' [mention: Middle English, of imitative origin], ***chuck*** 'to make a clucking sound' [mention: Middle English, of imitative origin], ***croak*** 'to utter a low, hoarse sound' [mention: Middle English, probably of imitative origin], ***hiss*** 'to make a sharp, sibilant sound' [Middle English, of imitative origin], ***hoot*** 'to utter the characteristic cry of an owl' [mention: Middle English, perhaps of imitative origin], ***neigh*** 'to utter the characteristic sound of a horse; whinny' [mention: Middle English, probably of imitative origin], ***quack*** 'to utter the characteristic sound of a duck' [mention: Middle English, of imitative origin], ***rattle*** 'to make or emit a quick succession of short percussive sounds' [mention: Middle English, probably of imitative origin], ***squeal*** 'to utter or produce with a squeal' [mention: Middle English, probably of imitative origin], ***tinkle*** 'to signal or call by tinkling' [mention: Middle English, perhaps of imitative origin], ***twitter*** 'to utter a succession of light chirping or tremulous sounds; chirrup' [mention: Middle English, ultimately of imitative origin];
- **Italian**: ***thrill / trill*** 'to produce or give forth a trill' [mention: Italian, ultimately probably of imitative origin];
- **Scandinavian**: ***wail*** 'to make a prolonged, high-pitched sound suggestive of a cry' [mention: probably of Scandinavian origin].

2. Unconventional English verbs of animal communication are 'human' or 'inanimate' verbs for which the English language dictionaries do not mention the animal(s) they are attributed to (since they are, normally, reserved to humans or to objects) but that specialists in animal science use frequently (Alcock 1989) – ***burble***, ***call***, ***cry***, ***rattle***, ***scream***, ***tinkle***, ***twitter***, ***utter***, ***wail***, ***warble***, and ***yodel / yodel*** – a phenomenon that also happens in verbs of human communication (Raţă 1997).

Conclusions

Our analysis of the 57 ***English verbs of animal communication*** shows that:

- 34 verbs come from the old Anglo-Saxon fund (Old English and Middle English), which attests an early consolidation of this type of verbs;
- 12 verbs (among which 2 are frequentative) are imitative, which attests the high degree of creativity of the English language;
- 14 verbs could be ranged in either of these two groups;
- 11 verbs are unconventional, which attests, again, the high degree of creativity of the English language.

As a conclusion, we can say that, despite the fact that 59% of the ***English verbs of animal communication*** have an Anglo-Saxon etymon, the rest of the verbs are the result of a recent renewal of the language by either imitation (21%) or borrowing from inventories that have nothing to do with the animal world (20%). (Figure 1-4)

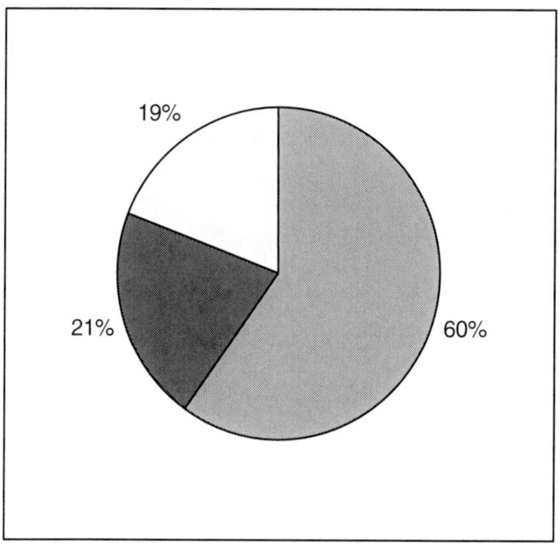

Figure 1-4. Share of the English verbs of animal communication: 60% - Anglo-Saxon etymon; 21% - imitative; 19% - borrowings

References

Adams, Valerie. (1987). *An Introduction to Modern English Word-formation*. London – New York: Longman.

Alcock, J. (1989). *Animal Behaviour: An Evolutionary Approach*. Sunderland, MA: Sinauer Associates, Inc.

Carstairs-McCarthy, A. (2002). *An Introduction to English Morphology. Words and Their Structure*. Edinburgh: Edinburgh University Press.

Dicţionarul explicativ al limbii române. [The Explanatory Dictionary of the Romanian Language]. (1998). Bucureşti: Editura Univers Enciclopedic.

Grzega, J. & Schöner, Marion. (2007). *English and General Historical Lexicology. Materials for Onomasiology Seminars*. Katholische Universität Eichstätt-Ingolstadt.

Plag, I. (2002). *Word-formation in English*. Cambridge: Cambridge University Press.

Raţă, Georgeta. (1997). Les verbes de communication non-conventionnels. Essai de classification [Unconventional Verbs of Communication. A Tentative Classification]. *Studia Universitatis Babeş-Bolyai. Philologia 4:* 67-71.

—. (2001). *Contribuţii la teoria comunicării*. [Contributions to the Theory of Communication]. Timişoara: Editura Mirton.

—. (2002). Verbele de comunicare animală în limba engleză. Încercare de clasificare. [Verbs of Animal Communication in English. A Tentative Classification]. *Lucrări ştiinţifice. Agricultură XXXIII*: 299-302.

Sheehan, M. J. (2000). *Word Parts Dictionary. Standard and Reverse Listings of Prefixes, Suffixes, Roots and Combining Forms*. Jefferson, NC – London: McFarland & Company, Inc., Publishers.

Štekauer, P. & Lieber, Rochelle. (2005). *Handbook of Word-Formation*. Dordrecht: Springer.

The American Heritage Dictionary of the English Language. (2008). Boston: Houghton Mifflin.

The Random House Dictionary of the English Language. (1968). New York: Random House.

Webster Comprehensive Dictionary. (1995). Chicago: J. G. Ferguson Publishing Company.

CHAPTER TWO

SYNTAX

AGRICULTURAL ENTOMOLOGY: *-ING* FORMS

GEORGETA RAȚĂ

Introduction

Teachers of English for Special Purposes are very much interested in finding teaching and learning algorithms that facilitate the study of this type of English. Thus, the learning of *-ing* words could be rendered simpler if students were aware of the fact that these words make up a relatively compact group of words.

Material and Methods

The method we have used here is quantitative.

We started from the hypothesis that the different *-ing* forms must share certain patterns, which are the result of a long process of creating words and/or phrases and of assimilating them.

Our goal was to inventory all the entries of the *-ing* type in Gordth and Hendrick's *Dictionary of Entomology* (2001) and see if they belong or not to the same category and if our findings could be of any help for our students (i.e. help them better and easier understand and learn the English terminology in the field of agricultural entomology).

We then compared the 3 basic types of **noun like uses of the *-ing* forms** (Chalker & Weiner 1994) and drew a few conclusions.

Results

We have inventoried a number of 20 entries in *-ing* in Gordth and Hendrick's dictionary (2001).

Below are these **noun like uses of the *-ing* forms** together with the definitions supplied by both Gordth and Hendrick's *Dictionary of Entomology* (2001): ***absconding*** n. 'Social insects: colony multiplication by the departure of a queen and a small group of workers from a larger colony'; ***alluring*** 'pertaining to the capacity to attract, entice, or tempt';

budding n. '1. Absconding. 2. Agamic reproduction by aphids.', ***catfacing*** n. 'injury to a plant which results from the feeding or piercing-sucking insects on developing fruit'; ***clearing*** 'replacing absolute ethanol in specimens being prepared for microscopical study'; ***converging*** 'approaching each other near the apex'; ***countershading*** n. 'a physical condition and aspect of mimicry in which an animal is dark dorsally and pale ventrally'; ***crosshatching*** n. 'hatching'; ***flaring*** 'expanding like the bell of a trumpet'; ***hatching***$_1$ n. 'surface marks, lines or scratches which are numerous, short and close-set'; ***hatching***$_2$ n. 'the breaking of the eggshell by an insect embryo during the process of emergence'; ***hearing*** n. 'the capacity of ability of perceiving sound, or vibrations in the air or substrate'; ***imprinting*** n. 'a form of learning in which a newborn recognises that an object has a particular function'; ***infringing*** 'encroaching upon'; ***learning*** n. 'a persistent change in insect behaviour caused by experience of association (conditioning), imprinting and habituation'; ***moulting*** n. 'a periodic process of loosening and discarding the cuticle, accompanied by formation of a new cuticle, and often by structural changes in body wall and other organ'; ***provisioning*** n. 'Solitary bees and wasps: the practice of feeding the larvae in open cells throughout larval development'; ***sembling*** 'assembling'; ***swarming*** n. '1. Social insects: the concerted, organised departure from a hive or nest by workers accompanied by reproductives. Collectively, the individuals form the nucleus of a new colony. 2. Solitary insects: aggregations of conspecific individuals characterised by periodic flight within a small region'; ***urticating*** 'nettling, causing a stinging or burning sensation of the skin'.

Discussion

Though 5 **noun like uses of the -*ing* forms** are not mentioned by English language dictionaries, they can be traced upon in their base verb: ***absconding*** < *to abscond*, ***converging*** < *to converge*, ***infringing*** < *to infringe*, ***provisioning*** < *to provision*, and ***swarming*** < *to swarm*.

Thirteen of the 20 entries (65%) are nouns – ***absconding***, ***budding***, ***catfacing***, ***countershading***, ***crosshatching***, ***hatching***$_1$, ***hatching***$_2$, ***hearing***, ***imprinting***, ***learning***, ***moulting***, ***provisioning***, and ***swarming***, while 7 (35%) are adjectives – ***alluring***, ***clearing***, ***converging***, ***flaring***, ***infringing***, ***sembling***, and ***urticating*** (Figure 2-1).

Contrary to the authors' mentions, the **-*ing* forms** ***hatching***$_2$, ***infringing***, ***provisioning***, and ***sembling*** are not verbs, and ***clearing*** is not an adjective, since **-*ing* forms** can be (*Webster Encyclopedic Unabridged Dictionary of the English Language*):

- nouns formed from verbs, expressing the actions of the verb or its result, product, material, etc. (*the art of* ***building***, *a new* ***building***, *cotton* ***wadding***) or nouns from words other than verbs (***offing***, ***shirting***);
- present participles of verbs, such participles being often used as participial adjectives (***warring*** *factions*);
- nouns meaning 'one belonging to', 'of the kind of', 'one descended from', or diminutive nouns (***bunting***, ***farthing***, ***gelding***, ***shilling***, and ***whiting***).

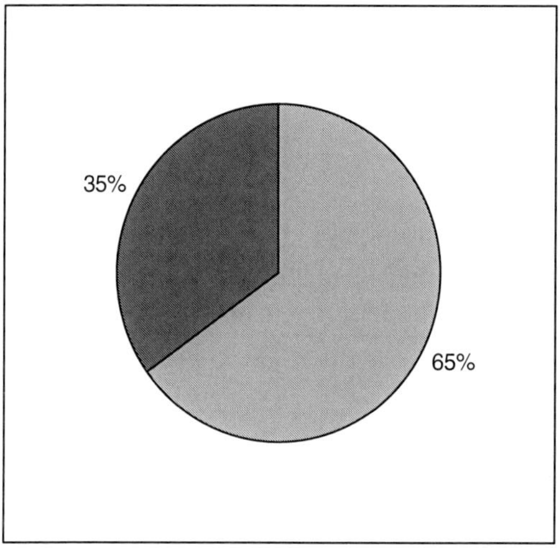

Figure 2-1. Distribution of *-ing* nouns and adjectives in agricultural entomology: 65% nouns, 35% adjectives

Conclusion

The terms not mentioned by English language dictionaries (***absconding***, ***converging***, ***infringing***, ***provisioning***, and ***swarming***) and the error of ascertaining grammatical values (***hatching***$_2$, ***infringing***, ***provisioning***, and ***sembling*** are erroneously considered verbs, and ***clearing*** is erroneously considered adjective) allow us to draw the conclusion that any specialised dictionary should be the collective work of a specialist (or of several specialists) in a certain field of knowledge and of a linguists (or of several linguists).

As far as professors, researchers, or students in agriculture, animal husbandry, biology, horticulture, etc. are concerned, we think that they should be warned against such misunderstandings and motivated to participate in the hard work of clarifying terminologies.

References

Chalker, Sylvia & Weiner, E. (1994). *The Oxford Dictionary of English Grammar*. London - New York - Sydney - Toronto: BCA.

Gordth, G. & Hendrick, D. H. (2001). *A Dictionary of Entomology*. Oxon: CABI Publishing.

Miller, J. (2002). *An Introduction to English Syntax*. Edinburgh: Edinburgh University Press.

Rață, Georgeta. (2004). Agricultural Entomology: *-ing* forms. *4th International Scientific Days of Land Management in the Great Hungarian Plain, October 21-22, 2004, Mezőtúr, Hungary*: 1-5.

Webster Encyclopedic Unabridged Dictionary of the English Language. (1996). New York: Gramercy Books.

'VERBAL N + N' COMPOUNDS IN AGRICULTURAL ENTOMOLOGY

GEORGETA RAȚĂ
AND LAURA-CONSTANTINA MICU

Introduction

Verbal nouns in *-ing* are often used attributively (*the* ***printing*** *trade*) and in forming **compounds** (***drinking*** *song*). In some **compounds** (***sewing*** *machine*), the first element might reasonably be regarded as the participial adjective, *-ing* (a suffix forming the present participle of verbs), the **compound** thus meaning 'a machine that sews', but is commonly taken as a verbal noun, the **compound** being explained as 'a machine for sewing'.

Material and Methods

We have inventoried a number of 79 'verbal N + N' **compounds** containing participial adjectives, verbal nouns, or having both values.

Results and Discussion

1. The following 58 **compounds** contain *-ing* forms that are participial adjectives, the **compounds** thus meaning 'a something that does something': ***advancing coloration*** 'any colour (e.g. yellow) or combination of colours that appears nearer the observer than other colours viewed in the same plane'; ***amplifying host*** 'a vertebrate host that is infected with pathogens at a level sufficiently high for a blood-feeding arthropod vector to be infected'; ***baiting sheet*** 'a device for collecting sedentary insects in an arboreal habitat [...] held beneath the branch or shrub which is struck sharply with a stick, insects fall onto the sheet and are collected with moistened brush, forceps or aspirator'; ***burrowing bugs*** 'a widespread, rather small Cydnidae family of heteropterous Hemiptera closely linked to stink bugs [whose] adults and nymphs [...] burrow into the soil where they feed upon roots'; ***burrowing cockroaches*** 'members of

the blaberid Genera *Macropanesthia* and *Geoscapheus* in Australia'; ***burrowing wasp*** '*Philanthus ventilabris* Fabricius [Hymenoptera: Sphecidae]: they are digging species'; ***burying beetles*** 'polyphagous Coleoptera assigned to Staphylinoidea and closely related to Staphylinidae'; ***clothing hairs*** 'the setae (chaetae) that invest (cover) the surface of the insect body or its appendages'; ***covering gall*** 'plant galls characterised by the gall inducer first acting externally to produce hypertrophy of surrounding tissue that quickly results in the inducer becoming enclosed within the gall'; ***creeping myasis*** 'a type of myasis [...] caused by species of bot flies that usually infest domesticated animals [...]. First-instar larvae of the bot fly penetrate human skin and burrow beneath surface [...]'; ***creeping waterbugs*** 'Family Naucoridae of heteropterous Hemiptera'; ***damping off*** 'the symptom of a fungal infection in plants characterised by the collapse and subsequent death of seedlings following the formation of stem lesions at soil level'; ***darning needle*** 'insect of the paleopterous Order Odonate [whose] naiads display elongate, scoop-like, prehensible labium which can be rapidly extended forward to grasp prey'; ***dayfeeding armyworm*** 'an insect of the cosmopolitan family of ditrysian Lepidoptera [whose adults are] nocturnal'; ***depluming mite*** '*Knemidokoptes gallinae* (Raillet) [Acas: Sarcoptidae]. *Neocnemidocoptes gallinae* (Raillet [Acari: Knemidokoptidae]. (Australia)'; ***devastating grasshopper*** '*Melanoplus devastator* Scudder [Orthoptera: Acrididae]'; ***dimpling bug*** '1. *Campylomma liebknechti* (Girault) [Hemiptera: Miridae]: a pest of apples [whose] bugs are attracted to apple blossoms and migrate into orchards during warm weather where they feed on developing fruit and cause dimpling of fruit which results in downgrading'; ***diving flight*** 'a form of passive flight in which the insect engages in a fast descent at a large angle between the flight direction and the horizontal'; ***flanking seta*** 'Psocoptera: a seta near the hyaline cone of the distal paraproctal margin'; ***flapping flight*** 'one of the two basic functional forms of insect flight distinguished by wingbeats that create active movement through air'; ***fluctuating variations*** 'the neo-Darwinian concept of unstable changes in the characteristics of a species, that appear or disappear from generation to generation, or which depend on genetic or somatic influences conditioned by known or unknown factors'; ***founding stage*** 'social insects: the earliest period in colony development, typified by the queen rearing the first brood'; ***fruiting body*** 'fungi: a general term for spore-bearing structures'; ***gliding flight*** 'a form of passive flight which follows full speed in fast flapping flight during which the insect engages prolonged forward movement without beating its wings'; ***glistening blue*** '*Sahuna scintillate*

(Lycas) [Lepidoptera: Lycaenidae]. (Australia)'; ***hovering flight***; ***hunting billbug*** '*Sphenophorus venatus vestitus* Chittenden [Coleoptera: Curculionidae]'; ***jumping bean*** 'seeds of Euphorbiaceae (*Sebastiania*, *Sapium*) which move or jump owing to the movement of insect larvae within the seeds'; ***jumping bristletails*** 'primitive Family Machilidae of Archaeognatha: Machilids are active and jump when disturbed'; ***jumping plantlice*** 'Family Psyllidae of the superfamily Psylloidea'; ***jumping spiders*** 'small to moderate-sized spiders of the Family Salticidae with large anteromedial eyes, good vision and the capacity to jump'; ***jumping tree bugs*** 'Heteropterous Hemiptera assigned to Miroidea and sometimes recognised as a subfamily of Miridae'; ***leaf feeding*** 'a specialised type of herbivory that is widespread among many groups of insects'; ***lightning bugs*** 'polyphagous Coleoptera assigned to Elateroidea: adults with luminescent organs on female abdominal sternum 5 and male abdominal sterna 5 or 6'; ***louping ill*** 'a tick-borne arboviral disease closely related to Central European Encephalitis: typically a disease of sheep in Scotland and Ireland that rarely occurs in humans'; ***measuring worms*** 'ditrysian Lepidoptera assigned to superfamily Geometroidea'; ***moulting fluid / liquid*** 'a fluid secreted abundantly by growing insects in the act of moulting, by dermal glands of the body surface'; ***parachuting flight*** 'a form of passive flight in which the insect engages in a slow, vertical descent'; ***relapsing fevers*** 'a group of diseases caused by spirochetes of the Genus Borrelia and vectored principally by argasid ticks […]'; ***sapling borer*** '*Sahyadrassus malabaricus* (Moore) [Lepidoptera: Hepialidae]'; ***scattering colour*** 'structural colour that is produced from the scattering of light from a surface, such as irregular and diffuse light reflected from ground glass, or the scattering of light within the tegument'; ***sleeping sickness*** 'an often fatal human disease caused by the parasitic protozoan *Trypanosoma brucei* and vectored by tsetse flies *Glossina* spp. [Diptera: Glossinidae]'; ***snapping beetles*** 'a species related to Lampyridae: posterolateral pronotal process corresponding to an anterolateral mesosternal sulcus that collectively form a click mechanism'; ***soaring flight*** 'a form of passive flight in which the insect engages ascending air currents'; ***spinning glands*** 'Arachnids; Lepidoptera larvae; Psocidae: the glands which produce the viscid secretion which forms the silk'; ***spiralling whitefly*** '*Aleurodicus disperses* Russell [Hemiptera: Aleyrodidae]: a pest of numerous fruits and vegetables in Central America, islands of the Pacific, Australia and New Guinea; common name from the habit of female laying eggs in a spiral pattern'; ***spitting spider*** '*Scytodes* sp. [Araneida: Scytodidae] (Australia)'; ***sucking lice*** 'lice of the Order Anoplura that complete three nymphal instars that feed by sucking blood from the capillaries within the skin of their hosts [...]'; ***swarming leaf-beetle*** '*Rhyaprida* spp. [Coleoptera: Chrysomelidae]: Australian species; adults prefer new leaves and young shoots and cause severe defoliation […]'; ***urticating anthelid*** '*Anthela nicothoe* (Boisduval) [Lepidoptera: Anthelidae] (Australia)'; ***urticating hairs*** 'in some caterpillars and adult insects, setae (chaetae) connected with dermal glands, through which the venom passes; barbed setae which cause discomfort presumably induced by mechanical irritation'; ***warning coloration*** 'conspicuous colours or patterns of colour which are frequently associated with qualities which render their possessor unpalatable, offensive or dangerous to predators'; ***webbing clothes-moth*** '*Tineola bisselliella* (Hummel) [Lepidoptera: Tineidae]: a cosmopolitan pest endemic in Africa: larva white with brown head capsule; spins silk webbing and forms feeding tube […]'; ***webbing coneworm*** '*Dioryctria disclusa* Heinrich [Lepidoptera: Pyralidae]'; ***whirling beetles*** 'aquatic, adephagous Coleoptera: adults swim on surface of water in tight circles aided by pygidial gland surfactant'; ***whirling mites*** 'Anystidae'; ***whistling spiders*** 'Selenocosmia spp. [Araneida: Theraphosidae] (Australia)'; ***wrapping flexure*** 'in Eucilidae (Hymenoptera): a form of wing bending in which the apical portion of the wing, in the flexed position, is notched with the region on either side of the notch wrapped downward over the dorsum of the metasoma'.

For 8 of these **compounds**, the dictionary supplies no information whatsoever allowing the establishment of a link between common name and features. Therefore, we can assume that the common name "describes" the insect and/or its behaviour: ***burrowing cockroaches***, ***burrowing wasp***, ***burying beetles***, ***jumping plantlice***, ***jumping tree bugs***, ***measuring worms***, ***spitting spider***, and ***webbing coneworm***. Similarly, we can also assume the same thing for 2 **compounds** mentioned in the dictionary, but without a special entry: ***hovering flight*** and ***whirling mites***.

2. The following 18 **compounds** contain *-ing* forms that are verbal nouns, the **compounds** thus meaning 'something for doing something': ***alluring coloration*** 'any colour, colour pattern or combination of colours that are attractive to some species of insects (prey) and displayed (used) by predaceous species as part of their feeding strategy: prey are attracted to predator with alluring coloration'; ***alluring glands*** 'glandular structures (frequently found on males) that diffuse odours attractive to the opposite sex'; ***breathing pore*** ' a pore, hole or aperture in the integument which serves as an adaptation to permit gas exchange between the body and the environment; [the] air passes through breathing pores, enters the atrium, then into tracheae and subsequently into tracheoles; CO_2 passes from tissue to tracheoles, then tracheae and exit *via* breathing pores'; ***collecting***

basket 'an arrangement of setae, bristles and/or spines on the forelegs of certain insects, in which they collect or hold food while devouring (Needham)'; ***feeding strategy*** 'the complex of anatomical, behavioural and physiological adaptations that are responsible for the acquisition of food and processing of nutrients by an organism'; ***feeding tube*** 'a cylindrical device created by adult female parasitic Hymenoptera [...] used to feed upon a host's ahemolymph in concealed habitats or places in which the body of the host cannot be contacted by mouthparts of the adult parasite'; ***flying-hairs*** 'very long, slender surface setae whose bases are set in cuticular punctures (pits)'; ***gathering hairs*** 'the flexible, flattened and often hookes setae that occur on the tongue of bees and other Hymenoptera'; ***jumping mechanism*** 'biomechanical devices which facilitate jumping by insects'; ***moulting glands*** 'epidermal glands that secrete moulting fluid which facilitates moulting'; ***moulting hormone*** 'a hormone produced by the prothoracic glands upon stimulation by prothoracicotropic hormones: its target tissue is the cellular layer of the integument, and the action is the initiation of apolysis: cell division, cell elongation and production of moulting fluid'; ***pulsating membranes*** 'small, muscular membranes found in the thorax, head, and appendages of various insects, the rhythmic contractions of which probably contribute to the circulation of the blood'; ***spinning glands*** 'arachnids [whose] glands produce the viscid secretion which forms the silk'; ***sucking pump*** 'the highly modified cibarium of insects which ingest liquid food (sap, blood, etc.)'; ***sucking spears*** 'the specialised mandibles and maxillae of hemerobiid larva, used for puncturing prey and sucking its juice'; ***sucking stomach*** 'a thin-walled muscular pouch connected with the end of the oesophagus; a food reservoir not commonly present except in some Lepidoptera'; ***swimming paddles*** 'terminal appendages of mosquito pupae'; ***triturating basket*** '1. Adapted for grinding or crushing. 2. A cuticular network in the foregut of bees adapted for filtering particulate material from nectar and other liquid nutrients'.

3. Other 3 **compounds** could be ranged in both groups: ***reducing sugar*** 'any sugar which contains the radical CO, which when heated with Fehling's solution (blue) reduces the copper compound in solution so that cuprous oxide (red) is formed'; ***spinning bristle*** 'Embiidae: a long hollow bristle on the plantar surface of the first and second tarsal segments of the fore leg, that occur in numbers and issue silk threads developed in spinning glands'; ***swarming flight*** 'a behavioural category for insect flight [in which] males gather in relatively high concentrations in flight and maintain a stationary position'.

Conclusion

Participial adjective **compounds** ('a something that does something') are the most numerous (73%), while participial adjectives + verbal nouns ('something for doing something') are sensibly less (23%). (Figure 2-2)

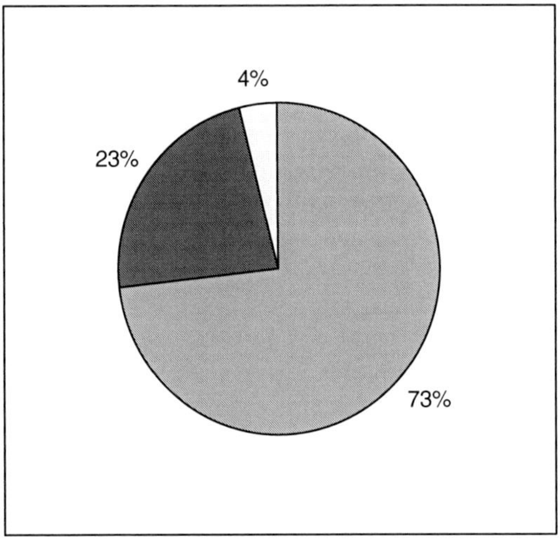

Figure 2-2. 'Verbal N + N' Compounds in Agricultural Entomology: 73% Participial Adjectives; 23% Verbal Nouns; 4% Participial Adjectives/Verbal Nouns.

References

Gordth, G. & Hendrick, D. H. (2001). *A Dictionary of Entomology.* Oxon: CABI Publishing.

Rață, Georgeta & Micu, Laura-Constantina. (2004). Verbal "Noun + Noun" Compounds in Agricultural Entomology. *Tradiție și modernitate în științele umaniste și sociale*. Timişoara: Editura Mirton. 121-126.

Webster Comprehensive Dictionary. (1995). Chicago: J. G. Ferguson Publishing Company.

Webster Encyclopedic Unabridged Dictionary of the English Language. (1996). New York: Gramercy Books.

'N + V-*ING* + N' Compounds in Agricultural Entomology

Anica Perković and Georgeta Rață

Introduction

Almost 15 years ago, Kennedy (1998) stated that **corpus linguistics** had "a tendency sometimes to focus on lexis and lexical grammar rather than pure syntax". The present paper shows that this is not the case of our research.

The purpose of the research was to show that **compound** nouns designating insect pests in English and having been built after the same pattern - 'Noun + Verbal nouns in *-ing* + Noun' - have a common general meaning, i.e. 'pest doing something'.

Our background information consisted of repeated observations of the fact that such insect pest compound names usually point to both 'the fact that a particular pest does something', and to the object of the action thus performed.

We used as sources of information Gordth & Hendrick's *Dictionary of Entomology* (Gordth & Hendrick 2001) for the inventory of insect pests and Chalker & Weiner's *Oxford Dictionary of English Grammar* (1994) as well as the *American Heritage Dictionary of the English Language* (2008) to define linguistic terms.

Material and Methods

We have inventoried a number of 57 insect pest compound names of the 'Noun + Verbal nouns in *-ing* + Noun' pattern, which we then analysed from the point of view of their meaning based on the following linguistic background:

- nouns ending in *-ing* are derived from verbs and express the action of the verb (*the art of building*) or its result (*a new building*), product, material (*cotton wadding*), etc. (Chalker & Weiner 1994);

- verbal nouns ending in *-ing* are often used attributively (*the printing trade*) and in **composition** (*drinking song*) (Chalker & Weiner 1994); that sews', but it is commonly taken as a verbal noun, the **compound** being explained as 'a machine for sewing.' (Chalker & Weiner 1994).

Results and Discussion

First, in our analysis of the insect pest compound names we focussed on 'the type of action' the pests perform. Thus, insect pests (Gordth & Hendrick 2001):

- ***bite*** (5 occurrences) *cattle* (1 occurrence): ***cattle-biting louse*** '*Bovicola bovis* (Linnaeus): Attacks base of tail, withers and shoulders', *dogs* (1 occurrence): ***dog-biting louse*** '*Trichodectes canis* (DeGeer): A pest of puppies', *goats* (1 occurrence): ***goat-biting louse*** '*Bovicola caprae* (Gurit)', *horses* (1 occurrence): ***horse-biting louse*** '*Bovicola equi* (Denny): Attacks host around neck and base of tail; *Werneckiella equi* (Denny)', and *sheep* (1 occurrence): ***sheep-biting louse*** '*Bovicola ovis* (Schrank): A pest of sheep';
- ***bore*** (2 occurrences) *shoots* (1 occurrence): ***balsam shoot-boring sawfly*** '*Pleroneura brunneicornis* (Tohwer)' and *wood* (1 occurrence): ***metallic wood-boring beetles*** 'Polyphagous Coleoptera: Larvae bore wood';
- ***cast*** (1 occurrence) *nets*: ***net-casting spiders*** 'Deinopids: When prey approach the spider, the spider's web is cast and ensnare the prey';
- ***cut*** (4 occurrences) *leaves* (4 occurrences): ***leaf-cutting ants*** '*Acromyrmex* sp. and *Atta* sp.', ***leaf-cutting bees*** 'Aculeate Hymenoptera nest in cells lined with leaf material cut from living plants', ***pale leaf-cutting bee*** '*Megachile concinna* (Smith)', ***Texas leaf-cutting ant*** '*Atta texana* (Buckley): Workers take leaves from plants to nest, malaxate them for substrate of fungus garden';
- ***eat*** (12 occurrences) *leaves* (6 occurrences): ***citrus leaf-eating cricket*** '*Tamborina australis* (Walker): Nymphs and adults feed nocturnally on foliage chewing margin of leaves, grazing on surface or chewing holes in leaves', ***citrus leaf-eating weevil*** '*Eutinophaea bicristata* (Lea): Adults feed on young foliage on the lower third of the tree; chew out sections on both leaf surfaces. Larvae feed on roots', ***coconut leaf-eating caterpillar*** '*Opisina arenosella* (Walker)', ***hairy leaf-eating caterpillar*** '*Xanthodes congenital* (Hampson)', ***large leaf-eating ladybird*** '*Epilachna guttatopustulata* (Fabricius): Adults and larvae eat leaf surface', ***lucerne leaf-eating beetle*** '*Colaspoides*

foveiventris (Lea)', *mites* (1 occurrence): ***mite-eating ladybirds*** '*Stethorus* spp.: Adults and larvae are important predators of various mite species', *pitches* (1 occurrence): ***pitch-eating weevil*** '*Pachylobius picivorus* (Germar)', *roots* (2 occurrences): ***root eating beetles*** 'Rhizophagidae: Biology poorly studied. Species predaceous on bark beetles, feeding on eggs and larvae of several species of scolytids. One species reported as carrion feeder', ***root eating flies*** 'Muscoid Diptera: Phytophagous habits' and *scales* (2 occurrences): ***scale-eating caterpillar*** '*Catoblemma dubia* (Butler)', ***scale-eating ladybird*** '*Rhyzobius lophanthae* (Blaisdell)';

- ***feed*** (1 occurrence) *on sap*: ***sap-feeding beetles*** 'Polyphagous Coleoptera: Phytophagous species';
- ***fold*** (1 occurrence) *leaves* (1 occurrence): ***lantana leaf-folding caterpillar*** '*Anania haemorrhoidalis* (Guenée): Larvae roll leaves';
- ***gnaw*** (1 occurrence) *bark* (1 occurrence): ***bark gnawing beetles*** 'Coleoptera: Found beneath bark in galleries';
- ***harvest*** (1 occurrence) *seeds* (1 occurrence): ***seed-harvesting ant*** '*Pheidole ampla* (Forel); *Pheidole anthracina* (Forel)';
- ***love*** (2 occurrences) *mud* (2 occurrences): ***minute mud-loving beetles*** 'Polyphagous Coleoptera: Adults inhabit mud', ***variegated mud loving beetles*** 'Polyphagous Coleoptera: Adults and larvae inhabit mud';
- ***make*** (2 occurrences) *cases* (1 occurrence): ***case-making clothes-moth*** 'A complex of species resembling *Tinea pellionella* (Linnaeus)' and *tubes* (1 occurrence): ***tube-making caddies-flies*** 'Trichoptera';
- ***mimic*** (1 occurrence) *ants* (1 occurrence): ***green-tree-ant mimicking spider*** '*Amyciaea albomaculata* (Cambridge)';
- ***mine*** (2 occurrences) *leaves* (2 occurrences): ***lantana leaf-mining beetle*** '*Octotoma scabripennis* (Guérin-Méneville): Larvae develop mine near main vein of leaf', ***lantana leaf-mining fly*** '*Calcomyza lantanae* (Frick): Larvae form blotch mines leaves';
- ***pierce*** (1 occurrence) *fruit* (1 occurrence): ***fruit-piercing moth*** '*Othreis fullonia* (Clerck): Adults use their proboscis to pierce rind of developing fruit';
- ***poison*** (1 occurrence) *cattle* (1 occurrence): ***cattle-poisoning sawfly*** '*Lophyrotoma interrupta* (Klug): Cattle eat mature larvae which accumulate at base of iron-bark trees before pupation';
- ***roll*** (4 occurrences) *balls* (1 occurrence): ***African ball-rolling dung beetle*** '*Kheper nigroaeneus* (Boheman): A large bodied, diurnal, ball-rolling dung beetle') and *leaves* (3 occurrences): ***leaf-rolling crickets*** 'Wingless long-horned grasshoppers (Gryllacrididae) conceal themselves in rolled leaves during the day', ***leaf-rolling sawflies***

'Holartic Symphyta', ***leaf-rolling weevils*** 'Polyphagous Coleoptera: Females roll leaves to form balls in which eggs are concealed';

- ***silver*** (1 occurrence) *bananas* (1 occurrence): ***banana silvering-thrips*** '*Hercinothrips bicinctus* (Bagnall)';
- ***spin*** (4 occurrences) *nets* (1 occurrence): ***net-spinning caddis-flies*** 'Hydropsychids construct a small cup-shaped net oriented to catch prey items' and *webs* (3 occurrences): ***plum web-spinning sawfly*** '*Neurotoma inconspicua* (Norton)', ***web-spinning sawflies*** 'Holartic Symphyta', ***web-spinning tree crickets*** 'Wingless long-horned grasshoppers';
- ***spot*** (2 occurrences) *bananas* (1 occurrence): ***banana spotting-bug*** '*Amplypelta lutescens lutescens* (Distant)' and *fruit* (1 occurrence): ***fruit-spotting bug*** '*Amblypelta nitida* (Stál): Damages most fruits and macadamia nuts by feeding activities';
- ***suck*** (5 occurrences) *blood* (1 occurrence): ***blood-sucking conenose*** '*Triatoma sanguisuga* (LeConte): A bloodsucking parasite', *dogs* (1 occurrence): ***dog sucking-louse*** '*Linognathus setosus* (Olfers)', *goats* (1 occurrence): ***goat-sucking louse*** '*Linognathus stenopsis* (Burmeister)', *horses* (1 occurrence): ***horse-sucking louse*** '*Haematopinus asini* (Linnaeus)' and *sheep* (1 occurrence): ***sheep sucking body-louse*** '*Linognathus ovillus*: A blood-sucking parasite of sheep.';
- ***weave*** (3 occurrences) *orbs* (3 occurrences): ***garden orb-weaving spider*** '*Eriophora biapicata* (L. Koch); *Eriophora transmarine* (Keyserling)', ***long-tailed orb-weaving spider*** '*Argiope protensa* (L. Koch)', ***orb-weaving spiders*** 'Spiders build large, planar webs of orb-shape';
- ***web*** (1 occurrence) *leaves* (1 occurrence): ***groundsel-bush leaf-webbing caterpillar*** '*Aristotelia* sp.'.

It is interesting to see that the number of verbs denoting "actions" of the pests is smaller compared to the number of the "objects" on which they act.

It is very interesting to notice the great difference in distribution of the 'objects' on which the pests mentioned above 'act'. Thus, if the number of occurrences of 'objects' on which insect pests perform an action is low:

- 1 occurrence: *ant*, *balls*, *blood*, *bark*, *cases*, *mites*, *pitches*, *sap*, *seeds*, *shoots*, *tubes*, and *wood*,
- 2 occurrences: *bananas*, *cattle*, *dogs*, *fruit*, *goats*, *horses*, *mud*, *nets*, *roots*, *scales*, and *sheep*,

- 3 occurrences: *orbs* and *webs*, there are 17 occurrences of the most important part of the plant – the *leaf* (Figure 2-3).

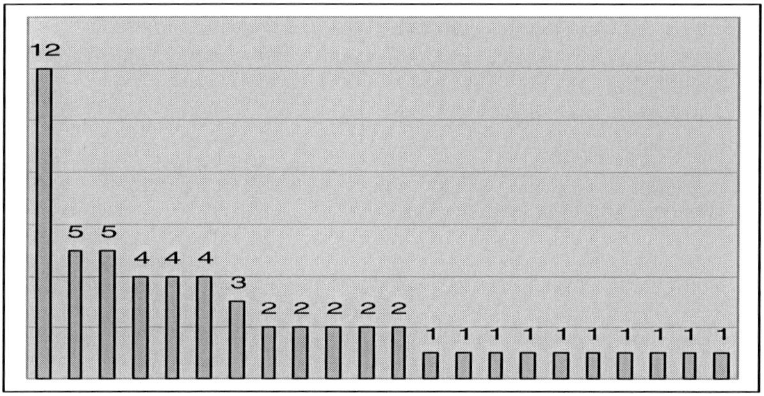

Figure 2-3. Number of occurrences of the verbs used attributively in pest nouns: 12 *eat*; 5 *bite* and *suck*; 4 *cut*, *roll*, and *spin*; 3 *weave*; 2 *bore*, *love*, *make*, *mine*, and *spot*; 1 *cast*, *feed*, *fold*, *gnaw*, *harvest*, *mimic*, *pierce*, *poison*, *silver*, and *web*

Conclusion

If there are lots of nouns of pests in Agricultural Zoology, whose names include participial adjectives of the V-*ing* form, they have nevertheless something in particular: they have always the same meaning ('an insect that does something'). In addition, the **derivatives** thus formed not only indicate 'the fact that a particular pest does something', it also indicates the object of the action thus performed – a pattern easy to understand and, therefore, to learn.

References

Chalker, Sylvia & Weiner, E. (1994). *The Oxford Dictionary of English Grammar*. London - New York - Sydney - Toronto: BCA.

Gordth, G. & Hendrick, D. H. (2001). *A Dictionary of Entomology.* Oxon: CABI Publishing.

Kennedy, G. (1998). *Introduction to Corpus Linguistics*. Harlow: Longmann.

Miller, J. (2002). *An Introduction to English Syntax*. Edinburgh: Edinburgh University Press.

Rață, Georgeta. (2004). Agricultural Entomology: *ing* forms. *4th International Scientific Days of Land Management in the Great Hungarian Plain, October 21-22, 2004, Mezötúr, Hungary*: 1-5.

Perković, Anica & Rață, Georgeta. (2010). Agricultural Zoology: Names of Pests of the '*N + v-ing + N*' Pattern. *The International Language Conference on the Importance of Learning Professional Foreign Languages for Communication between Cultures 2010*. Online: http://fl.uni-mb.si/.

The American Heritage Dictionary of the English Language. (2008). Boston: Houghton Mifflin.

Chapter Three

Lexis

PRECISION AGRICULTURE TERMINOLOGY

GEORGETA RAŢĂ, FLORIN SALA
AND ANICA PERKOVIĆ

Introduction

Gaining a general introduction to Precision Agriculture and to its underlying philosophy as a main objective in learning about it asks, among others, for gaining a general understanding of the Global Positioning System (GPS) terminology. But how difficult is this? Are language dictionaries of any help in approaching this special terminology or do we need special linguistic tools to do it? Our expertise as teachers of English for Special Purposes shows that things are rather blurred in this area, as it has always happened when new scientific or technical branches immerged. However, as it is often the case, there must be a small number of terms belonging to the general vocabulary, and a very large one belonging only to GPS terminology.

Material and Method

To test this hypothesis, we have picked up the shortest corpus (42 entries) of the kind we could find on the Internet and checked the GPS terminology in both Ryan Larsen's *GPS Terminology. Glossary and Terminology* (2006) and in *Webster Comprehensive Dictionary* (1995).

Results and Discussion

We thus found out the following:

- only 2 (5%) of the terms have the same meaning (though put in different words) in both inventories: ***celestial navigation*** 'Navigating by use of traditional methods having to do with observations of stars, the sun, and moon' (Larsen 2006) and 'Navigation by means of observations made of the apparent position of heavenly bodies. Also called astronavigation, cello-navigation. [1935-1940] (WEUDEL);

LCD 'Liquid Crystal Display. A type of screen commonly found in small electronic devices' (Larsen 2006) and ' liquid-crystal display: a method of displaying readings continuously, as on digital watches, portable computers, and calculators, using a liquid-crystal film, sealed between glass plates, that changes its optical properties when a voltage is applied' (WEUDEL);

- 5 (12%) terms occurring with the same meaning (though put in different words) in both inventories, retain in Larsen's *Glossary* only one of the several meanings in *Webster Comprehensive Dictionary*: ***altimeter*** 'A device that measures barometric pressure to calculate altitude' (Larsen 2006) and '1. *a sensitive aneroid barometer that is graduated and calibrated, used chiefly in aircraft for finding distance above sea level, terrain, or some other reference point by a comparison of air pressures*. 2. Any device used for the same purpose that operates by some other means, as by radio waves. [1820-1830; *alti- + meter*]' (WEUDEL); ***frequency*** 'The rate at which a cycle is repeated. Measurements based on low frequency signals are less accurate than those based on high frequencies. Low frequency signals can, however, travel greater distances' (Larsen 2006) and '1. The state or fact of being frequent; frequent occurrence. 2. Rate of occurrence. 3. Physics. a. the number of periods or regularly occurring events of any given kind in unit of time, usually in one second. b. *the number of cycles or completed alternations per unit of time of a wave or oscillation*. 4. Math. The number of times a value recurs in a unit change of the independent variable of a given function. 5. Statistics. The number of items occurring in a given category. [1545-1555; < L *frequentia* 'assembly, multitude, crowd' (WEUDEL); ***latitude*** 'Polar coordinates used in geographic coordinate systems. The angular distance from the Earth's equatorial plane. Latitude ranges from $+90^0$ at the north pole to -90^0 at the south pole' (Larsen 2006) and '1. Geog. a. *the angular distance north or south from the equator of a point on the earth's surface, measured on the meridian of the point*. b. a place or region as marked by this distance. 2. Freedom from narrow restrictions; freedom of action, opinion, etc. 3. *Astro*n. a. see celestial latitude. b. see galactic altitude. 4. *Photog*. The ability of an emulsion to record the brightness values of a subject in their true proportion to one another, expressed as the ratio of the amount of brightness in the darkest possible value to the amount of brightness in the brightest. [1350-1400; ME < L *lātītūdo* 'breadth') (WEUDEL); ***longitude*** 'Polar coordinates used in geographic coordinate systems. The angular distance from the adopted reference point located in Greenwich,

England. Latitude ranges from 180^0 west of Greenwich to 180^0 east of Greenwich' (Larsen 2006) and '1. Geog. *angular distance east or west on the earth's surface, measured by the angle contained between the meridian of a particular place and some prime meridian, as that of Greenwich, England, and expressed either in degrees or by some corresponding difference in time*. 2. Astron. A. See celestial longitude. B. See galactic longitude. [1350-1400; ME < L *longitūdō* 'length'. See *longi-* + *-tude*) (WEUDEL); ***satellite*** (?) 'A body, natural or manmade, that orbits a planet. The term is commonly used to describe manmade objects that orbit the Earth' (Larsen 2006) and '1. Astron. *A natural body that revolves around a planet; a moon*. 2. A country under the domination or influence of another. 3. Something as a branch office or an off-campus facility of a university, that depends on, accompanies, or serves something else. 4. An attendant or follower of another person, often subservient or obsequious in manner. 5. *A device designated to be launched into orbit around the earth, another planet, the sun, etc*. [1540-1550; < L *satellite* 'attendant, member of the bodyguard or retinue'] (WEUDEL);

- 1 (2%) term has partially the same meaning in both inventories: ***transducer*** 'A device which transmits sound pulses and listens for echoes' (Larsen 2006) and 'a device that receives a signal in the form of one type of energy and converts it to a signal in another form. [1920-1925; < L *trānsdūc(ere)* 'to transfer' + *-er*) (WEUDEL);
- 5 (12%) terms have rather different meanings in the two inventories: ***basemap*** 'A map or GS that is pre-loaded in a GPS receiver' (Larsen 2006) and ***base map*** 'an outline map of an area to which specific information is added for any of various purposes' (WEUDEL); ***cache*** 'A small container hidden at a particular location by a geocacher' (Larsen 2006) and '1. A hidden place, especially one in the ground, for ammunition, food, treasures, etc. 2. Anything so hidden. 3. Alaska and Northern Canada. A small shed elevated on poles above the reach of animals and used for storing food, equipment, etc. [1585-1595; < F *cacher* 'to hide' < VL **coācticāre* 'to stow away, to pack together'] (WEUDEL); ***route*** 'A series of points that allows one to navigate from one location to another' (Larsen 2006) and '1. A course, way, or road for passage or travel. 2. A customary or regular line of passage or travel. 3. A specific itinerary, round, or number of stops regularly visited by a person in the performance of his or her work or duty. [1175-1225; ME *route* 'way, course' < OF < L *rupta (via)* 'broken (road)']' (WEUDEL); ***sounder*** 'A device that determines water depth by measuring sound echoes through water' (Larsen 2006) and

sounder$_2$ 'a person or thing that sounds depth, as of water. [1565-1575; < *sound* 'to measure or try the depth of (water, a deep hole, etc.) by letting down a lead of plummet at the end of a line, or by some equivalent means' + *-er*]' (WEUDEL); ***waypoint*** 'A geographic coordinate that is stored in a GPS receiver or computer' (Larsen 2006) and way point '1. A place or point between major points on a route. [1875-1880; Amer.]' (WEUDEL);

- 29 (69%) terms occur only in Larsen's *Glossary*: ***12-channel*** 'A GPS that can track a maximum of 12 GPS satellites at any one time'; ***automatic routing*** 'A feature commonly found in high-end automotive GPS receivers. Automatic routing is the process by which a GIS calculates an optimal driving between two locations'; ***bluechart*** 'Garmin's most detailed mapping system. Bluechart data can be loaded on compatible GPS receivers and includes detailed nautical chart information'; ***chartplotters*** 'Marine GPS receiver that features detailed mapping capabilities'; ***DGPS*** 'Differentiated GPS. A method of improving the accuracy of GPS positions by receiving 'differential' signals that are used to improve observed GPS ranges'; ***differential y-corrected*** 'DGPS'; electronic compass 'A compass that outputs a digital heading. Used in GPS receiver to calculate a heading without satellites or when not moving'; ***etrex*** 'A series of GPS receivers manufactured by Garmin. Etrex receivers are small, highly portable and intended for outdoor use'; ***fishfinder*** 'A sounder that uses sonar technology to locate fish'; ***fixed mount*** 'A GPS receiver that is permanently or semi-permanently mounted on a vehicle (usually a boat)'; ***FRS*** 'Family Service Radio Band. A public radio band that consists of 14separate channels'; ***geocaching*** 'An emerging sport based on GPS positions. Geocachers share the locations of caches which can be found by other participants with GPS receivers'; ***geographic coordinates*** 'A polar coordinate system used to determine positions on the globe. Geographic positions are typically given in terms of latitude and longitude'; ***geographic information system*** 'A database system that is used to reference information based on real-world locations'; ***GPS engine*** 'An algorithm used to calculate geographic coordinates from GPS satellite observations'; ***GPS receiver*** 'An electronic device used to measure ranges to GPS satellites and determine geographic coordinates'; **IPX7** waterproof standard 'A standard indicating that a device can be submerged 1 meter for up to 30 minutes'; ***MapSource*** 'A detailed mapping programme that can load map information into a Garmin GPS receiver. The MapSource software includes functionality to backup personal waypoints and routes'; ***nautical charts*** 'A series of detailed maps for the purpose of aiding safe marine navigation'; nautical charts 'A series of detailed maps for the purpose of aiding safe marine navigation'; ***NMEA*** 'A U.S. standards committee that defines data protocols used in GPS receivers. NMEA standard protocols allow the GPS receivers to communicate with other equipment or software'; ***position reporting*** 'A system used in Rino GPS receivers whereby one receiver can report its position to another receiver'; ***Rino*** 'Radio Integrated with Navigation for the Outdoors. Rino GPS receivers, made by Garmin, are robust outdoor receivers with the ability to communicate with each other by radio'; ***Selective Availability*** 'A process that made GPS positioning less accurate for non-military GPS. Selective Availability was removed on May 1, 2000'; ***sound pulse*** 'A short burst of sound energy. Sound pulses are used in sonar devices, which can interpret the pulse's echo'; ***squelch code*** 'A code entered into public band radio devices that is used to eliminate undesirable radio traffic'; ***TracBack*** 'The proprietary Garmin feature that converts a track log and converts it into a route to guide you back to a starting position'; ***tracklog*** 'A series of points recorded in a GPS receiver as the user moves. Often called a breadcrumb trail'; ***Voice Enabled*** 'An automotive GPS unit with the capability of transmitting route directions by voice'; ***WAAS*** 'Wide Area Augmentation System. A system of satellites and ground stations that provide GPS signal corrections for better position accuracy'.

First, there is no consistency in presenting acronyms: ***DGPS***, on the one hand, and ***FRS*** (misspelled, as it should have been ***FRSB***, from Family Radio Service Band) or ***NMEA*** (the author should have supplied the whole name for the acronym, i.e. National Marine Electronics Association), on the other hand. Then, there are also misspelled words, such as ***differential y-corrected*** (it should have been ***differentially-corrected***).

Second, *Webster Comprehensive Dictionary* (1995) supplies more detailed and technical explanations (it is a comprehensive dictionary!). In addition, it also supplies figurative meanings, details concerning the time the different terms appeared, etymons, synonyms, etc.

Conclusions

The large share of terms (67%, i.e. two thirds) not being recorded in one of the most popular English language dictionaries emphasises the necessity of developing specialised GPS dictionaries that really help students, scholars, and specialists in Precision Agriculture willing to study it.

References

Larsen, R. (2006). *GPS Terminology. Glossary and Terminology*. Online: http://www.studentcomputers.co.uk/garmin/gps-product-review/gps-terminology.htm.

Rață, Georgeta & Perković, Anica. (2007). Precision Agriculture: Terminological Issues. *Scientific Papers. Faculty of Agriculture XXXIX (2)*: 671-674.

Webster Comprehensive Dictionary. (1995). Chicago: J. G. Ferguson Publishing Company. (W)

NAMES OF WEEDS

ANICA PERKOVIĆ AND GEORGETA RAŢĂ

Introduction

The purpose of this paper is to supply useful tools for the study of the English of weed science by our students in agriculture, be they Croatian or Romanian. The main argument in doing so is the fact that common weed names can be misleading because of their structure, i.e. they may contain the word *weed* and designate weeds, they may contain the word *weed* and not designate weeds, or they may not contain the word *weed* but designate weeds. Our hypothesis is that this inconvenience can be overcome by getting to reach the proper meaning of these words. Getting to know the precise meanings of common weed names depends on our knowledge of the world (for instance, knowing weeds by their scientific names rather than by guessing their nature on the ground of their name alone) rather than on purely linguistic knowledge. The hypothesis of the present research is that there are linguistic algorithms in the study of English for Special Purposes that should be identified, studied, and presented in a manner that allows our students instantaneous understanding of a specialised vocabulary such as that of pest control, in general, and of weed control, in particular. As for the background information, there is no corpus of *-weed* ending plant names that allows instant identification of the weeds. Thus, we have searched one of the best English language dictionaries available nowadays, *The American Heritage Dictionary of the English Language* (2008), containing over 200,000 entries, which we corroborated with similar Croatian (*Hrvatski enciklopedijski rječnik* 2003) and Romanian (*Dicţionarul explicativ al limbii române* 1998) comprehensive language dictionaries.

Material and Method

The material we present and analyse below consists of a corpus of 114 plant names containing the word ***weed*** and their definitions. In everything that follows, we understand by *definition* 'the dictionary or lexical definition of a word, i.e. a definition that reports the meaning of a word as it is normally used, usually by supplying an approximately equivalent expression in which the original word does not occur'. This is not always the case here since, as we shall see, there are situations in which the base word of the common weed name, i.e. *weed*, often reappears in the definition. The research methodology is a linguistic one: it consists of the semantic analysis (i.e. the analysis of the meaning) of the common weed names in English.

Results

We have identified a number of 114 English common weed names (a ***weed*** being defined as 'a plant considered undesirable, unattractive, or troublesome, especially one growing where it is not wanted, as in a garden'). Common weed names including the word ***weed*** have been grouped into two categories:

The first category includes 14 common weed names containing the word *weed* and designating weeds (12%): ***bindweed*** '1. Any of various trailing or twining, often weedy plants of the genera *Calystegia* and *Convulvulus*, having white, pink, or purple bell-shaped or funnel-shaped flowers. 2. Any of various similar trailing or twining plants, such as the *black bindweed*'; ***black(-)bindweed*** 'A twining annual vine (*Polygonum convolvulus*), native to Eurasia but widespread as a weed and having heart-shaped leaves and clusters of small, greenish-white flowers'; ***blueweed*** 'A biennial Eurasian plant (*Echium vulgare*) naturalised as a weed in eastern North America and having usually blue flowers'; ***butterweed*** '1. A succulent annual or biennial plant (*Senecio glabellus*), native to the eastern United States and having pinnately divided leaves and bright yellow, radiate flower heads. 2. *Horseweed*'; ***carpetweed*** 'A prostrate, mat-forming annual plant (*Mollugo verticillata*) widespread as a weed throughout North America and having whorled leaves and small, greenish-white flowers'; ***chickweed*** 'Any of various herbs of the genera *Cerastium* and *Stellaria*, especially *S. media*, a European weed naturalised worldwide. The herb has small white flowers, petals with two deep lobes, and opposite leaves'; ***fireweed*** '1. (also *willow herb*) any of various plants of the genus *Epilobium*, especially *E. angustifolium*, having long, terminal, spike-like clusters of pinkish-purple flowers. 2. Any of several weedy North American plants of the genus *Erechtites*, having small white or greenish flowers grouped in discoid heads; etc.'.

The second category includes 45 common weed names containing the word *weed* but not designating weeds explicitly (39%): ***beetleweed*** (also *coltsfoot*, *galax*, *wandflower*) 'A stemless, evergreen, perennial plant

(*Galax urceolata*) of the eastern United States, having a rosette of glossy, heart-shaped leaves and small, white flowers in spike-like clusters'; ***bishop's weed*** '1. *Goutweed.* 2. A chiefly Mediterranean annual plant (*Ammi magus*) in the parsley family, grown as a source of psoralens used in medicine and as an ornamental for its compound umbels of small white flowers'; ***bugleweed*** (also *bugle*) 'Any of several herbs of the genera *Ajuga* and *Lycopus* in the mint family, having opposite eaves, square stems, and axillary clusters of purplish to white flowers'; etc.

A third category includes 55 common weed names not containing the word *weed* but designating weeds (49%): ***air(-)potato*** (also *aerial yam*, *potato yam*) 'A tropical Old World yam (*Dioscorea bulbifera*) having axillary potato-like tubers, some of which are edible after cooking. It is a weed in the tropics and Florida'; ***alfilaria / alfileria*** (also *filaree*, *pin clover*) 'An annual Mediterranean plant (*Erodium cicutarium*) having pinnately dissected leaves and small pink or purple flowers. It is a widespread weed and is used for spring forage in the western United States'; ***alyssum*** '1. (also *sweet alyssum*) A widely cultivated annual or perennial herb (*Lobularia maritime*) of the mustard family, native to the Mediterranean region, having racemes of long-lasting flowers varying in size and colour. 2. (also *madwort*) Any of various chiefly Mediterranean weeds or ornamentals of the genus *Alyssum* in the mustard family, having racemes of white or yellow flowers. 3. (also *hoary alyssum*) An annual European herb (*Berteroa incana*) of plants of the mustard family, having silvery foliage, oblong fruits, and white, deeply notched flowers. It is naturalised in Eastern North America'; etc. (Figure 3-1).

Discussion

The problem with some of these entries is that they are sometimes defined in a rather confusing manner, i.e. for a single entry there may be two definitions one of which may not be that of a weed. Thus, ***bindweed*** is defined as 'Any of various similar trailing or twining plants, such as the black bindweed' (a tricky definition, since one needs to look up for black bindweed to see what it is), while ***butterweed*** is defined as '1. A succulent annual or biennial plant (*Senecio glabellus*), native to the eastern United States and having pinnately divided leaves and bright yellow, radiate flower heads', etc. The definitions of these common weed names are based on criteria lacking consistency (frequency, place of origin, posture): common weeds (***pigweed***), cosmopolitan weeds (***pigweed***), European weeds (***orange hawkweed***), European weeds naturalised worldwide (***chickweed***), often weedy plants (***bindweed***), plants naturalised as weeds (***blueweed***). Another source of confusion is the fact that ***weedy*** also means 'full of or consisting of weeds (*a* ***weedy*** *lawn*)', 'resembling or characteristic of a weed (*a* ***weedy*** *plant*)', and 'of a scrawny build; spindly or gawky'. Therefore, plants designated as 'weedy' (***fireweed***, ***hogweed***, ***horseweed***) or as 'often weedy plant' (***bindweed***) should be looked up in a specialised dictionary to determine what it really is.

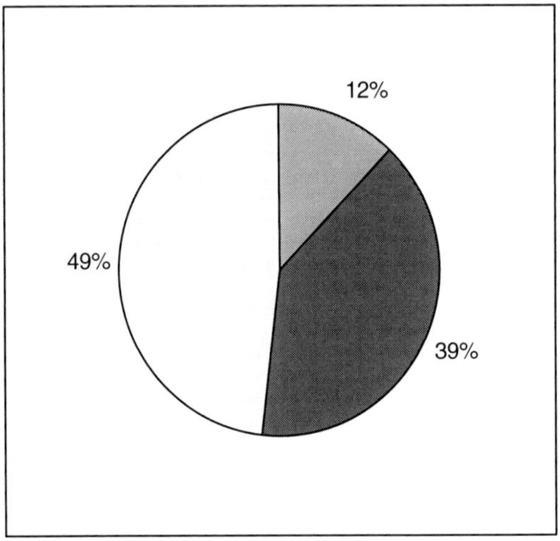

Figure 3-1. Names of weeds: 12% *weed* names designating weeds; 39% *weed* names not designating weeds; 49% no-*weed* names designating weeds

The second category of common weed names shares three weed names with the first category, as shown above. Thus, ***butterweed*** '1. A succulent annual or biennial plant (*Senecio glabellus*), native to the eastern United States and having pinnately divided leaves and bright yellow, radiate flower heads', ***fireweed*** '1. (also *willow herb*) any of various plants of the genus *Epilobium*, especially *E. angustifolium*, having long, terminal, spike-like clusters of pinkish-purple flowers. 2. Any of several weedy North American plants of the genus *Erechtites*, having small white or greenish flowers grouped in discoid heads; etc.' and ***ragweed*** '2. *Chiefly British*. *Ragwort*. Any of several plants of the very large genus *Senecio* in the composite family, having yellow flower heads, especially *S. aureus* of eastern North America and *S. jacobea* of Europe'. Another common weed name is defined in a manner that may designate both weeds and non-weeds: ***stickweed*** 'Any of various plants having clinging seeds or fruit, especially ragweed'. As for ***stinkweed*** 'Any of various plants that have flowers or foliage with an unpleasant odour', it is defined in the most confusing way, with no scientific name whatsoever.

The third category of common weed names is, perhaps, the trickiest. They all designate weeds but are not designated as such by their name. Another problem with these entries is that they are sometimes defined in a confusing manner, i.e. for a single entry there may be two or even three definitions, one or two of which may not be that of a weed: ***alyssum*** 1 and 3, ***beggar('s) ticks*** 2, and ***goat('s)(-)beard*** 2. In other cases, the definition of the weed is ambiguous because of such formulations as: and also some weeds (***bluegrass / blue grass***), including several weeds (***brome***, ***convolvulus***), or any of several / various similar or related plants (***dandelion*** 2, thistle 2). Again, a considerable number of weeds in this category are defined as often weedy (***bracken***), usually weedy (***bent grass / bentgrass***), or weedy (***bedstraw***, ***black mustard***, ***bladder campion***, ***bur cucumber***, ***burdock***, ***butter-and-eggs***, ***cockle***, ***corn cockle***, ***dead nettle***, ***knawe(l)***, ***knotgrass***, ***mercury***, ***ribgrass***, ***spur(e)y***, ***strawberry blite***, ***tare***, ***thistle***, ***yard grass***). As for ***dog fennel***, it designates both a weed and a weedy plant.

Conclusions

Our hypothesis that arriving at the precise meanings of the common weed names depends on our knowledge of the world rather than on purely linguistic knowledge is, thus, supported. However, it supposes a deep knowledge of the world, in general, and of weed science, in particular. Our analysis shows that common weed names that need to be clarified are more numerous than those explained by language and/or specialised dictionaries. This means that both teachers and students – no matter the level – should build up their own inventories of terms when specialising in a certain field of knowledge. Common weed names make up a special vocabulary that needs special skills to be instantly and properly understood since almost half of it (49%) consists of common weed names whose meaning is difficult to catch on the ground of their structure alone. The implications of the research and results are huge: they point to the need for both specialists (academics, lexicologists) and students (undergraduates, MSc students, PhD students, and post-graduates) in the field of weed science to build up their own specialised inventories of terms. Joint efforts of specialists in weed science, on the one hand, and of specialists in linguistics, on the other hand, would be ideal.

References

Četverojezični rječnik: hrvatsko-englesko-latinski. [Multilingual Dictionary: Croatian – English – Latin]. (2011). Online: prevodi.freehostia.com/glosari.htm.

Dicționarul explicativ al limbii române. [The Explanatory Dictionary of the Romanian Language]. (1998). București: Editura Univers Enciclopedic.

Herren, R. V. & Donahue, R. L. (1991). *The Agricultural Dictionary.* Albany, NY: Delmar Publishers Inc.

Hrvatski enciklopedijski rječnik. [Croatian Encyclopaedic Dictionary]. (2003). Zagreb: Novi liber.

Perković, A. & Rață, G. (2008). Names of Weeds in Contemporary English: A Terminological Approach. *Poljoprivreda 2*: 52-54.

Rață, Georgeta, Samfira, I., Diana-Andreea Boc-Sânmărghițan & Butnariu, Monica. (2012). *Compendium of Forage Technical Terms in English, French and Romanian.* Newcastle upon Tyne: Cambridge Scholars Publishing.

The American Heritage Dictionary of the English Language. (2008). Boston: Houghton Mifflin.

THE VOCABULARY OF IRRIGATION

GEORGETA RAŢĂ, CORNELIA PETROMAN AND IOAN PETROMAN

Introduction

Words of different origin abound in contemporary English, including the **vocabulary of irrigation**. English has taken over a great number of words belonging to the field of irrigation from other languages, words which have nevertheless become somehow a permanent part of it. Most of them have been modified and brought into line with the phonological rules of English, helping native English-speaking people better understand and use them. Other words, though plain English ones, are confusing even to specialists: they are terms belonging to different technical fields related more or less to the field of irrigation and gathered in encyclopaedias that abound on the Internet. We present some of the terms belonging to the field of irrigation pointing out the trends in this type of specialised vocabulary.

Material and Method

We have inventoried a number of 63 entries belonging to the field of "irrigation", as follows: ***acequia***, ***Arthur Thomas Cotton***, ***aryk***, ***atmospheric vacuum breaker***, ***backflow prevention device***, ***Banaue Rice Terraces***, ***Centre Pivot Irrigation in Saudi Arabia***, ***centre pivot irrigation***, ***chemigation valve***, ***Columbia Basin Project***, ***disk filter***, ***distribution uniformity***, ***double check valve***, ***drip irrigation***, ***drip tape***, ***dripperline***, ***Dujiangyan Irrigation System***, ***emitter***, ***Espada Acequia***, ***Gezira Scheme***, ***Godavari River Basin Irrigation Projects***, ***Great Manmade River***, ***Hazen-Williams equation***, ***hydrozoning***, ***irrigation***, ***Irrigation Association***, ***irrigation controller***, ***irrigation scheduling***, ***irrigation sprinkler***, ***Kallanai***, ***Kokee Ditch***, ***Laramie-Poudre Tunnel***, ***Majalgaon***, ***media filter***, ***microtubing***, ***moisture stress***, ***Natural Resources Conservation Service***, ***Orbit Irrigation Products, Inc.***, ***permeability***, ***pipe***, ***pressure vacuum breaker***, ***qanat***, ***Rain Bird***

Corporation, ***rain sensor***, ***rainwater harvesting***, ***runoff***, ***Salt River Project***, ***sand separator***, ***screen filter***, ***settling basin***, ***subirrigation***, ***submersible pump***, ***Sukkur***, ***Sukkur barrage***, ***tensiometer***, ***Truckee-Carson Irrigation District***, ***Vaalharts Irrigation Scheme***, ***Valmont Industries, Inc.***, ***water filter***, ***water hammer***, ***water resources***, ***water table***, and ***water tank***.

We then compared the way all these terms are defined on the Internet and by English language dictionaries and encyclopaedias.

Results and Discussion

The 63 entries designate a large variety of terms related to "irrigation", as follows:

- 21 types of "devices" (32%): ***atmospheric vacuum breaker*** 'a backflow prevention device used in plumbing to prevent backflow of non-potable liquids into the potable (drinking water) system' (I), ***backflow prevention device*** 'a device used to protect water supplies from contamination' (I), ***chemigation valve*** 'an apparatus designed to protect water supplies from agricultural chemicals used during chemigation (the application of chemicals such as fertilizers and pesticides through irrigation water)' (I), ***disk filter*** 'a type of water filter' (I), ***double check valve*** 'a backflow prevention device designed to protect water supplies from contamination' (I), ***drip tape*** 'a type of thin walled dripperline used in drip irrigation' (I), ***dripperline*** 'a type of drip irrigation tubing with emitters pre- installed at the factory' (I), ***emitter*** 'a device used in drip irrigation to transfer water from a pipe or tube to the area to be irrigated' (I), ***irrigation controller*** 'a device to operate automatic irrigation systems such as lawn sprinkler systems and drip irrigation systems' (I), ***irrigation sprinkler*** 'a device used on farms, golf courses, residential yards, and other places to water crops, lawns, gardens or other plants in the event of drought' (I), ***media filter*** 'a type of filter utilizing a bed of sand, crushed granite or other material to filter water for drinking, swimming pools, aquaculture, irrigation, and other applications' (I), ***microtubing*** 'a very fine plastic tubing used in drip irrigation, typically in greenhouses' (I), ***pipe*** 'a hollow cylinder of material' (I) and 'a hollow cylinder of metal, wood, or other material, used for the conveyance of water, gas, steam, petroleum, etc. [before 1000] (WEUDEL), ***pressure vacuum breaker*** 'a backflow prevention device' (I), ***rain sensor*** 'a switching device activated by rainfall' (I), ***sand separator*** 'a device which utilizes centrifugal force to separate sand or other heavy particles out of water' (I), ***screen filter*** 'a type of filter utilizing a rigid or flexible screen to separate sand and other fine particles out of water for irrigation or other applications' (I), ***submersible pump*** 'a pump which has a hermetically sealed motor close-coupled to the pump body' (I), ***tensiometer*** 'a device used to determine soil moisture tension, an indirect measure of soil moisture content' (I) and 'an instrument for measuring the surface tension of a liquid [1910-1915]' (WEUDEL), ***water filter*** 'a device which removes impurities from water by means of a fine physical barrier and/or chemical processes' (I), and ***water tank*** 'a tank used to store water' (I);

- 14 irrigation systems (22%): ***acequia*** 'a waterway used for irrigation in the USA' (I) and '*Southwestern U.S.* an irrigation ditch [1835-1845]' (WEUDEL), ***aryk*** '(Turkic *arık*, Russian *Арык*) an irrigation canal widely used throughout Central Asia' (I), ***Banaue Rice Terraces*** '2000-year-old terraces carved into the mountains of Ifugao in the Philippines' (I), ***Dujiangyan Irrigation System*** 'an irrigation system in China)' (I), ***Espada Acequia*** 'an irrigation system in the USA' (I), ***Great Manmade River*** 'a network of water in Libya' (I), ***Kallanai*** 'a dam in India' (I), ***Kokee Ditch*** 'an irrigation canal in the USA' (I), ***Laramie-Poudre Tunnel*** 'an irrigation tunnel in the USA' (I), ***Majalgaon*** 'an irrigation area in India' (I), ***qanat*** 'a water management system in ancient Persia' (I), ***Sukkur*** 'a dam in Pakistan' (I) and 'a city in Pakistan' (WEUDEL), ***Sukkur barrage*** 'a barrage in Pakistan' (I), and ***Vaalharts Irrigation Scheme*** 'an irrigation scheme in South Africa' (I);
- 7 "organisations" (11%): ***Irrigation Association*** 'a worldwide industry trade group for irrigation professionals and manufacturers' (I), ***Natural Resources Conservation Service*** 'a service in the USA' (I), ***Orbit Irrigation Products, Inc.*** 'a manufacturer and supplier of irrigation products for residential and commercial markets' (I), ***Rain Bird Corporation*** 'a large privately held manufacturer and provider of irrigation products and services in the USA' (I), ***Salt River Project*** 'the Salt River Valley Water Users' Association' (I), ***Truckee-Carson Irrigation District*** 'a political subdivision of the State of Nevada (USA), which operates dams, canals, and drains' (I), and ***Valmont Industries, Inc.*** 'a manufacturer of Valley centre pivot irrigation systems and steel utility poles in the USA' (I);
- 5 irrigation projects (8%): ***Centre Pivot Irrigation in Saudi Arabia*** 'an irrigation project in Saudi Arabia' (I), ***Columbia Basin Project*** 'an irrigation project in the USA' (I), ***Gezira Scheme*** 'an irrigation project

in Sudan' (I), ***Godavari River Basin Irrigation Projects*** 'irrigation projects in India' (I), and ***Salt River Project*** 'the Salt River Project Agricultural Improvement and Power District (a political subdivision of the state of Arizona), and any improvement project along the Salt River' (I);

- 3 methods of irrigation (5%): ***centre pivot irrigation*** 'a method of crop irrigation in which equipment rotates around a pivot' (I) and 'a method of irrigation, used mainly in the USA, in which water is dispersed through a long, segmented arm that revolves about a deep well and covers a circular area [1970-1975]' (WEUDEL), ***drip irrigation*** 'an irrigation method that applies water slowly to the roots of plants, by depositing the water either on the soil surface or directly to the root zone, through a network of valves, pipes, tubing, and emitters' (I) and 'a system of crop irrigation involving the controlled delivery of water directly to individual plants through a network of tubes and pipes [1970-1975]' (WEUDEL), and ***subirrigation*** 'a method of irrigation where water is delivered to the plant root zone from below the soil surface and absorbed upwards' (I) 'irrigation beneath the surface of the ground, as with water passing through a system of underground porous pipes or transmitted through the subsoil from ditches, etc. [1900-1905] (WEUDEL);
- 14 other cases (22%): ***Arthur Thomas Cotton*** 'British general and irrigation engineer' (I), ***distribution uniformity*** 'a measure of how uniformly water is applied to the area being watered, expressed as a percentage' (I), ***Hazen-Williams equation*** 'a formula that can be used to calculate the pressure loss in a length of pipe due to friction dependent on the flow' (I), ***hydrozoning*** 'a landscape practice that groups plants with similar water requirements together in an effort to conserve water' (I), ***irrigation*** 'the replacement or supplementation of rainfall with water from another source in order to grow crops' (I) and 'the artificial application of water to land to assist in the production of crops [1605-1615]' (WEUDEL), ***irrigation scheduling*** 'the process used by irrigation system managers to determine the correct frequency and duration of watering' (I), ***moisture stress*** 'stress occurring when the water in a plant's cells is reduced to less than normal levels' (I), ***permeability*** 'in soil science, a measure of the infiltration rate of precipitation into the soil' (I) and '*Geol.* the capability of a porous rock or sediment to permit the flow of fluids through its pore spaces [1750-1760] (WEUDEL), ***rainwater harvesting*** 'the collection and storage of rain from roofs or a surface catchment for future productive use' (I), ***runoff*** 'water from rain, snowmelt, or elsewhere that does not

percolate into the ground, but flows along the surface of the earth and eventually drains into rivers, lakes, and/or oceans' (I) and 'something that drains or flows off, as rain that flows off from the land in streams [1850-1855]' (WEUDEL), ***settling basin*** 'a method of removing very fine particles from water by means of gravity' (I), ***water hammer*** 'a pressure surge or wave caused by the kinetic energy of a fluid in motion when it is forced to stop or change direction suddenly' (I) and 'the concussion and accompanying noise that result when a volume of water moving in a pipe suddenly stops or loses momentum [1795-1805]' (WEUDEL), ***water resources*** 'sources of water that are useful or potentially useful to humans' (I), and ***water table*** 'the upper limit of abundant groundwater' (I) and 'the planar, underground surface beneath which earth materials, as soil or rock, are saturated with water [1400-1450]' (WEUDEL).

Of the 63 entries, only 14 (22%) can also be found in general language dictionaries, which emphasises the need for specialised dictionaries developed by both linguists and specialists in irrigation: ***acequia*** 'a community operated waterway used in the USA for irrigation' (I) and '*Southwestern U.S.* an irrigation ditch [1835-1845]' (WEUDEL); ***centre(-)pivot irrigation*** 'a method of crop irrigation in which equipment rotates around a pivot' (I) and 'a method of irrigation, used mainly in the USA, in which water is dispersed through a long, segmented arm that revolves about a deep well and covers a circular area [1970-1975]' (WEUDEL); ***drip irrigation*** 'an irrigation method that applies water slowly to the roots of plants, by depositing the water either on the soil surface or directly to the root zone, through a network of valves, pipes, tubing, and emitters' (I) and 'a system of crop irrigation involving the controlled delivery of water directly to individual plants through a network of tubes and pipes [1970-1975]' (WEUDEL); ***emitter*** 'a device used in drip irrigation to transfer water from a pipe or tube to the area to be irrigated' (I) and 'a person or thing that emits' (WEUDEL); ***irrigation*** 'the replacement or supplementation of rainfall with water from another source in order to grow crops' (I) and 'the artificial application of water to land to assist in the production of crops [1605-1615]' (WEUDEL); ***permeability*** 'in soil science, a measure of the infiltration rate of precipitation into the soil' (I) and '*Geol.* the capability of a porous rock or sediment to permit the flow of fluids through its pore spaces [1750-1760] (WEUDEL); ***pipe*** 'a hollow cylinder of material' (I) and 'a hollow cylinder of metal, wood, or other material, used for the conveyance of water, gas, steam, petroleum, etc. [bef. 1000] (WEUDEL); ***runoff*** 'water from rain, snowmelt, or elsewhere that does not percolate into the ground, but flows along the surface of the earth and eventually drains into rivers, lakes, and/or oceans' (I) and 'something that drains or flows off, as rain that flows off from the land in streams [1850-1855]' (WEUDEL); ***subirrigation*** 'a method of irrigation where water is delivered to the plant root zone from below the soil surface and absorbed upwards' (I) 'irrigation beneath the surface of the ground, as with water passing through a system of underground porous pipes or transmitted through the subsoil from ditches, etc. [1900-1905] (WEUDEL); ***Sukkur*** 'a dam in Pakistan' (I) and 'a city in Pakistan' (WEUDEL); ***tensiometer*** 'a device used to determine soil moisture tension, an indirect measure of soil moisture content' (I) and 'an instrument for measuring the surface tension of a liquid [1910-1915]' (WEUDEL); ***water hammer*** 'a pressure surge or wave caused by the kinetic energy of a fluid in motion when it is forced to stop or change direction suddenly' (I) and 'the concussion and accompanying noise that result when a volume of water moving in a pipe suddenly stops or loses momentum [1795-1805]' (WEUDEL); ***water table*** 'the upper limit of abundant groundwater' (I) and 'the planar, underground surface beneath which earth materials, as soil or rock, are saturated with water [1400-1450]' (WEUDEL); ***water tank*** 'a tank used to store water' (I) and 'the planar, underground surface beneath which earth materials, as soil or rock, are saturated with water [1400-1450]' (WEUDEL).

Conclusions

It is interesting to note that, of the 63 entries, only 75% are strictly related to the vocabulary of irrigation (the 21 types of devices, the 14 irrigation systems, the 12 of the 14 'other cases', and the 3 irrigation methods), while the remaining 25% are dedicated to 'organisations' or to irrigation projects. This pragmatic trend, the direct result of the fact that the Internet is "for" the people and "by" the people, speaks not only of sponsorship, but also of the need for more than just 'words'. (Figure 3-2)

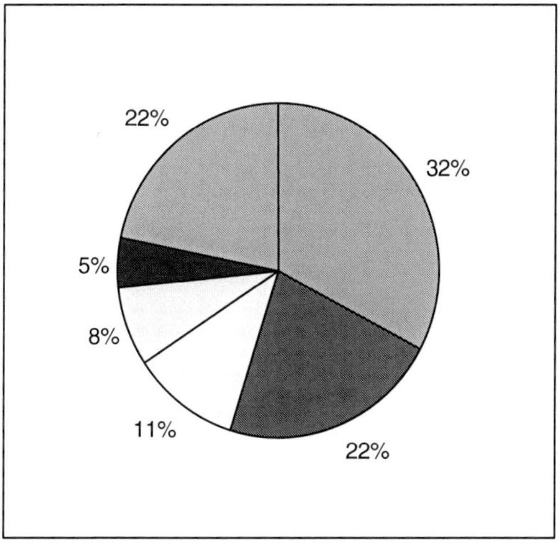

Figure 3-2. Share of the terms specific to the English of irrigation: 32% types of "devices", 22% irrigation systems, 11% "organisations", 8% irrigation projects, 5% methods of irrigation, 22% other cases

References

Benson, M., Benson, Evelyn & Ilson, R. (1990). *The Combinatory Dictionary of English*. Amsterdam – Philadelphia: John Benjamins Publishing Company.

Chalker, Sylvia & Weiner, E. (1994). *The Oxford Dictionary of English Grammar*. London - New York - Sydney - Toronto: BCA.

Irrigation. Online: http://www.betterworldbooks.com/irrigation-id-1156506948.aspx (I)

Leviţchi, L. & Bantaş, A. (1995). *Dicţionar englez-român*. [English-Romanian Dictionary]. Bucureşti: Ed. Teora

Leviţchi, L. (Ed.) (2004). *Dicţionar englez-român*. [English-Romanian Dictionary]. Bucureşti: Ed. Univers Enciclopedic.

Leviţchi, L. (Ed.). (1974). *Dicţionar englez-român*. [English-Romanian Dictionary]. Bucureşti: Ed. Academiei R.S.R.

Longman Dictionary of English Language and Culture (2003). London: Longman.

New Standard Encyclopedia (1995). Chicago: Standard Educational Corporation.

Perković, Anica & Raţă, Georgeta. (2006). Presentation of Water Importance to Tertiary Level Agriculture Students. *Buletinul*

Universităţii de Ştiinţe Agricole şi Medicină Veterinară – Cluj-Napoca, Seria Horticultură 63: 459.

Perković, Anica & Raţă, Georgeta. (2007). Misleading Combining Forms: *Hydro-*. *Scientific Papers. Faculty of Agriculture* XXXIX (2): 591-594.

Raţă, Georgeta & Proca, Maria-Ana. (2006). On Greek Loanwords in the English Vocabulary of Hydrology (A Lexicological Approach). *Conference on Water Observation and Information System for Decision Support, May 23-26, 2006, Ohrid, Republic of Macedonia*: 1-6.

Raţă, Georgeta, Petroman, Cornelia & Petroman, I. (2008). The Vocabulary of Irrigation: A Typological Approach. *Conference on Water Observation and Information System for Decision Support, 27-31 May 2008, Ohrid, Republic of Macedonia*: 381.

The New Encyclopaedia Britannica. (1992). Chicago: Encyclopaedia Britannica, Inc.

The Random House Dictionary of the English Language. (1968). New York: Random House. (RHDEL)

Webster Comprehensive Dictionary. (1995). Chicago: J. G. Ferguson Publishing Company. (WCD)

Webster Encyclopedic Unabridged Dictionary of the English Language. (1996). New York: Gramercy Books. (WEUDEL)

Webster's Universal Dictionary & Thesaurus (2003). New Lanark: Geddes & Grosset.

‘BERRY’ OR ‘BERRYLIKE’

GEORGETA RAȚĂ AND ANICA PERKOVIĆ

Introduction

The word ***berry*** has two meanings: one is based on a *botanical definition*, indicating a *true berry* (‘fleshy fruit in which the entire ovary wall ripens into an edible pericarp’: NPDB, W): *avocado*, *blackcurrant*, *chilli pepper*, *eggplant*, *goose**berry***, *lychee*, *plantain*, *redcurrant*, *tomato*, and *uchuva*; the other one is based on *common identification*, indicating a *false berry* (‘any small, sweet, juicy and brightly-coloured fruit’: NPDB, W): *black**berry***, *blue**berry***, *boysen**berry***, *cran**berry***, *mul**berry***, *rasp**berry***, and *straw**berry***. We have shown (Rață 2005, 2006) that, because of this polysemantism, plant names containing the word ***berry*** are a nuisance for undergraduates specializing in agriculture, horticulture, or food processing.

We think we should also add to these **compound words** (words formed by combining two or more bases, i.e. words – Chalker & Weiner 1994) the names of plants claimed to bear ***berries*** or of ***berrylike*** fruits, as is the case with some authoritative English language dictionaries. Romanian makes no linguistic difference whatsoever between *true* and *false berries*; moreover, finding the proper equivalent for English **compounds** with ***berry*** in Romanian is an ordeal. Therefore, here, more than anywhere else, academic study is, among others, about forgetting common parlance and acquiring scientific language.

Material and Method

We have inventoried a rather limited number of plant nouns collected from general English language and from specialized dictionaries, claimed to either bear ***berries*** or be ***berrylike***.

We then analysed them from a lexicographical point of view.

Results

Forty plants are said to have '***berries***' ('indehiscent fruits derived from a single ovary and having the whole wall fleshy, such as the grape or tomato; small, juicy, fleshy fruits, such as a blackberry or raspberry, regardless of their botanical structure', AHDEL). The following plants bear 'berries': ***allspice / pimento*** 'a tropical American evergreen tree (*Pimenta dioica*) having [...] berries used as a spice, especially in baking', ***bay*** or ***bay laurel*** or ***laurel*** or ***sweet bay*** 'a Mediterranean evergreen tree (*Laurus nobilis*) having [...] small blackish berries', ***boxthorn*** or ***matrimony vine*** 'any of various often thorny shrubs of the genus *Lycium* [having] purplish flowers and brightly coloured berries', ***bryony*** 'any of various Eurasian tendril-bearing vines of the genus *Bryonia*, having red or black berries', ***butcher's broom*** 'an evergreen shrub (*Ruscus aculeatus*) native to Europe and the Mediterranean region, having [...] usually red berries', ***cassis*** 'a Eurasian currant (*Ribes nigrum*) bearing black berries', ***catbrier / greenbrier / smilax*** 'any of several woody, usually prickly delicious vines of the genus *Smilax,* having [...] usually bluish to black berries', ***Chinese lantern plant*** or ***winter cherry*** 'a frequently cultivated Eurasian plant (*Physalis alkekengi*) having small red berries', ***clintonia*** 'any of various perennial herbs of the genus *Clintonia* in the lily family, native to North America and eastern Asia and having [...] blue or black berries', ***cuckoopint / lords-and-ladies*** 'a European plant (*Arum maculatum*) having [...] scarlet berries', ***currant*** 'any of various deciduous, spineless shrubs of the genus *Ribes,* native chiefly to the Northern Hemisphere and having [...] edible, variously coloured berries', ***fire thorn*** 'any of various thorny shrubs of the genus *Pyracantha,* native to Asia and [having] showy reddish or orange berries', ***fox grape / skunk grape*** 'a wild grape (*Vitis labrusca*) of the eastern US that bears purplish-black berries', ***golden club*** 'an aquatic plant (*Orontium aquaticum*) of the eastern US, having [...] small blue-green berries', ***grape*** 'any of numerous woody vines of the genus *Vitis*, bearing clusters of edible berries', ***holly*** 'any of numerous trees or shrubs of the genus *Ilex*, usually having bright red berries', ***honeysuckle*** 'any of various shrubs or vines of the genus *Lonicera,* having [...] small berries', ***horse nettle*** 'a prickly-stemmed plant (*Solanum carolinense*) of eastern and central North America, having [...] yellowish berries', ***juniper*** berries (*choucroute*, *gin*), ***madroña*** 'an evergreen tree (*Arbutus menziesii*) native to Pacific North America, having [...] orange or red edible berries', ***mangosteen*** 'a Malaysian evergreen tree (*Garcinia mangostana*) having [...] large edible berries', ***maqui*** 'a Chilean evergreen shrub (*Aristotelia chilensis*) bearing edible purple berries', ***mistletoe*** 'a Eurasian parasitic shrub (*Viscum album*) having [...] waxy white berries', ***mountain ash*** 'any of various deciduous trees of the genus *Sorbus* having [...] bright orange-red berries', ***myrtle*** 'any of several evergreen shrubs or trees of the genus *Myrtus*, especially *M. communis,* an aromatic shrub native to the Mediterranean region and western Asia, having [...] blue-black berries', ***Natal plum*** 'a South African evergreen shrub (*Carissa grandiflora*) having [...] an edible scarlet berry', ***Oregon grape*** 'any of various evergreen shrubs of the genus *Mahonia*, especially *M. aquifolium* of northwest North America, having [...] black berries', ***pepper*** 'any of several tropical American, cultivated forms of *Capsicum frutescens* or *C. annuum*, having podlike, many-seeded, variously coloured berries', ***peppercorn*** 'a dried berry of the pepper vine *Piper nigrum*', ***plant species of the genus Rhus*** (*Japan wax*)', ***poison elder / sumac*** 'a swamp shrub (*Rhus vernix*) of the southeast US, having [...] greenish-white berries', ***poison ivy / oak*** 'a North American shrub or vine (*Rhus radicans*) that has [...] whitish berries', ***queen cup*** 'a perennial stemless plant (*Clintonia uniflora*) of Pacific North America, having [...] a blue berry', ***red cedar*** (*cedar waxwing*), ***rowan*** 'a small deciduous European tree (*Sorbus aucuparia*) of the rose family, having [...] orange-red berries', ***umbrella leaf*** 'a perennial herb (*Diphylleia cymosa*) of the southeast US, having [...] small blue berries', ***Virginia creeper*** or ***woodbine*** 'a North American climbing vine (*Parthenocissus quinquefolia*) having [...] bluish-black berries', ***wayfaring tree*** 'a deciduous Eurasian shrub (*Viburnum lantana*) having [...] berries that turn from red to black', and ***white bryony*** 'a climbing Eurasian vine (*Bryonia alba*) having [...] blackish berries'.

Other four plants are said to have ***berrylike*** fruits, that is fruits that resemble ***berries*** (< *berry* 'an indehiscent fruit derived from a single ovary and having the whole wall fleshy, such as the grape or tomato; a small, juicy, fleshy fruit, such as a blackberry or raspberry, regardless of its botanical structure' + suffix *-like* 'resembling or characteristic of', AHDEL). The plants indicated as bearing ***berrylike*** fruits are: ***cubeb*** 'a tropical southeast Asian shrubby vine (*Piper cubeba*) having spicy, berrylike fruits [...] used in perfumery, pharmaceuticals, and commercial flavourings', ***elder*** 'any of various shrubs or small trees of the genus *Sambucus,* having [...] red or purplish-black berrylike fruit', ***manzanita*** 'any of several evergreen shrubs or small trees of the genus *Arctostaphylos* of the Pacific coast of North America, especially *A. manzanita*, [...] producing red berrylike drupes', and ***wax myrtle*** 'an evergreen shrub (*Myrica cerifera*) of the southeast US, having [...] small, berrylike fruit' (Figure 3-3).

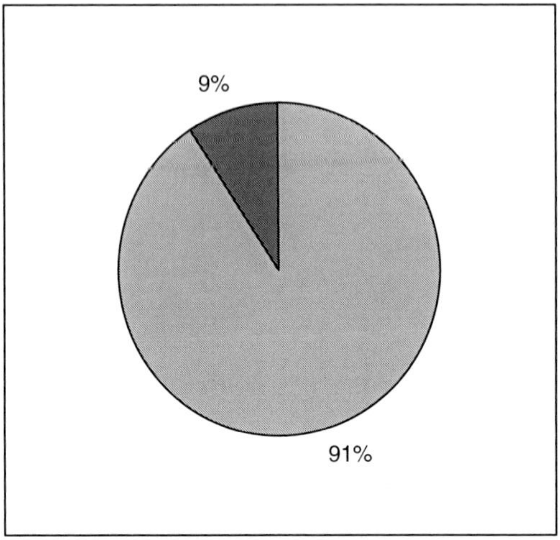

Figure 3-3. True and false berries: 91% true berry plants, 9% berry-like plants

Discussion

As far as plants bearing / having ***berries*** are concerned, there seems to be a problem with English language dictionaries. Thus, three plants are indicated indirectly as producing ***berries***, while in reality there are no ***berries*** at all:

- ***juniper berries*** are claimed to be added in both *choucroute* (defined as 'an Alsatian dish of sauerkraut with wine, sausages, pork, and *juniper berries*', AHDEL) and *gin* (defined as 'a strong, colourless alcoholic beverage made by distilling or redistilling rye or other grain spirits and adding *juniper berries* or aromatics such as anise, caraway seeds, or angelica root as flavouring', AHDEL); or, ***juniper*** is defined as 'any of various evergreen trees or shrubs of the genus *Juniperus, having [...] aromatic, bluish-grey,* ***berrylike****, seed-bearing* ***cones***' (AHDEL);
- ***red cedar berries*** are claimed to be eaten by a cedarbird (defined as 'a North American bird (*Bombycilla cedrorum*) having a crested head, a yellow-tipped tail, and predominantly brown plumage', 'probably so called because it eats the *berries of the red cedar*', *cedar waxwing*); or, ***red cedar*** is defined as 'an evergreen, coniferous, eastern North American tree (*Juniperus virginiana*) *having fleshy, purplish-black*

seed ***cones***; a tall, evergreen, Pacific North American tree (*Thuja plicata*) *having [...] small, ovoid seed-bearing* ***cones***' (AHDEL);
- ***Rhus berries*** are claimed to be the source of *Japanese wax* (defined as 'a pale yellow solid wax obtained from the berries of certain plant species of the genus *Rhus* and used in wax matches, soaps, and food packaging and as a substitute for beeswax', AHDEL); investigation of other sources shows that genus *Rhus* is represented by 250 species of flowering plants among which the ***poison elder / sumac***.

Among plants bearing ***berrylike*** fruits, ***manzanita*** is defined as 'any of several evergreen shrubs or small trees of the genus *Arctostaphylos* of the Pacific coast of North America, especially *A. manzanita*, [...] producing red berrylike drupes', a definition which associates, in the most outrageous manner, two words that suggest two different types of fruits from a botanical point of view: ***berry*** (***berrylike*** 'resembling or being characteristic of berries') and ***drupe*** ('a fleshy fruit, such as a peach, plum, or cherry, usually having a single hard stone that encloses a seed').

Conclusions

English-language dictionaries:

- are not always accurate when it comes to identifying plants (***juniper*** is said to have both ***berries*** and ***berrylike*** *cones*; red cedar is said to have both ***berries*** and ***seed cones*** or ***seed-bearing cones***);
- should be completed by other information tools such as specialized dictionaries, encyclopaedias, etc. to allow proper identification of plants (to trace certain plant species of the genus *Rhus*, for instance);
- should avoid such misleading word associations as ***berrylike*** and ***drupe***.

Therefore, authors of English language dictionaries should do one of the following: either correct the definitions to meet scientific requirements, or add a third meaning to the entry ***berry*** to comply with the rest of the apparently faulty definitions including this word.

References

Abercrombie, M., Hickman, M., Johnson, M. L. & Thain, M. (1990). *The New Penguin Dictionary of Biology*. London: Penguin Books. (NPDB)

Berry. Online: http://en.wikipedia.org/wiki/Berry. (W)

Carstairs-McCarthy, A. (2002). *An Introduction to English Morphology. Words and Their Structure*. Edinburgh University Press.

Chalker, Sylvia & Weiner, E. (1994). *The Oxford Dictionary of English Grammar*. London - New York - Sydney - Toronto: BCA.

Rață, Georgeta & Perković, Anica. (2008). Being 'a Berry' or 'Berrylike': A Lexicographical Approach. *Journal of Linguistic Studies 1*: 51-54.

Rață, G. (2005). True and False 'Berries' in English and Romanian: A Contrastive Approach. *Scientific Works L (6)*: 411-414.

Rață, G. (2006). A Case of Polysemy in the English of Horticulture: True and False 'Berries'. *Language and the media – One language: many worlds*. Zagreb - Split: HDLP. 603-612.

The American Heritage Dictionary of the English Language. (2008). Boston: Houghton Mifflin. (AHDEL)

SPECIAL TYPES OF TOURISM: TOURISM IN THE COUNTRYSIDE

GEORGETA RAȚĂ, ANICA PERKOVIĆ AND IOAN PETROMAN

Introduction

Nothing could be more disconcerting than **tourism nomenclature** nowadays – a field in which the *different types of tourism* related to *countryside and/or nature* interfere or overlap resulting in noun phrases such as ***agrarian tourism***, ***agricultural tourism***, ***agritourism***, ***country(side) tourism***, ***farm tourism***, ***rural tourism***, ***village tourism***, etc. for which not even the World Tourism Organisation supplies proper definitions.

The list could be completed with other noun phrases strictly related to the types of tourism above, and that enrich tourism vocabulary: ***country vacations***, ***dude ranches***, ***farm holidays***, ***farm house holidays***, ***farm recreation***, ***ranch recreation***, ***ranch vacations***, and ***vacation farms***.

Some other specialists consider there are more terms for **tourism in the countryside**, including ***rural tourism***, ***agricultural tourism***, ***ecotourism***, ***green tourism*** and ***agritourism***.

All the above types of tourism could be grouped under **alternative tourism**, a type of tourism that:

- gives emphasis to the contact and understanding between the hosts and the tourist, as well as the environment;
- is consistent with the natural, social and community values and allows a positive relationship among locals and tourists;
- includes micro- and small companies of local inhabitants' property;
- has smaller impacts in the natural and social environments, links with other sectors (agriculture, craft) of the local economy and retention of earnings in the region.

The classifications that can be included under the concept of **alternative tourism** can be Natural, Cultural, Events and Others:

- the "Natural" (tourism that you can undertake in natural places, about the nature, and/or for the preservation of the natural environment) includes: adventure tourism, ecotourism, and nature tourism;
- the "Cultural" (tourism that involves contact and learning about a culture) includes archaeological tourism, ***rural tourism***, religious tourism, and ethnic tourism;
- "Events" (tourism interested in experiencing characteristic events of an area or important annual events) include sports, carnivals and festivals for example;
- "Others" include what we cannot include in the other classifications: volunteering, farm stays, educational tourism, etc.

Material and Method

We have inventoried 7 noun phrases designating different types of tourism related to the countryside and very in fashion nowadays: ***agrarian tourism***, ***agricultural tourism***, ***agritourism***, ***country(side) tourism***, ***farm tourism***, ***rural tourism***, and ***village tourism***. For each of these terms, we looked for possible definitions and for the different relationships they may have with one another (association, identification, inclusion, opposition) in an attempt to define them more clearly for our students.

Results and Discussion

This is what we have found out after searching over 40 Internet sites.

1. ***Agrarian Tourism*** (or ***Agrarian-Tourism***), whose main ingredients are simple attractions, a tot of relaxation and natural foods, tourism and nature, is considered an ***alternative type of tourism***. It is defined as '*a farm stay holiday*', '*a special holiday formula to spend in the genuine world of the farmer*', and as '*visiting farms with crops of fine fruits, cherry trees and vineyards, [...] tulip fields*'. It is identified with ***Agritourism*** and ***Country Tourism***.

2. ***Agricultural Tourism*** (or ***AgriCultural Tourism***), whose main ingredients are farm-based accommodations, meals, activities, farm festivals / events, and retail activities where the travelling public interacts directly with the farm family / workers, food, lodging, activities for people who want to experience farm life and nostalgia, farm festivals, activities, fun, affordable, family-oriented recreational and educational activities and opportunities to learn about the production of food and agricultural products and the state's rich farming heritage, fairs, festivals, is considered an alternative for improving the incomes and potential economic viability of small farms and rural communities, a way to practice sustainable tourism. It is defined as '*a growing trend in agriculture which merges the world of travel with experiences of farming and our food system*', '*a tourist activity organized and ran by the family agricultural farms*', '*exploring art roads and farm trails*', '*the act of visiting a working farm or any agricultural, horticultural or agribusiness operation for the purpose of enjoyment, education, or active involvement in the activities of the farm or operation*', '*the practice of visiting an agribusiness, horticultural, or agricultural operation, including a farm or winery or a companion animal or livestock*', and as '*visiting agricultural operations that throw their doors wide open for visitors*'. It is identified with ***Agritourism***.

3. ***Agritourism*** (or ***Agri-Tourism*** or ***Agro Tourism*** or ***Agrotourism***), whose main ingredients are watching or taking part in traditional agricultural practices, without disturbing the ecosystem or the productivity of host areas, assisting with farming tasks during the visit, picking fruits and vegetables, riding horses, tasting honey, learning about wine, shopping in gift shops and farm stands for local and regional produce or hand-crafted gifts, farm based accommodations, meals, activities, farm festivals - events, and retail activities where the travelling public interacts directly with the farm family / workers, and assisting with farming tasks while on vacation, is considered the strategy of using the farm to attract visitors for the purpose of education, enjoyment, or active involvement in the activities of the farm, an alternative for improving the incomes and potential economic viability of small farms and rural communities, and a means to prop up a local agricultural economy when local producers are no longer economically competitive otherwise. Its main point of reference is the village. It is defined as '*a commercial enterprise at a working farm, ranch, or agricultural plant conducted for the enjoyment of visitors that generates supplemental income for the owner*', '*a style of vacation in which hospitality is offered on farms*', '*any business conducted by a farmer for the enjoyment or education of the public*', '*any farm-based business offered for the enjoyment and education of the public*', '*farm based tourism*', '*the act of visiting a working farm or any agricultural, horticultural or agribusiness operation for the purpose of enjoyment, education, or active involvement in the activities of the farm or operation*', and as '*tourism in which tourists board at farms or in rural villages and experience farming at close hand*'. It is identified with ***Agrarian Tourism***, with ***Agricultural Tourism***, with ***Farm Tourism*** (on the East Coast), and with a subset of a larger industry called ***Rural Tourism*** that includes resorts, off-site farmers' markets, non-profit agricultural tours, and other leisure and hospitality businesses that attract visitors to the countryside. It is associated with ***Rural Tourism***. It is considered a subsct of ***Cultural Tourism*** or of ***Rural Tourism***. An ***agritourism business*** is a farm enterprise operated for the enjoyment and education of the public that may also generate additional farm income by promoting farm products.

4. ***Country Tourism*** (or ***Countryside Tourism***), whose main ingredients are nature, culture, tradition, gastronomy - concepts that our people have learnt to preserve and encourage -, is considered more and more popular. It is not defined anywhere. It is identified with ***Agrarian Tourism*** and with ***Rural Tourism***. It is associated with ***Farm Tourism*** and with ***Rural Tourism***. Practiced in the countryside, in the counties, and in the regions, it is opposed to ***Town Tourism***.

5. ***Farm tourism*** (or ***On-Farm Tourism***), whose main ingredients are farm visits with retail sales of locally-grown products, activities conducted on private agricultural lands, which might include fee-hunting and fishing, overnight stays, educational activities, etc., is considered one mechanism by which non-farmers can learn about agriculture. Its main point of reference is the village. It is defined as '*travel associated with farming*'. It is identified with ***Agritourism***. It is associated with ***Country Tourism***, ***Rural Tourism***. It is considered a subset of ***Cultural Tourism***.

6. ***Rural Tourism***, whose main ingredients are appreciating culture and rural activities in farmhouses (with tourists hosted and guided by the local people), attaching great significance to local values and to local cultural identity, making a tortilla, milking a cow, relying on the natural and heritage resources and people of rural communities, seeking rural peace, showcasing the rural life, art, culture and heritage at rural locations, travelling to and staying in rural areas (without mass tourism) other than those of their usual environment for less than one consecutive year for leisure, business and other purposes (excluding the exercise of an activity remunerated from within the place visited), witnessing or participating in activities that form the core of country life such as farming, is considered a little staid and school-trip-ish, a lot less sexy than ***Adventure Tourism***, a magic wand to stop rural decay, a tourism strategy in developed countries. It is not easy to give a definition of ***Rural Tourism*** because a generic model does not exist. The definition of ***Rural Tourism*** has been subject of much debate in the literature without arriving at any firm consensus. Most definitions tend to focus on the types of activities visitors engage with in a rural area, this leads to labelling of different tourism types. However, it is defined as '*any activity that takes place in a non-urban, populated area*', and as '*tourism away from the mainstream, away from areas of intensive tourism activity, engaged in by visitors who wish to interact with the rural environment and the host community in a meaningful and authentic way*'.

It is identified with *any tourism that is not urban tourism*, including ***Agritourism*** and ***Country Tourism***. It is associated with ***Agritourism*** and ***Countryside Tourism***, both forms closely associated with the basic requirements of sustainable development, and with ***Farm Tourism***. ***Rural Tourism*** differs from ***Agritourism*** in two ways: first, ***Rural Tourism*** enterprises do not necessarily occur on a farm or ranch, or at an agricultural plant; second, they do not generate supplemental income for the agricultural enterprise. Practiced by people who are mainly from the middle or upper classes, it is opposed to *Urban Tourism*. It encloses ***Agritourism***. It is also seen as a component of *Cultural Tourism* or of *Ecotourism* (although the reciprocal is also valid).

7. ***Village Tourism*** is not defined anywhere, but the phrase occurs in association with ***Country Tourism***.

Conclusion

From the point of view of their definition, the only element that seems to occur with the highest frequency is '*visiting*': the act of visiting a working farm (***Agricultural Tourism*** and ***Agritourism***), the practice of visiting an agribusiness (***Agricultural Tourism***), visiting agricultural operations (***Agricultural Tourism***), and visiting farms with crops' (***Agrarian Tourism***), the rest of definitions being singular in content.

Of the 7 'types of tourism' related to the rural area, ***Agritourism*** occurs 4 times as a synonym (for ***Agrarian Tourism***, ***Agricultural Tourism***, ***Farm Tourism***, and ***Rural Tourism***), which makes it a favourite denomination for countryside-related tourism, followed by ***Agrarian Tourism***, ***Country Tourism***, and ***Rural Tourism*** (2 occurrences each), and by ***Agricultural Tourism*** and ***Farm Tourism*** (1 occurrence each).

The term most associated with other 'types of tourism' related to the countryside is ***Rural Tourism*** (3 occurrences), followed by ***Country Tourism*** and ***Farm Tourism*** (2 occurrences each), and by ***Agritourism*** (1 occurrence).

The only clear antonymy is between ***Country Tourism*** and ***Rural Tourism*** on the one hand, and *Town Tourism* and *Urban Tourism* (obviously, synonyms), on the other hand.

The only clear inclusive relationship is between ***Rural Tourism*** and ***Agritourism***, the former including the latter.

It seems that the different **tourism types** are created from the experiences that tourists want to experience (nature tourism, cultural tourism, adventure tourism, etc.). Thus, each type of tourism is a way to give a denomination to a new market niche for a different experience.

References

Benson, M., Benson, Evelyn & Ilson, R. (1990). *The Combinatory Dictionary of English*. Amsterdam – Philadelphia: John Benjamins Publishing Company.

Carstairs-McCarthy, A. (2002). *An Introduction to English Morphology. Words and Their Structure*. Edinburgh: Edinburgh University Press.

Chalker, Sylvia & Weiner, E. (1994). *The Oxford Dictionary of English Grammar*. London - New York - Sydney - Toronto: BCA.

Chiţoran, D. (1973). *Elements of English Structural Semantics*. Bucureşti: E.D.P.

Collins English Dictionary. (2003). Online: http://www.thefreedictionary.com.

Columbia Electronic Encyclopedia. (2004). Online: http://www.reference.com.

Crystal, D. (1999). *The Penguin Dictionary of Language*. London: Penguin Books.

Dicţionarul explicativ al limbii române. [The Explanatory Dictionary of the Romanian Language]. (1998). Bucureşti: Editura U.E.

Dictionary of American English. (1997). London: Longman.

Harper, D. (2009). *Online Etymology Dictionary.* Online: http://www.etymonline.com.

Leviţchi, L. & Bantaş, A. (1995). *Dicţionar englez-român*. [English-Romanian Dictionary]. Bucureşti: Ed. Teora.

Leviţchi, L. (Ed.). (2004). *Dicţionar englez-român*. [English-Romanian Dictionary]. Bucureşti: Ed. Univers Enciclopedic.

Leviţchi, L. D. (1970). *Limba engleză contemporană. Lexicologie.* [Contemporary English. Lexicology]. Bucureşti: E.D.P.

Merriam-Webster Dictionary. (2011). Online: www.merriam-webster.com/dictionary.

Microsoft® Encarta® Encyclopedia. (2002). ©1993 – 2001 Microsoft Corporation.

New Standard Encyclopedia. (1995). Chicago: Standard Educational Corporation.

Raţă, G., & Perković, A. & Petroman, I. (2006). Special Types of Tourism: Tourism in the Countryside (A Terminological Approach). *5th International Scientific Days of Land Management in the Great Hungarian Plain, October 26-27, 2006, Mezőtúr, Hungary*: 1-4.

The American Heritage Dictionary of the English Language. (2008). Boston: Houghton Mifflin.

The New Encyclopaedia Britannica. (1992). Chicago: Encyclopaedia

Britannica, Inc.
The Random House Dictionary of the English Language. (1968). New York: Random House.
Webster Comprehensive Dictionary. (1995). Chicago: J. G. Ferguson Publishing Company.
Webster Encyclopaedic Unabridged Dictionary of the English Language. (1996). New York: Gramercy Books.
Webster's Universal Dictionary & Thesaurus (2003). New Lanark: Geddes & Grosset.
www.agnet.org
www.agritourismitaly.com
www.backyardnature.com
www.bartleby.com
www.bulgariatravel.org
www.ceres.ca.gov
www.ces.ncsu.edu
www.country-vacations.com
www.craigmarlatt.com
www.crctourism.com.au
www.csl.doc.govt.nz
www.ctahr.hawaii.edu
www.deba.net
www.dekust.org
www.duderanches.com
www.ec.europa.eu
www.ecoturismolatino.com
www.ecoturismorural.com
www.en.wikipedia.org
www.ersa.org
www.eupronet.cz
www.exploreruralindia.org
www.farmhouse-holidays.com
www.fleurop.zce.szczecin.pl
www.forum.europa.eu.int
www.greek-ecotourism.gr
www.greentour.rrdi.go.kr
www.health-tourism.info
www.homegrownhandmade.com
www.ilcontadino.it
www.invitationtotuscany.com
www.johnstonshomestyle.pe.ca

www.maisontourisme.nassogne.marche.be
www.mass.gov
www.nal.usda.gov
www.satglobal.com
www.senate.michigan.gov
www.sfc.ucdavis.edu
www.state.nj.us
www.terrazasdeesquel.com
www.tourisme-gaspesie.com
www.trrworld.org
www.tuscany.nucl.be
www.uwadmnweb.uwyo.edu
www.villa-branko
www.visitkankakeecounty.com
www.wikipedia.org
www.wkycorp.org
www.zum.lt

Chapter Four

Semantics

LANGUAGE TRAPS: VEGETABLE OR FRUIT?

OANA BOLDEA

Introduction

When learning a foreign language, the first thing one is taught is the vocabulary considered basic, and among the first words to introduce, irrespective of the learner's age, are those referring to vegetables and fruits, together with those expressing names of animals. Thus, ***pear***, ***potato***, ***dog***, ***tomato***, ***orange***, ***tiger*** are among the first words in English for any learner of this language.

Things get a little bit more complicated when the learners are students at an agricultural university. Sometimes things are not what they seem to be, and language can be misleading in certain ways. The problem in teaching vegetables and fruits in this context has a cultural nature: it arises because of the difference in classification between the traditional context and the scientific one. Thus, many parts of plants traditionally considered vegetables are fruits for a botanist.

Intrigued by this aspect, we decided to try and see whether the language can help us in separating one category from another. In other words, this paper presents an attempt to identify a linguistic logarithm (if any), which enables the speaker to categorize a part of a plant as being a vegetable or a fruit. We were interested in possible suffixes or prefixes, or in any other way of determining the classification of a plant taking its name into consideration.

Material and Method

In the attempt to solve the problem of vegetables and fruits in English (and Romanian, for that matter, because the difference between the traditional use of the terms and the scientific one is the same in this language), we first turned to the books which could provide the proper definitions of these terms. Thus, we focused our attention first on those books which could enlighten us regarding the scientific classification of plants. In this respect, we consulted books and courses in botany.

The next step was to turn towards the books considered to be basic for all linguists: dictionaries. There were two types of dictionaries we took into consideration: the so-called "general" language dictionaries and encyclopaedias, i.e. dictionaries displaying information on all (or at least most) aspects of human interest, such as the *Webster's Encyclopedic Unabridged Dictionary of the English Language* and the *Microsoft® Encarta® Encyclopedia*.

The second type of dictionaries consulted were "specialised" dictionaries, i.e. dictionaries which give information in a certain field. This is how we came across Borza's *Dicționar etnobotanic* (1968) and Vaczy's *Dicționar botanic polyglot* (1980), but we also looked into special dictionaries like Usher's *Wordsworth Dictionary of Botany* (1996), which also proved to be extremely helpful for our research.

The method adopted in solving the problem already presented in the introduction was that of lexical analysis. We looked at the vernacular and scientific names of various plants or parts of plants traditionally considered to be vegetables, and to names of plants traditionally treated as fruits. What we wanted to find was a possible lexical feature which should make the distinction between vegetables and fruits. If we had found any, then we would have made our students' task of learning new words in this field a lot easier.

Results and Discussion

We started from the definitions given to "vegetable" and "fruit", and what we noticed was that the meanings they offered were different from one dictionary to another. The *Microsoft® Encarta® Encyclopedia*'s definition for the word ***vegetable*** reads 'the edible product of a herbaceous plant – that is, a plant with a soft stem, as distinguished from the edible nuts and fruits produced by plants with woody stems such as shrubs and trees'. This signifies that the fruits of herbaceous plants are vegetables. Although this first definition of the word was not satisfactory for us, it helped us move along: at least we had the evidence that the matter under discussion was a real issue, since it left the reader still wondering.

The *Wordsworth Dictionary of Botany* gives another definition: a ***vegetable*** is 'any food plant, especially leafy plants, roots and tubers, legumes and even flowers'. Although it seems to shed a bit more light on the matter, as it gives some examples, the definition is still not enough to make the reader understand the term.

The most complex definition seems to be the one provided by the *Webster's Encyclopedic Unabridged Dictionary of the English Language* (1996): ***vegetable***, 'noun 1. any herbaceous plant whose fruit, seeds, roots, tubers, bulbs, stems, leaves or flower parts are used as food; 2. the edible part of such a plant; 3. any member of the vegetable kingdom; plant; 4. a dull, spiritless, and uninteresting person'. And it goes on with the definition of the same word when used as an adjective. Since only the meaning of the noun was of interest to us, this is the only one we have cited here. Also, because of the fact that definition (4) is a figurative meaning, it is of no interest to us.

We can see from this last definition that the word ***vegetable*** actually seems to have at least three layers of understanding: the largest one, which includes a smaller one, which, in its turn, includes the smallest of all three, as we have tried to represent in Figure 4-1.

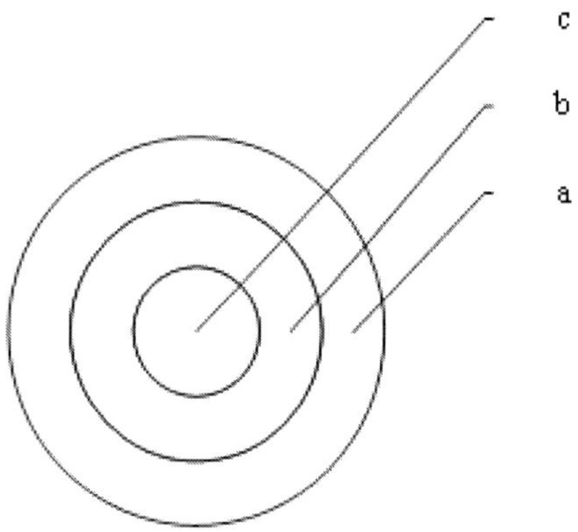

Figure 4-1. The concentric meanings of the word "vegetable": a – plant, in general; b – (herbaceous) plant whose parts are used as food; c – plant part

In what the term ***fruit*** is concerned, the definitions are all alike though more or less detailed: they all indicate that, in botany, a ***fruit*** is the ripened ovary in flowering plants, with its contents and accessory parts. Some dictionaries go further in saying that a ***fruit*** is the edible part of a plant developed by a flower. This last definition is, actually, a mixture between the scientific definition of the term and the cuisine one, which gives, for the same word, the meaning of the sweet and fleshy part of a plant that can be eaten by people, examples being the apples, oranges or plums.

When consulting scientific books in order to find more about the term ***fruit***, we could see that many of the plants or parts of plants commonly treated as vegetables are, in fact, fruits. A good example is provided by one of the most used vegetables, the common ***tomato***. What we eat from the plant *Lycopersicon esculentum* is exactly its edible fruit. Cut in half, a ***tomato*** displays distinct sections, each representing a separate carpel with many seeds. This makes it a true "berry", a type of fleshy simple fruit. In the case of berries, the entire pericarp, with the exception, in some cases, of the rind, becomes soft and fleshy, while the seeds are embedded in the pulp, so in the fleshy portion. Coming from the same genus as the tomato, namely Solanaceae, the ***eggplant*** is another perfect example of a true berry, with the seeds embedded in the greyish pulp. In the same way, the ***cucumber*** we eat is, actually, the unripe fruit of the plant botanically named *Cucumis sativus*. The botanical name of the type of fruit represented by the ***cucumber*** is a "pepo", which is a kind of berry, but having a hard rind. The same holds also true for the ***squash***, which is the fruit of one of the plants pertaining to genus *Cucurbita*.

What seemed to be extremely interesting for us was that, while some parts of plants popularly considered vegetables proved to be berries, some others, considered berries, were not berries at all. This is the case of the ***raspberry*** or ***strawberry***, which develop differently from the flower and are not really berries, although they misleadingly contain the word "berry" in their names. ***Raspberries*** actually belong to the class of "aggregate fruits", namely fruits consisting of many ripened ovaries produced by a single flower. At the same time, ***strawberries*** are "accessory fruits". An accessory fruit contains a mature ovary, but it consists mainly of tissues other than the ovary itself, such as other flower and stem parts. A ***strawberry*** consists of many small achenes on a fleshy, enlarged receptacle.

The "drupe" is the other type of fleshy fruit besides the "berry". It differs from the latter in that the seed is contained in a hard stone placed in the centre of the fleshy fruit. For this class, too, we found examples of fruits eaten as vegetables. It is the case of the ***olives*** and of the ***avocado***.

Simple fruits can be fleshy or dry. We have already dealt with the examples of vegetable-like fleshy fruits. But we have also found such examples of dried fruits. First, we should make clear that "dry fruits" are called "dry" because, when they mature, the entire wall becomes dry instead of fleshy, as for the ones mentioned before. ***Beans***, ***lentils***, ***peas***, and ***soybeans*** produce pods which are considered by botanists to be "dehiscent dry fruits" because, when they mature, they split open for the seeds to emerge. So, when eating ***green beans***, we do not eat a vegetable, but a fruit, and when consuming ***peas***, we actually eat the seeds coming from such a fruit.

Another class of vegetables which are actually fruits is made up of the cereals. ***Wheat***, ***maize***, ***oats***, ***barley***, ***rye***, and ***rice*** are all plants from which people consume the fruit. They make up an entire class of fruits, called "caryopses", or "grains", belonging to the bigger class of "indehiscent dry fruits". A fruit is indehiscent if it does not split when ripe, and grains are one-seeded fruits in which the outer husk is tightly attached to the entire seed.

Table 4-1 contains a coarse classification of fruits, with examples of botanical fruits popularly considered to be vegetables. For the other three main classes of fruits occurring in botanical courses (aggregate, multiple and accessory) together with the simple fruits, we did not find examples of fruits considered vegetables. That is why we did not include them in the table, as the aim is not to present the entire classification of fruits, but only the part which is of interest to us, namely that for which we have found examples of fruits commonly treated as vegetables.

Table 4-1. Fruits traditionally considered vegetables

Type of fruit				**Fruits traditionally considered vegetables**
Simple	**Fleshy**	**Berry**	**true berry**	*eggplant, green pepper, pimento, red pepper, tomato*
			pepo	*cucumber, gourd, pumpkin, squash*
		Drupe		*avocado, olive*
	Dry	**Dehiscent**	**legume**	*pods of bean, lentil, pea, soybean*
			silique	*pods of mustard*
		Indehiscent	**caryopsis** or **grain**	*barley, corn, oats, rice, rye, wheat*
			schizocarp	*dill*

For the Romanian learners of English, there is another problem, represented by the word ***legume***. This word is a false friend for the Romanians, as it very much resembles the Romanian word ***legumă*** 'vegetable'. Thus, when reading that a pod of ***pea***, for instance, is a "legume", they will tend to interpret the information as 'a pod of pea is a vegetable'. Actually, a "legume" designates a type of dry fruit.

Another very interesting fact is that popularly vegetables include not only fruits, but also organisms which are not plants at all. For instance, ***algae***, which although resemble plants, are not considered by botanists to be plants, since they lack the roots, leaves and other structures typical of true plants. ***Mushrooms*** are another example. They are actually "fungi", and not "plants", but still are treated as "vegetables".

We were interested in seeing whether **word derivation** or **word composition** helped building up the names of the plants analyzed. **Derivation** consists in adding affixes to the stems of different words in order to form new words. Affixes may be classified from various points of view: according to their origin, to the parts of speech they serve to form, to the meaning they render to the parts of speech, to their productivity, etc. We have been particularly interested in the classification according to the part of the word they are added to (the beginning or the end, thus being called prefixes and suffixes, respectively), and the meaning (if any) they bring to the word. Unfortunately, we could not find any affix to help us in reaching at a special "formula" telling us whether a "vegetable" is a fruit or not. This means that the names of vegetables and fruits have to be learned as such.

Composition also proved to be useless. There are no words which, placed next to others, indicate whether the object named is a vegetable or a fruit. Moreover, as we have seen, **composition** can be misleading, as in the already-mentioned case of "berry". Many 'berry' **compounds** do not refer to "berries" (in a botanical sense) at all. The same holds true for the **compounds** of "nut". A "nut" is a type of "indehiscent dry fruit" whose single seed is contained in a woody shell. That is how, although called ***peanut***, the fruit bearing this name is actually a legume (it has a pod with many seeds which splits along two seams). Much in the same way, the ***coconut*** and the ***walnut*** are not true nuts: botanically, they fall into the category of "drupes".

Other findings of our research were that some ***spices*** are also fruits. It is the case of the ***allspice*** and of the ***nutmeg***.

And, to our amazement, we found that, sometimes, some vegetables which are not botanically "fruits" are considered to be so, not by specialists, nor by common people, but by institutions. It is true that it happens extremely rarely, and for various reasons. This is the case with the ***carrot***, although everybody knows that it is actually a root. Indeed, under European Union trade rules, a ***carrot*** is defined as a "fruit", presumably because fruits are taxed at a higher duty and carrot jam is a popular Portuguese dish (according to www.gardenerskitchen.co.uk).

Conclusions

In every-day life, the plants or plant parts people eat are roughly separated into two large classes: vegetables and fruits. Usually, people consider that fruits are the sweet parts of plants which can be eaten raw or made into jams while vegetables are usually those we cook for eating, or at least which are not sweet. Our attempt in this paper was to clarify which of the commonly-called ***vegetables*** were botanically ***fruits***. The problem occurs especially when teaching English to students at an agricultural university. Here the matter does not seem as insignificant as it would seem to anybody else: accuracy regarding plants is of great importance in this field.

Then, the aim was to track any lexical sign which would help the students separate vegetables from fruits in what their names are concerned. We found that many of the so-called vegetables are, actually, fruits. Such is, for instance, the ***tomato***, which is a true berry botanically, as opposed to ***strawberry*** or ***raspberry***, which, although containing this word, are fruits of totally different kinds: aggregate and accessory, respectively. ***Cucumbers***, ***pumpkins*** and ***squashes*** are also fruits. ***Cereals*** fall into the category of dry fruits, although this is known only to botanists. And there are other examples of vegetable-like fruits we discussed in the paper.

We also clarified the problem of the false friend "legume", which has by no means the Romanian meaning: it designates a sub-class of dry fruits.

If for the first problem (identifying vegetables from fruits), we had encouraging results, for the second we did not. We could not find any lexical pattern to help us tell vegetable from fruits.

References

Boldea, Oana. (2005). Language Traps: Vegetable or Fruit? *Proceedings of the Jubilee Scientific Conference with International Participation "State-of-the-art and Problems of Agricultural Science and Education", Plovdiv, Bulgaria*: 239-244.

Borza, Al. (1968). *Dicţionar etnobotanic*. [An Ethnobotanical Dictionary]. Bucureşti: Editura Academiei.

Gardeners Kitchen. Online: www.gardenerskitchen.co.uk.

Leech, G. (1989). *An A-Z of English Usage*. London: Longman.

Matthews, P. H. (1991). *Morphology*. Cambridge: Cambridge University Press.

Microsoft® Encarta® Encyclopedia. (2002). ©1993 – 2001 Microsoft Corporation.

Raţă, Georgeta. (1999). *English for Horticulturists*. Timişoara: Editura Mirton.

—. (1999). *Curs practic de limba engleză – profilul agricol (anul al II-lea de studiu)*. [A Practical Course of English – Agriculture (2nd year)]. Timişoara: Editura Mirton.

Usher, G. (1996). *The Wordsworth Dictionary of Botany*. Reston, VA: Wordsworth Editions Ltd.

Vaczy, C. (1980). *Dicţionar botanic poliglot*. [A Multilingual Botanical Dictionary]. Bucureşti: Editura Ştiinţifică şi Enciclopedică.

Webster Encyclopedic Unabridged Dictionary of the English Language. (1996). New York: Gramercy Books.

HOMONYMY AND SYNONYMY

OANA BOLDEA

Introduction

The present paper is meant to be a lexical analysis of some aspects of the English vocabulary related to water used in agriculture. It deals with some words which prove to be, in a way, more difficult to understand or use by non-native English speaking students. This difficulty comes exactly from the fact that words are either homonymous or polysemous, i.e. the students already know the word, but with another meaning, usually the primary meaning used in everyday speech. The paper tackles some of these words in an attempt to show that as far as **homonymy** is concerned, students have a hard time coping with the new meaning; in the case of **polysemy**, things may be different in the sense that if the new meaning is, in a way or another, related to the one already known by the student, it is easier to keep in mind and to use.

When learning a foreign language, it is quite difficult to keep in mind the new words one needs for conveying a message. This task is even more complicated when another process interferes, namely getting used to more meanings of the same word. This is an understandable phenomenon: it is only natural to find it difficult to get used to new contexts and meanings for a word the "meaning" of which one has "taken for granted" for some time.

Students in agricultural sciences have to go through this experience when they find themselves in the second year of university. In the first year, they study basic, common, everyday English, just enough to help them sustain a conversation on general topics. But when they are sophomores, they pass on to a different level, namely that of in-taking the specialized vocabulary they need to understand articles or books written in English on specific topics related to their future profession, or to write such works themselves.

This paper is based upon our observations in class on what words seem to really be new for the students and for which words they only learn new meanings. Naturally, as in English (or in any other language for that matter) the number of meanings is greater than that of words, a lot of words prove to be polysemantic. The actual number of meanings of the commonly used words ranges from five to about a hundred (Matthews 1993). The commoner the word, the more meanings it has. In the following pages, some of these words will be looked at more closely.

Very important to mention is the fact that it is not the intention of this paper to deal with all the possible meanings of the words analyzed. As stated before, we will focus on the one meaning found in relation to agriculture, and one, two or a maximum of three of the most representative meanings of the word in everyday English.

Material and Method

The approach used is that of semantic analysis. Each word taken into consideration was analyzed from the point of view of **homonymy** and **polysemy**. We checked whether the word had more meanings than that with which it occurred in the source and, if so, whether those meanings were more commonly used than the one occurring there. The premise was that the "agricultural" meaning was less common or newer than others.

For the analysis, sometimes the etymological approach proved useful. Therefore, we looked at the origin of words to see whether the meanings had a common root, started from the same etymon or not, because if so, then, at least at the beginning, they had something in common. If they had something in common, then they were not **homonyms**, but rather different meanings of the same word, that is a **polysemantic word**.

The corpus we have used in this study was provided by the book *Water and Agriculture* (Raţă *et al.* 2004), a collection of terms related to water and agriculture. The book addresses students in agriculture. We have chosen this book because it supplies a limited number of entries, as opposed to very extensive, comprehensive dictionaries. Although the book comprises terminology in French, German, and Romanian besides English, only the latter was taken into account in the present study. Nevertheless, when the equivalent in another language was relevant for the analysis, it did get some attention. Thus, 3,143 terms were analyzed from the perspective of multiple meanings. Of these, only those 100 most commonly met with in everyday life found their way into the present paper as illustrations.

The theoretical background on **homonymy** and **polysemy** can be traced back to books written by Romanian researchers (Salapina 1974; Leviţchi 1975; Bădescu 1984), as well as in works of well-known foreign scholars (Leech 1989; Matthews 1993; Swan 1995; Crystal 1999). For the practical side of the paper, the *Webster Encyclopaedic Unabridged*

Dictionary of the English Language and the *Dictionary of American English* were the tools used in finding the meanings of words, while the *Dictionary of Word Origins* was of great help, together with the *Webster Encyclopaedic Unabridged Dictionary of the English Language*, for establishing the etymologies of certain terms. Other relevant sources are mentioned in References.

Results and Discussion

Both **homonymy** and **polysemy** are terms coming from Greek, and both deal with the meaning of words. While ***homonymy*** comes from *homo*- ('the same') and -*onym* ('name'), thus having the significance 'having the same name', ***polysemy*** is built from *poly*- ('many') and *sema* ('sign'), meaning 'with many significations'. The problem of criteria distinguishing **homonymy** from **polysemy**, i.e. between the different meanings of one word and the meanings of **homonyms**, is one of long-debated problems in semasiology. As a matter of fact, there is not as yet a universal criterion for the distinction between the two phenomena (Salapina 1974). Below is a description of the two as applied in the present analysis.

We only dealt with the so-called **perfect (absolute) homonyms**, or **homonyms proper**, which are identical both in pronunciation and spelling, different in meaning only, leaving aside the **homographs** (identical in spelling, but different both in sound and meaning), the **homophones** (identical in sound, but different in spelling and meaning) and the **homoforms** (different in meaning but identical in some of their grammatical forms). In other words, we looked for **full lexical homonyms**.

The sources of **homonymy** are exactly the problem in differentiating between **homonyms** and **polysemantic words**: they can have come into being either through *convergent sound development* (words of different origin accidentally coincide in sound and, consequently, in graphic form, mostly after losing the inflections), or through *divergent sense development* (also called **split polysemy**). The second source is, actually, the cause of the problem which occurs when trying to establish whether a word belongs to a class or to another. The break-up of **polysemy** is due to the fact that it is natural for **polysemantic words** to develop meanings which, in the course of time, may deviate very far from the original one. When the intermediate links happen to fall out, some of these meanings lose all connection with the rest of the structure and start a separate existence (Salapina 1974).

As there are many cases when the demarcation line between **polysemy** and **homonymy** is rather unclear (Matthews 1993), in the present paper those words were considered to be **homonyms** which did not have a common etymology, or those which presented no connection between various meanings; if the meanings were felt as being connected, they were considered as belonging to the same **polysemantic word**.

From a synchronic point of view, researchers have noted different degrees in the importance or frequency of various meanings. Some meanings are called "basic" and others "secondary", or "minor". It is the basic meaning that is given first in language dictionaries; it is followed by the other(s), not so frequently used. What the present study aimed at was establishing whether any of the terms gathered in the "water and agriculture" compendium were used in their primary meaning, starting from the hypothesis that there were very few of those, if any.

Conclusions

The study revealed that there were much more **homonyms** and **polysemantic words** than expected at the beginning.

The **homonyms** we found can be separated into different classes. One such class of terms is made up of the words the students already know in the form of proper names. Such is, for instance, the word ***Eddy***, which written without capital letter becomes a common noun having the meaning 'whirlpool'. The same happens with ***Ford***, the brand name of one of the best-known vehicles in the world: when written without capital letter, it denotes 'a place where a river or other body of water is shallow enough to be crossed by wading'. The same is true for ***Dutch***, which the students know to denote nationality, but which, when spelt as a common noun, means 'socket'. Things are somehow different when we deal with the word ***Peg***. When capitalized, it is a woman's name, and when not, it becomes a **common homonym**. The complication arises here, because the **homonym** of ***Peg*** is a **polysemantic word**, the meaning already known to the students being that of 'hanger', 'rack'. Among the other many meanings this word displays, the one the students have to learn is 'a pin of wood or other material driven or fitted into something, as to fasten parts together or to stop a hole'.

Then there is the class of the **homonyms** some of the students may know, as they are more commonly used in both senses. Such is, for instance, ***bank***, which denotes both 'an institution for receiving, lending, exchanging and safeguarding money', and 'the slope immediately bordering a stream course along which the water normally runs'. In the case of ***bank***, etymology also helps us prove that we deal with two **homonyms** and not with a **polysemantic word**. While the first comes from the Italian *banca* 'counter', the second has its origins in the Old Danish ***bake*** 'elevation, hill'. Another pair of words likely to be known by students is ***ram***, which may mean either 'the he-sheep' or 'a strike with great force'. Yet another word pertaining to this class is ***beam***, which means 'any of various relatively long pieces of metal, wood, stone, etc., manufactured or shaped especially for use as rigid members or parts of structures or machines' and also 'a ray of light'. Another example is ***shed***, which is best known to the students as the verb meaning 'to weep, to express one's regret, sorrow, grief or other overpowering emotion through tears', but which also, as a noun, means 'a rude structure built for shelter, storage, etc.'.

Another class is that of **debatable homonyms**: in this case, the meanings seem to have something in common, so they might be taken as **polysemantic words**, they are in-between. Thus, students might guess what they mean even if they have not encountered that specific meaning before. Such is the case of the expression ***water table***, denoting 'the depth below which the ground is saturated with water', the meaning of which can be arrived at if one thinks of the fifth meaning given for the word ***table*** in the dictionary, namely 'a flat or plane surface; a level area'. The same holds true for ***weep hole***, which denotes 'a hole for draining off accumulated moisture', as opposed to the word ***weep*** meaning 'shedding tears'. ***Water pocket*** is another such case. Although ***pocket*** denotes 'a shaped piece of fabric attached inside or outside a garment and forming a pouch used especially for carrying small articles', if one keeps in mind that it has come to denote any pouch-like cavity, then it is not too hard to understand that ***water pocket*** means 'plunge basin'. In much the same way, if one knows that ***crab*** denotes a crustacean with a more or less flattened body, catching its prey with the help of claws, then one might understand that, in certain contexts, it may come to mean 'pipe grab'. ***Chinese hat*** is the denomination for a 'shutter', the conic shape of which is easy to guess from its name.

Polysemantic words are present in great numbers too: a ***blind drainage area*** is 'an area closed at one end' (as the students probably know from the expression ***blind alley***), a ***bed*** can be 'a piece or area of ground in a garden or lawn where flowers are grown'. An ***arm*** is not only 'the upper limb of the human body', but also 'any arm-like part or attachment, as of a lever on a machine', thus ***an overhanging arm*** may belong to someone who has fallen asleep half out of bed, but also to a machine. A ***knee joint*** in agriculture is easily understood after the students are reminded of the primary sense of the word ***joint***, that of 'the place at which two things or two separate parts of the same thing, are united, either rigidly, or in such a way as to admit motion'. Unfortunately, watching too many American movies, students are much more exposed to the **homonyms** of the word, which in slang means either 'cigarette with marijuana' or 'a cheap bar with a bad reputation'.

The word ***mole*** is, in a way, one of the most interesting found in the database, because it is part of a large "family" of **homonyms**: it has no less than 5 meanings! The first one to be learned in school is that of 'a small insectivorous mammal living chiefly underground and having velvety fur, very small eyes and strong, fossorial forefeet'. Then, one finds out that it also means 'a small, congenital spot or blemish on the human skin'. ***Mole*** is also used in chemistry, where it denotes 'the molecular weight of a substance expressed in grams'. Students in agriculture, then, learn the meaning of 'a massive structure, especially of stone, set up in the water, as for a breakwater or a pier'. And the fifth meaning is a medical term denoting 'a fleshy mass in the uterus formed by a hemorrhagic dead ovum'.

Another word which calls for our attention is ***head***, which has many varied meanings, even in agriculture. It may mean 'pressure', so the expression ***loss of head*** actually reads 'loss of pressure', and has nothing to do with 'to lose one's head over somebody'. ***Working head*** does not denote somebody who is thinking hard, but rather 'the height a pump can bring water to'. Even in a phrase containing the word *water*, it may have different meanings. Thus, ***head of water*** expresses 'the height of a water spout / column', while ***headwater*** denotes 'the upper tributaries of a river'.

There are also those **homonyms** that were not included in the other classes presented above. They form a distinct class and are the most numerous of all. Generally, students do not know the ones related to agriculture when they come into the second year and there is no way they can guess the meaning from the **homonym** used in the everyday language. Such a word is ***dolphin***, which, besides the 'mammal living in the ocean', also means 'a cluster of piles used as a fender'. A ***borehole*** is not a hole which annoys people to death, as one might guess knowing only the meaning of the **homonym**, but 'a hole drilled in the earth for the purpose of getting at water'. A ***riddle*** is, in everyday English, a question or statement so framed as to exercise one's ingenuity in answering it. Little do the students know when coming in the second year that the same word denotes 'a coarse sieve, as one for sifting sand'. As for ***race***, the only meaning they know is that related to sports, but they soon find out that it also means 'an artificial channel leading water to or from a place where its energy is put to use', or even 'the current of water in such a channel'. ***Prime*** means 'to pour or admit liquid into a pump, in order to expel air and make it ready for action', as opposed to the well-known meaning in the television-related expression ***prime time***. ***Sounding*** does not only mean 'producing sounds' or 'pompous', but also 'the act of measuring'.

As we could see, much of the new vocabulary meant to be introduced to students is, actually, based on already-known words to which new meanings are attached. If the meanings are related, then we are dealing with **polysemantic words**. If not, then the words are **homonyms**. They can be then parted into different subclasses, according to different criteria. But in the end, the context, the domain we tackle helps us understand which of the meanings the right one to use or to interpret is.

References

Bădescu, Alice. (1984). *Gramatica limbii engleze*. [English Grammar]. Bucureşti: Editura ştiintifică şi enciclopedică.

Boldea, Oana. (2004). On the Phenomenon of Homonymy and Polysemy Regarding the Agricultural Terminology. *Proceedings of the 4th International Scientific Days of Land Management in the Great Hungarian Plain, Mezőtúr, Hungary*: 5 p.

Crystal, D. (1999). *The Penguin Dictionary of Language*. London: Penguin Books.

Dictionary of American English. (1997). London: Longman.

Flavell, Linda & Flavell, R. (1995). *Dictionary of Word Origins*. London: Kyle Cathie Limited.

Leech, G. (1989). *An A-Z of English Usage*. London: Longman.

Leviţchi, L. (1975). *Limba engleză contemporană - Lexicologia*. [Contemporary English: Lexicology]. Bucureşti: Editura Academiei R.S.R..

Matthews, P. H. (1993). *Morphology*. Cambridge: Cambridge University Press.

Raţă, Georgeta, Moisuc, Al., Samfira, I., Silion, Raluca-Felicia Silion & Boldea, Oana. (2004). *Water and Agriculture*. Timişoara: Editura Mirton.

Raţă, Georgeta.(1999). *Curs practic de limba engleză – profilul agricol (anul al II-lea de studiu)*. [A Practical Course of English – Agriculture (2nd year)]. Timişoara: Editura Mirton.

Salapina, Galina. (1974). *Limba engleză contemporană*. [Contemporary English]. Tipografia Universităţii din Timişoara.

Swan, M. (1995). *Practical English Usage*. Oxford: Oxford University Press.

Webster Encyclopedic Unabridged Dictionary of the English Language. (1996). New York: Gramercy Books.

Defining Concepts and Practices Associated with Sustainable Agricultural Systems

Georgeta Rață and Anica Perković

Introduction

Understanding agricultural terminology at academic level is of utmost importance for researchers, teachers, and undergraduates in agricultural science that need to classify, describe, discuss, explain, identify, locate, paraphrase, recognize, report, select, and translate ideas or concepts. Understanding this terminology also means:

- thinking about it and using concepts to deal adequately with it;
- being able to apply it to situations likely to be encountered, to recognize significant deviations, and to carry out the research necessary to arrive at reasonable solutions;
- using concepts and bringing individuals under them;
- reasoning and making judgements relating to its applicability.

In this context, we need to point out that specialised terminologists have noticed that most of the **concepts and practices associated with sustainable agricultural systems** defy *definition*, i.e. they do not delimit or describe the meaning of a concept or term by stating the essential properties of the entities or objects denoted by the concept.

Material and Method

We have analysed the definitions of 18 different **concepts and practices associated with sustainable agricultural systems** among the 79 ones supplied by Mary V. Gold's in her reference compilation *Sustainable Agriculture: Definitions and Terms* (1999). These definitions were then analysed from the point of view of their meaning as suggested by Simpson & Weiner in their *Oxford English Dictionary* (1989).

Results

We have inventoried 79 **concepts and practices associated with sustainable agricultural systems** in usage (and abusage) nowadays in Gold's reference compilation (1999). Eighteen of these **concepts and practices** not labelled as *conventional modern farming* are given tentative definitions. We have analysed these definitions in an attempt to see why they are misunderstood by our undergraduates (and, sometimes, even by our colleagues teaching agricultural subjects).

Sustainable agriculture is similar to the type of agriculture that was practiced in the early 1900's and involved purchasing few inputs and marketing little of the outputs. "If nothing else, the term 'sustainable agriculture' has provided 'talking points', a sense of direction, and an urgency that has sparked much excitement and innovative thinking in the agricultural world." says Gold in the *Introduction* to her compilation (1999). She could not be more right, and the analysis below is the best example.

A. Of the 18 definitions, 16 (88%) are *definitions by genus and difference* or *genus-differentia definitions*, i.e. *intensional definitions* (definitions that specify all and only the properties required of something in order that falls under the term defined, i.e. its necessary and sufficient conditions), *in which a word or concept that indicates a species* (a specific type of item) *is described first by a broader category it belongs to, the genus, and then distinguished from other items in that category by differentia* (i.e. specific properties): ***alternative agriculture / farming*** is '[a *farming method* encompassing] a vast *array of [agricultural] practices and enterprises*, all of which are considered different from *prevailing or conventional agricultural activities* […]: nontraditional crops, livestock, and other farm products; service, recreation, tourism, food processing, forest / woodlot, and other enterprises based on farm and natural resources (ancillary enterprises); unconventional production systems such as *organic farming* or *aquaculture*; or direct marketing and other entrepreneurial marketing strategies' (Grudens Shuck *et al.* in Gold 1999); ***biodynamic agriculture / farming*** is '[a *farming method* encompassing] *specific [agricultural] practices and preparations* that enable the farmer or gardener to work in concert with [the forces within living nature]' (Gold 1999); ***biointensive gardening*** or ***mini-farming*** is 'a *production system* that makes it possible for one person to grow all of his or her family's food using truly sustainable methods that maintain the fertility of the soil without relying on non-renewable resources like petrochemicals or imported organic matter' (Jeavons in Gold 1999); ***biological / ecological***

farming is '[a *farming method* encompassing] *various and more specific practices and techniques of farming sustainability*, e.g., *organic, biodynamic, holistic, natural*' (Gold 1999); ***biological farming*** is 'a *system of crop production* in which the producer tries to minimize the use of 'chemicals' for control of crop pests' (Pesek in Gold 1999); ***do-nothing farming*** or ***natural farming*** is 'a *farming method* involving no tillage, no fertilizer, no pesticides, no weeding, no pruning, and remarkably little labour […] by careful timing of his seeding and careful combinations of plants (polyculture)' (Gold 1999); ***Kyusei nature farming*** '[is a *farming method* that] often employs technology involving beneficial microorganisms as inoculants to increase the microbial diversity of agricultural soils, which, in turn, can enhance the growth, health, and yield of crops' (*Idem*); ***low input agriculture*** '[is a *low input farming system* that] seeks to optimize the management and use of internal production inputs (i.e., on-farm resources) and to minimize the use of production inputs (i.e., off-farm resources) such as purchased fertilizers and pesticides, wherever and whenever feasible and practicable, to lower production costs, to avoid pollution of surface and groundwater, to reduce pesticide residues in food, to reduce a farmer's overall risk, and to increase both short- and long-term farm profitability' (*Idem*). The term is 'somewhat *misleading* [we underline] and indeed unfortunate. For some it implied that farmers should starve their crops, let the weeds choke them out, and let insects clean up what was left. In fact, the term low-input referred to purchasing few off-farm inputs (usually fertilizers and pesticides), while increasing on-farm inputs (i.e. manures, cover crops, and especially management). Thus, a more accurate term would be *different input [agriculture]* or *low external input [agriculture]* rather than *low-input [agriculture].*' (Norman 1997); ***nature farming*** is a *farming system* utilizing 'the inherent power of the soil' and focusing on 'analyzing and building soil through composting, green manuring, mulch, and various other soil management techniques' (Gold 1999); ***organic farming*** is 'a *production system* which avoids or largely excludes the use of synthetically compounded fertilizers, pesticides, growth regulators, and livestock feed additives. [It relies] upon crop rotations, crop residues, animal manures, legumes, green manures, off-farm organic wastes, mechanical cultivation, mineral-bearing rocks, and aspects of biological pest control to maintain soil productivity and tilth, to supply plant nutrients, and to control insects, weeds and other pests.' (*Idem*); ***permaculture*** (from *perma*nent agri*culture*) is 'one of the many alternative *agriculture systems* described as sustainable [emphasising] on design; that is, the location of each element in a landscape, and the evolution of landscape over time, [whose goal is] to produce an efficient, low-maintenance integration of plants, animals, people and structure applied at the scale of a home garden, all the way through to a large farm' (*Idem*); ***precision agriculture / farming*** or ***prescription farming*** or ***site-specific management*** is 'a MANAGEMENT STRATEGY that employs detailed, site-specific information to precisely manage production inputs' (*Idem*); ***regenerative / sustainable agriculture*** or ***low-input sustainable agriculture*** is defined as 'a *farming system* that is capable of maintaining its productivity and usefulness to society indefinitely. Such systems must be resource-conserving, socially supportive, commercially competitive, and environmentally sound.' (*Idem*); or as 'an *integrated system of plant and animal production practices* having a site-specific application' (*Idem*); or as 'a *way of practicing agriculture* which seeks to optimize skills and technology to achieve long-term stability of the agricultural enterprise, environmental protection, and consumer safety' (*Idem*); or as 'a *whole-systems approach* to food, feed, and other fibber production that balances environmental soundness, social equity, and economic viability among all sectors of the public, including international and intergenerational peoples' (*Idem*).

B. Only 1 (6%) of the 18 definitions is a *dictionary* or *lexical definition*, i.e. *a definition that reports the meaning of a word or expression as it is normally used, usually by supplying an approximately equivalent expression in which the original word does not occur*: *biological farming* 'often refers to *organic farming* (in Europe, e.g. in the Netherlands)' (Norman 1997).

C. Only 1 (6%) of the 18 definitions is an *extensional definition*, i.e. *a definition formulating the meaning of a concept or term by specifying its extension, that is, every object that falls under the definition of the concept or term in question*: *ecological agriculture / farming* 'refers to *organic farming* plus environmental considerations such as on-farm wildlife management (i.e., the relationships between parts of the agroecosystem)' (*Idem*).

Discussion

From a teacher's point of view, it is good that 88% of the definitions of **concepts and practices associated with sustainable agricultural systems** are *definitions by genus and difference* or *genus-differentia definitions*, since *this is the natural thing to do if you are to explain the meaning of a particular word to someone*. Again, the fact that 6% of the definitions of **concepts and practices** represent a *dictionary* or *lexical* *definition* is not a problem, since it relies on reporting the meaning of a word or expression as it is normally used by supplying an approximately equivalent expression in which the original word does not occur. A *lexical definition* is usually *the type expected from a request for definition, and it is generally expected that such a definition will be stated as simply as possible in order to convey information to the widest audience.*

As for the only *extensional definition* representing other 6% of the definitions of **concepts and practices associated with sustainable agricultural systems**, it should not be perceived as a burden, since they address undergraduates in agriculture. *Extensional definitions* are used *when listing examples would give more applicable information than other types of definition, and where listing the members of a set tells the questioner enough about the nature of that set.*

The problem is that 15 (i.e. 83%) of these definitions are not pure definitions. Thus:

Of these 15 definitions, 14 (13 *definitions by genus and difference* and 1 *extensional definition*) could also be ranged among *precising definitions*, i.e. *definitions that extend the dictionary / lexical definition of a term for a specific purpose by including additional criteria that narrow down the set of things meeting the definition*: *biodynamic agriculture / farming* is '[a *farming method* encompassing] *specific [agricultural] practices and preparations that enable the farmer or gardener to work in concert with [the forces within living nature]*' (Gold 1999); ***biointensive gardening*** or ***mini-farming*** is 'a *production system* that makes it possible for one person to grow all of his or her family's food *using truly sustainable methods that maintain the fertility of the soil without relying on non-renewable resources like petrochemicals or imported organic matter*' (Jeavons in Gold 1999); ***biological / ecological farming*** is '[a *farming method* encompassing] *various and more specific practices and techniques of farming sustainability, e.g., organic, biodynamic, holistic, natural*' (Gold 1999); ***biological farming*** is 'a *system of crop production* in which the producer tries to minimize the use of 'chemicals' *for control of crop pests*' (Pesek in Gold 1999); ***do-nothing farming*** or ***natural farming*** is 'a *farming method* involving no tillage, no fertilizer, no pesticides, no weeding, no pruning, and remarkably little labour […] *by careful timing of his seeding and careful combinations of plants (polyculture)*' (Gold 1999); ***ecological agriculture / farming*** 'refers to *organic farming* plus environmental considerations *such as on-farm wildlife management (i.e., the relationships between parts of the agroecosystem)*' (Norman 1997); ***Kyusei nature farming*** '[is a *farming method* that] often employs technology involving beneficial microorganisms as inoculants to increase the microbial diversity of agricultural soils, *which, in turn, can enhance the growth, health, and yield of crops*' (Gold 1999); ***low input agriculture*** '[is a *low input farming system* that] seeks to optimize the management and use of internal production inputs *(i.e., on-farm resources)* and to minimize the use of production inputs *(i.e., off-farm resources) such as purchased fertilizers and pesticides, wherever and whenever feasible and practicable,* to lower production costs, to avoid pollution of surface and groundwater, to reduce pesticide residues in food, to reduce a farmer's overall risk, and to increase both short- and long-term farm profitability' (*Idem*); ***nature farming*** is a *farming system* utilizing 'the inherent power of the soil' *and focusing on 'analyzing and building soil through composting, green manuring, mulch, and various other soil management techniques*' (*Idem*); ***organic farming*** is 'a *production system* which avoids or largely excludes the use of synthetically compounded fertilizers, pesticides, growth regulators, and livestock feed additives. *[It relies] upon crop rotations, crop residues, animal manures, legumes, green manures, off-farm organic wastes, mechanical cultivation, mineral-bearing rocks, and aspects of biological pest control to maintain soil productivity and tilth, to supply plant nutrients, and to control insects, weeds and other pests.*' (*Idem*); ***permaculture*** (from *perma*nent agri*culture*) is 'one of the many alternative *agriculture systems* described as sustainable [emphasising] on design; *that is, the location of each element in a landscape, and the evolution of landscape over time, [whose goal is] to produce an efficient, low-maintenance integration of plants, animals, people and structure applied at the scale of a home garden, all the way through to a large farm*' (*Idem*); ***precision agriculture / farming*** or ***prescription farming*** or ***site-specific management*** is 'a MANAGEMENT STRATEGY that employs detailed, site-specific information *to precisely manage production inputs*' (*Idem*); ***regenerative / sustainable agriculture*** or ***low-input sustainable agriculture*** is 'a *farming system* that is capable of maintaining its productivity and usefulness to society indefinitely. *Such systems must be resource-conserving, socially supportive, commercially competitive, and environmentally sound.*' (*Idem*); ***regenerative / sustainable agriculture*** or ***low-input sustainable agriculture*** a '*whole-systems approach* to food, feed, and other fibber production that balances environmental soundness, social equity, and economic viability among all sectors of the public, *including international and intergenerational peoples*' (*Idem*).

Of these 15 definitions, 1 *definition by genus and difference* could also be ranged among *enumerative definitions*, i.e. *definitions that formulates its meaning by specifying its extension, that is, finite sets of objects that fall under the definition of the concept or term in question*: *alternative agriculture / farming* is '[a *farming method* encompassing] a vast *array of [agricultural] practices and enterprises*, all of which are considered different from prevailing or conventional agricultural activities [...]: *nontraditional crops, livestock, and other farm products; service, recreation, tourism, food processing, forest / woodlot, and other enterprises based on farm and natural resources (ancillary enterprises); unconventional production systems such as organic farming or aquaculture; or direct marketing and other entrepreneurial marketing strategies*' (Grudens Shuck *et al.* in Gold 1999)

In fact, this is what confuses our undergraduates in their approach of specialised texts on sustainable agriculture: too many terms to define types of agriculture and/or sustainable agriculture **concepts and practices** that are sometimes very difficult to differentiate; there is lack of consequence in defining these **concepts and practices associated with sustainable agricultural systems**, identified as *farming methods* (1), *farming systems* (2), *production systems* (2), *agriculture systems* (1), *arrays of agricultural practices and enterprises* (1), *integrated systems of plant and animal production practices* (1), *low input farming systems* (1), *management strategies* (1), *systems of crop production* (1), *ways of practicing agriculture* (1), or *whole-system approaches* (1); and 83% of the definitions are *hybrid ones*: 13 definitions are 50% *definition by genus and difference* + 50% *precising definition*, 1 definition is 50% *definition by genus and difference* + 50% *enumerative definition*, and 1 definition is 50% *extensional definition* + 50% *precising definition*.

Conclusions

Though *specialised definitions* should be rigorous, unambiguous, conventional, prescriptive or stipulating (Bidu-Vrânceanu 1968), we wish to stress one point: since "sustainable agriculture is a dynamic rather than static concept" (Norman *et al.* 1997), we have to admit that, at any point in time, in any society, the definition of any of the **concepts and practices associated with sustainable agricultural systems** above is going to be a compromise among differing world views, sets of values, etc., no one of which has any way to prove the other wrong, or illegitimate. Therefore, despite any interdisciplinary (specialists in agricultural systems and linguists) effort, the "definition" of something like the **concepts and practices** mentioned above is going to remain fluid (poly-semantic and poly-referential), driven by changes in community values, ideology, politics, science, etc.

The question we always ask ourselves is: Do we really need to spend much more time and effort attempting to define all these **concepts and practices associated with sustainable agricultural systems**? And here is a possible answer: Maybe agriculturists have sufficient commonality among their different understandings of these **concepts and practices** to continue moving in the right general direction, even if they are not yet all moving toward precisely the same destination by the same route: they may never have generally accepted definitions of these **concepts and practices**, and perhaps, they don't need any, but people involved in teaching others (this is our case) certainly do.

References

Bidu-Vrânceanu, A. (Ed.) (2002). *Lexic comun, lexic specializat* [*Common Vocabulary, Specialised Vocabulary*]. Universitatea din Bucureşti. Online: http://ebooks.unibuc.ro/filologie/vranceanu.

Gold, M. V. (1999). *Sustainable Agriculture: Definitions and Terms*. Online: http://www.nal.usda.gov.

Norman, D., Janke, R., Freyenberger, S., Schurle, B. & Kok, H. (1997). *Defining and Implementing Sustainable Agriculture*. Online: http://www.kansassustainableag.org.

Perković, Anica & Raţă, Georgeta. (2006). On Word Formation in the English of Agriculture: Practices Associated with Sustainable Agriculture, *Lucrări ştiinţifice. Facultatea de Agricultură XXXVIII*: 475-480.

Raţă, Georgeta & Perković, Anica. (2008). Understanding Agricultural Terminology: Defining Concepts and Practices Associated with Sustainable Agricultural Systems. *Journal of Linguistic Studies 1*: 31-36.

Raţă, Georgeta, Petroman, Cornelia & Petroman, I. (2009). Understanding Agricultural Terminology. Communication: Understanding / Misunderstanding. *Acta Semiotica Fennica XXXIV (3)*: 1489-1496.

Raţă, Georgeta. (2007). Understanding Agricultural Terminology: Defining concepts and Practices Associated with Sustainable Agricultural Systems. *Communication: Understanding Misunderstanding. IASS-AIS 9th World Congress of Semiotics, Helsinki, 11-17 June 2007*. Helsinki: Hakapaino. 367-368.

Simpson, J. & Weiner, E. (1989). *Oxford English Dictionary*. Oxford: Oxford University Press.

HOMONYMY: *CORN*

GEORGETA RAȚĂ

Introduction

There are two **homonyms** having yielded two long series of **compounds** and **derivatives** in English: U.S. ***corn*** < ME, OE; c. D *Koren*, Icel, G *Korn*, Goth *kaúrn*; akin to L *grānum*, Russ *zerno* GRAIN, and ***corn*** < late ME *corne* < MF < L *corn(ū)* HORN. This is quite a challenge for anyone learning English for Special Purposes or documenting on English literature in the field of agriculture. Therefore, knowing if there is any algorithm in acquiring new agricultural vocabulary is of extreme importance for both students and researchers in agriculture.

Material and Methods

In our analysis, we used some of the most reputed English languages dictionaries in an attempt to grouping together all the **compounds** and **derivatives** of ***corn*** in both British and American English. The method we used is the quantitative one. First, we left aside all the proper names (**compounds** and **derivatives**) containing ***corn*** but having nothing to do with either 'grain' or 'horn'. We have inventoried 125 entries that are **compounds** (but not also **compounds** of corn **derivatives**) or **derivatives** of ***corn***. We then grouped these words depending on their etymology, i.e. depending on the root word they have developed from ***corn*** 'grain' and ***corn*** 'horn'

Results and Discussion

The 125 entries containing corn were grouped into four smaller groups, as follows:

1. The first group contains 63 entries (**compounds** and **derivatives**) related to ***corn*** 'maize': ***cornball*** 'popcorn rolled into a ball and flavoured with molasses or caramel'; ***Corn Belt*** 'a region in the Midwest U.S.,

especially Iowa, Illinois, and Indiana, excellent for raising corn and cornfed livestock'; ***corn borer*** 'any of the several pyralid moths, as *Pyrausta (ostrinia) nubialis* (***European corn borer***), the larvae of which bore into the stem and crown of corn and other plants'; ***corn bread*** 'a bread made of cornmeal; (especially in north-eastern U.S.) a sour-dough rye bread, moist and heavy in texture'; ***corn cake*** 'Midland and Southern U.S., a flat corn bread baked on a griddle'; ***corn chip*** 'a thin, crisp piece of snack food made from cornmeal'; ***corn circle*** 'crop circle'; ***corn cob / corncob*** 'the elongated woody core in which the grains of an ear of corn are embedded'; ***corncob (pipe)*** 'a tobacco pipe with a bowl made from a corncob'; ***corn cockle*** 'a plant, *Agrostemma githago*, of the pink family, having magenta-purple flowers and occurring commonly as a weed among crops of grain; cornflower'; ***corn colour*** ' light yellow'; ***corn coloured*** 'coloured in light yellow'; ***corn-cracker*** 'Slang (disparaging and offensive) one of a poor class of white people in the southern U.S., cracker'; ***corn crake*** 'a short-billed Eurasian rail, *Crex crex*, frequenting grain fields, land rail'; ***corncrib*** 'a ventilated structure for the storage of unhusked corn'; ***corn dodger / corndodger*** 'South Midland and Southern U.S., a small, usually oval cake made of corn bread and baked or fried hard in a skillet; Chiefly South Atlantic States and Eastern Virginia, a boiled dumpling made of cornmeal'; ***corn dog*** 'a sandwich consisting of a frankfurter baked or fried in corn bread and usually spread with mustard before eating; often served on a stick'; ***corn dolly*** 'a figure made from straw, used for decoration'; ***corn earworm*** 'the larva of a cosmopolitan noctuid moth, *Heliotis zea*, that is highly destructive to crops, especially corn, cotton, and tomato, (cotton) bollworm, tomato fruit-worm'; ***corn exchange*** 'a place where corn is, or was formerly bought and sold'; ***corn-fed / cornfed*** 'fed on corn; having a well-fed, healthy, and guileless appearance'; ***corn field / cornfield*** 'a field in which corn is grown'; ***corn flakes / cornflakes*** 'a packaged breakfast cereal in the from of small toasted flakes made from corn, fro serving cold with milk, sugar, etc.'; ***corn flour*** 'flour made from corn; Brit., cornstarch'; ***corn flower / cornflower*** 'a European composite plant, *Centaurea cyanus*, growing in grain fields, having blue to white flower heads, often cultivated as an ornamental; strawflower'; ***corn gluten*** 'gluten separated from corn, during milling, used primarily as a livestock feed'; ***corn grit(s)*** 'hominy grits'; ***corn house*** 'New England and South Atlantic States, a corncrib'; ***cornhusk*** 'the husk of an ear of corn'; ***corn husker / corn-husker / cornhusker*** 'a person or thing that husks corn; (cap.) a Nebraskan (used as a nickname)'; ***cornhusking*** 'U.S., the removing of the husks from corn'; ***(the) Corn Law(s)*** 'English History, any of the laws regulating domestic and foreign trading of grain, the last of which was repealed in 1846'; ***corn liquor*** 'corn whiskey'; ***corn meal / cornmeal*** 'meal made of corn; Scottish, oatmeal'; ***corn muffin*** 'a muffin, often shaped like a cupcake, made from cornmeal'; ***corn oil*** 'a pale-yellow, water-insoluble liquid obtained by expressing the germs of corn kernels, used in the preparation of foodstuffs, lubricants, soaps, and hair dressings'; ***corn on the cob*** 'the woody central part of an ear of corn and the eatable pieces of corn which grow in rows along it'; ***corn picker*** 'a machine for picking the ears of corn from standing stalks and removing the husks'; ***corn picking*** 'the picking of the ears of corn from standing stalks and the removing of the husks'; ***corn pone*** 'Southern U.S., corn bread, especially of a plain or simple kind'; ***cornpone*** 'Usually Disparaging, of or characteristic of an unsophisticated rural person, especially from the South; hick'; ***corn poppy*** 'a common Old World poppy, *Papaver rhoeas*, having bright-red flowers, so called from its growing in grain fields'; ***corn-root aphid*** 'an aphid, *Anuraphis maidiradicis*, that lives as a symbiont in colonies of cornfield ants and feeds on the roots of corn: an agricultural pest'; ***corn rootworm*** 'the larva of any of several leaf beetles of the genus *Diabrotica* that feeds on roots and underground stems: an agricultural pest, especially of corn'; ***cornrow*** 'a type of braid, originating in Africa, in which a narrow strip of hair is plaited tightly against the scalp from front to back or from side to side; a hair style consisting of such braids in close parallel rows'; ***corn shock*** 'a stack of upright cornstalks'; ***corn silk*** 'the long, thread-like styles on an ear of corn'; ***corn smut*** 'a disease of corn caused by a fungus, *Ustilago maydis*, and characterised by blackish, powdery masses of spores on the affected parts of the plant'; ***corn snake*** 'a large, harmless rat snake, *Elaphe guttata guttata*, of the south-eastern U.S., having yellow, tan, or grey scales with dark-red blotches: once common in cornfields but now an endangered species'; ***corn snow*** 'Skiing, snow in the form of small pellets or grains produced by the alternate melting and freezing of a snow layer'; ***corn stack*** 'Delmarva Peninsula, corncrib'; ***corn stalk / cornstalk*** 'the stalk or stem of corn, especially Indian corn'; ***cornstarch*** 'a starch or a starchy flour made from corn and used for thickening gravies and sauces, making puddings, etc.; especially British, corn flour'; ***cornstick*** 'Southern Cookery, a corn muffin baked in the form of a small ear of corn'; ***corn sugar*** 'dextrose'; ***corn syrup*** 'syrup prepared from corn'; ***corn whiskey*** 'whiskey made from a mash having at least 80% corn; corn, corn liquor'; ***cornily*** 'in a corny way'; ***corniness*** 'the condition of being corny'; ***corny*** 'of or abounding in corn'; ***Indian corn*** 'corn'; ***sweet corn*** 'corn; especially American English, (the tender young seed of) a sweet type of maize'. Other 25 entries with ***corn*** 'grain' **compounds** occur only in English-

Romanian dictionaries: ***corn binder***, ***corn chandler***, ***corn cleaner***, ***corn cutter***, ***corn dealer***, ***corn factor***, ***corn failure***, ***corn floor***, ***corn grass***, ***corn harvester***, ***corning***, ***corn lift***, ***corn loft***, ***corn market***, ***corn middlings***, ***corn mill***, ***corn mint***, ***corn pimpernel***, ***corn-pipe***, ***corn planter***, ***corn reaper***, ***corn rent***, ***corn sieve***, ***corn van***, and ***corn weevil***. Other 3 **compounds** with ***corn*** have produced, in their turn, other **compounds**: ***cornfield ant*** 'a small, brown ant, *Lasius alienus*, that lives in cornfields and feeds on honeydew of the corn-root aphid'; ***cornfield bramble*** 'a rosaceous plant, *Rubus caesius*'; ***cornflower blue*** 'a deep, vivid blue'; ***Cornhusker State*** 'Nebraska (used as a nickname)'. It is worth mentioning that, though there is no ***corn leaf*** **compound**, there is ***corn-leaf aphid*** 'a green aphid, *Rhopalosiphum maidis*, widely distributed in the U.S.: a pest of corn and other grasses'.

2. The second group is that of 29 **derivatives** and **compounds** of **derivatives** of ***corn*** 'horn': ***cornea*** 'the transparent anterior part of the external coat of the eye covering the iris and the pupil and continuous with the sclera'; ***corneal*** 'related to the cornea'; ***corneous*** 'consisting of a horny substance; horny'; ***corner*** 'the place at which two converging lines or surfaces meet'; ***corner back***; ***corner cabinet***; ***corner chair***; ***corner kick***; ***corner shop***; ***corner stone / cornerstone***; ***corner table***; ***cornerwise / cornerways***; ***cornet*** 'a valved wind instrument of the trumpet family; a small cone of paper twisted at the end and used for holding candy, nuts, etc.; a pastry cone, usually filled with whipped cream; British, a conical wafer, as for ice cream, cone; a large, white, winged headdress formerly worn by the members of the Sisters of Charity; a woman's headdress, often cone-shaped, usually of delicate fabrics and having lappets of lace or other material, worn by women from the 14th to the 18th century; a pennant or flag used for signalling in a navy; (formerly) the officer who carried the colours in a troop of cavalry'; ***cornet(t)ist***; ***cornet-à-pistons***; ***cornetfish***; ***cornhole*** 'Slang (vulgar), to have anal intercourse with'; ***cornice*** 'any prominent, continuous, horizontally projecting feature surmounting a wall or other construction, or dividing it horizontally for compositional purposes; any of various other ornamental horizontal mouldings or bands, as for concealing hooks or rods from which curtains are hung or for supporting picture hooks; a mass of snow, ice, etc., projecting over a mountain ridge; ***cornicle*** 'any of various, small, horn-shaped processes, especially one of a pair of tubes at the posterior end of the abdomen of aphids, from which a waxy fluid is emitted'; ***corniculate***; ***cornification*** 'the formation of a horny layer of skin, or horny skin structures, as hair, nails, or scales, from squamous epithelial cells'; ***cornu*** 'a horn, especially a bony part that resembles a horn'; ***cornucopia*** 'Classical Mythology, a

horn containing food, drink, etc., in endless supply, said to have been a horn of the goat Amalthaea; a representation of this horn, used as a symbol of abundance; an abundant, overflowing supply; a horn-shaped or conical receptacle or ornament'; ***cornucopia leg***; ***cornucopian***; ***cornute*** 'to cuckold; cornuted'; ***cornuted***; ***cornuto*** 'a cuckold, one who is horned'; ***corny*** 'pertaining to or affected with corns of the feet'. Other 15 entries with ***corn*** 'horn' **compounds** occur only in English-Romanian dictionaries: ***corneous lead***; ***corneous silver***; ***corner box***; ***corner boy***; ***corner house***; ***corner iron***; ***corner man***; ***corner pillar***; ***corner plate***; ***corner post***; ***cornered***; ***cornice plane***; ***corniced***; ***cornon***; and ***cornopean***. To also note ***-corn*** 'a combining form meaning 'having a horn', of the kind specified by the first element'.

3. The third group is that of 7 ***corn*** **compounds** and **derivatives** for which we could not trace any link with either 'grain' or 'horn': ***corn beef*** 'corned beef'; ***corn broom*** 'North-eastern U.S. older use, a broom made from the panicles of broomcorn'; ***corned*** 'marinated in brine, often containing garlic, peppercorns, cloves, etc.; preserved or cured with salt'; ***cornel*** 'any tree or shrub of the genus *Cornus*; dogwood'; ***corn flag*** 'a Mediterranean plant, *Gadiolus segetum*, of the iris family, having loose, one-sided spikes or pinkish-purple flowers'; ***corn lily*** 'any of the several plants of the genus Ixia, of the iris family, native to southern Africa, having spikes of flowers and grown as an ornamental; a woodland plant, *Clintonia borealis*, of the lily family, native to eastern and mid-western North America, having broad leaves, nodding yellowish-green flowers, and dark-blue berries'; ***corn marigold*** 'a composite plant, *Chrysanthemum segetum*, of Eurasia, having daisy-like, solitary yellow flowers'; ***corn plant*** 'any of several tree-like tropical plants of the genus *Dracaena*, especially *D. fragrans massangeana*, widely cultivated as a houseplant'; ***corn salad*** 'any of several plants of the genus *Valerianella*, of the valerian family, especially *V. locusta* or *V. olitoria*, having small light blue flowers and tender, narrow leaves eaten in salads'. Other 3 entries with ***corn*** **compounds** and **derivatives** occur only in English-Romanian dictionaries: ***cornaceous***; ***corned leather***; and ***cornel berry***.

4. The fourth group is that of 1 apparent ***corn*** **derivative** that has nothing to do with either 'grain' or 'horn': ***cornelian*** 'carnelian'.

Conclusions

Corn **compounds** and **derivatives** are represented, quantitatively, as follows: 'grain'-meaning entries account for 63%, 'horn'-meaning ones account for 29%, the 'unknown-origin' ones for 7%, and the 'definitely no-connection' ones for 1%. The large number of ***corn*** 'grain' **compounds** points to the widespread use of grains, while the large number of **derivatives** points the fact that ***corn*** 'horn' has acquired a solid status in English. It is obvious that, while ***corn*** 'grain' entries are mainly **compounds** (93.66%), ***corn*** 'horn' entries are 100% **derivatives**. The low percentage of ***corn*** 'grain' **derivatives** (6.34%) allows us to draw the conclusion that one can assume a **compound** of ***corn*** refers rather to 'grain' than to 'horn', an observation which could be of help for anyone interested in the agricultural English.

References

Benson, M., Benson, Evelyn & Ilson, R. (1990). *The Combinatory Dictionary of English*. Amsterdam – Philadelphia: John Benjamins.

Leviţchi, L. & A. Bantaş. (1995). *Dicţionar englez-român*. [English-Romanian Dictionary]. Bucureşti: Ed. Teora.

Leviţchi, L. (Ed.). (2004). *Dicţionar englez-român*. [English-Romanian Dictionary]. Bucureşti: Ed. Univers Enciclopedic.

Leviţchi, L. D. (1970). *Limba engleză contemporană. Lexicologie*. [Contemporary English. Lexicology]. Bucureşti: E.D.P.

Longman Dictionary of English Language and Culture (2003). London: Longman.

Raţă, Georgeta. (2005). A Case of Homonymy in the English of Agriculture: *Corn*. *Proceedings. Linguistics and Theory of Literature. History, Ethnology and Folklore 44*. Rousse, Bulgaria: 34-38.

Raţă, G. (2004). Combining and Derivative Forms of *Corn*: A Lexicographical Approach. *Buletinul Universităţii de Ştiinţe Agricole şi Medicină Veterinară – Cluj-Napoca, Seria Horticultură* 60: 366-370.

The American Heritage Dictionary of the English Language. (2008). Boston: Houghton Mifflin.

The Random House Dictionary of the English Language. (1968). New York: Random House.

Webster Comprehensive Dictionary. (1995). Chicago: J. G. Ferguson Publishing Company.

Webster Encyclopedic Unabridged Dictionary of the English Language. (1996). New York: Gramercy Books.

Webster's Universal Dictionary & Thesaurus (2003). New Lanark: Geddes & Grosset.

COFFEE

CORNELIA PETROMAN, IOAN PETROMAN
AND SNJEŽANA TOLIĆ

Introduction

The purpose of the research is to show that the proper understanding of the meaning of **compounds** containing the word ***coffee*** is a matter of purely linguistic knowledge (to be taught by the teacher) rather than of knowledge of the world (since they have almost no practice in the field of food services). To our knowledge, no research has so far been done in this field and on this particular problem. The hypothesis of the research was that, since compound definitions cannot always help understanding phrases similar in structure, undergraduates should learn them as such and/or develop meaning inference skills if they want to avoid confusion. The background information was collected from different English language dictionaries, specialized dictionaries, and specialised encyclopaedias.

Material and Methods

We inventoried 102 phrases containing the word ***coffee***, leaving aside all **derivatives** such as: ***cafeteria***, ***cafetiere*** / ***cafetière***, ***cafetorium***, ***caffeine***, ***caffeinism***, ***caffeol***, ***decaf***, ***decaffeinated***, which could be subjected to a separate analysis. We then analysed them to see if the meaning of the phrases with no definition at all could be inferred from the meaning of the phrases defined by their users.

Results

We have identified a number of 94 (52 + 42) occurrences in which the noun ***coffee*** is a noun (*black* ***coffee***) or a noun modifier (***coffee*** *break*), other 6 in which it is used attributively, and other 2 with special uses (**calque** and adjective).

The Word *Coffee* as a Noun

Of the 52 occurrences in which the noun ***coffee*** occurs as a noun – *acorn* ***coffee***, *adulterate* ***coffee***, *Arabic* ***coffee***, *Arabica* ***coffee***, *aromatic* ***coffee***, *black* ***coffee***, *brewed* ***coffee***, *camp* ***coffee***, *chagga* ***coffee***, *chicory* ***coffee***, *Colombian* ***coffee***, *dandelion root* ***coffee***, *decaffeinated* ***coffee***, *drip* ***coffee***, *ersatz* ***coffee***, *espresso* ***coffee***, *filter* ***coffee***, *full-bodied* ***coffee***, *full-flavoured* ***coffee***, *German* ***coffee***, *ground* ***coffee***, *hot* ***coffee***, *iced* ***coffee***, *instant* ***coffee***, *Irish* ***coffee***, *Java* ***coffee***, *Kenya* ***coffee***, *mature* ***coffee***, *midmorning* ***coffee***, *muddy* ***coffee***, *Mysore* ***coffee***, *'naturally' decaffeinated* ***coffee***, *organic* ***coffee***, *potted* ***coffee***, *pulverized* ***coffee***, *regular* ***coffee***, *rich* ***coffee***, *roasted (grain)* ***coffee***, *Robusta* ***coffee***, *root* ***coffee***, *sharp* ***coffee***, *strong* ***coffee***, *sweetened* ***coffee***, *traditional* ***coffee***, *Turkish* ***coffee***, *unadulterated* ***coffee***, *unna* ***coffee***, *Vienna* ***coffee***, *Viennese* ***coffee***, and *weak* ***coffee*** – only 20 (38%) are defined in literature, as follows: ***Arabic coffee*** 'coffee flavoured with ground cardamom seeds' (Bender & Bender 1999); ***Arabica coffee*** 'a type of coffee from the bush *Coffea arabica*, with a finer flavour than the higher yielding Robusta varieties' (Sinclair 2005); ***camp coffee*** 'a liquid extract of coffee and chicory once used as an instant coffee' (Sinclair 2005); ***chagga coffee*** 'a full-bodied coffee from the slopes of Mount Kilimanjaro in Tanzania, named after the Wachagga people who grow and process it' (Sinclair 2005); ***Colombian coffee*** 'a smooth strong coffee from Colombia' (Sinclair 2005); ***decaffeinated coffee*** 'coffee made in the usual way from green beans which have been solvent treated to selectively remove the caffeine [with] similar flavour to, but […] not as stimulating as coffee' (Sinclair 2005), 'coffee beans (or instant coffee) from which the caffeine has been extracted with solvent (e.g. methylene or ethylene chloride), carbon dioxide under pressure (supercritical CO_2) or water' (Bender & Bender 1999); ***filter coffee*** 'coffee made by letting near boiling water flow through ground coffee held in a filter paper supported on a metal or plastic mesh or perforated container' (Sinclair 2005); ***German coffee*** 'a mixture of kirsch and hot sweetened coffee topped with whipped cream and drunk through the cream' (Sinclair 2005); ***green coffee bean*** 'coffee beans which are ready to be roasted [that] keep indefinitely and some […] improve with age' (Sinclair 2005); ***instant coffee*** 'dried coffee extract which can be used to make a beverage by adding hot water or milk. It may be manufactured by spray drying or freeze drying' (Bender & Bender 1999); ***Irish coffee*** 'hot coffee, served in a glass, with Irish whiskey added to it and whipped cream poured on top' (DFSN); ***Java coffee*** 'a mature coffee from Indonesia with a subtle mellow flavour' (Sinclair 2005); ***Kentucky***

coffee tree 'a deciduous North American tree *(Gymnocladus dioica)* having bipinnately compound leaves and flat, pulpy pods with large seeds formerly used as a coffee substitute' (AHDEL); ***Kenya coffee*** 'a sharp, aromatic coffee from Kenya' (Sinclair 2005); ***Mysore coffee*** 'a rich full-flavoured coffee from Mysore in S. India' (Sinclair 2005); ***'naturally' decaffeinated coffee*** 'coffee decaffeinated by water extraction' (Bender & Bender 1999); ***Robusta coffee*** 'a high yielding variety of coffee (*Coffea canephora*) but with a flavour inferior to that of the arabica variety' (Sinclair 2005); ***Turkish coffee*** 'a sweetened brew of pulverized coffee' (AHDEL); ***unna coffee*** '(E. Africa) the traditional coffee of Ethiopia which is roasted, ground and brewed at the table' (Sinclair 2005); ***Vienna coffee*** 'a particular blend of coffee beans favoured in Vienna' (Sinclair 2005); ***Viennese coffee*** '(Austria) 1. A mocha coffee 2. Ground coffee mixed with dried figs' (Sinclair 2005), 'ground coffee containing dried figs' (Bender & Bender 1999); ***weak coffee*** (AHDEL).

To also note ***kaffebröd*** '(Sweden) coffee cake' (Bender & Bender 1999) and ***Kaffeekuchen*** '(Germany) coffee cake' (Bender & Bender 1999), words that have entered cookbooks all over the world.

The Word *Coffee* as a Noun Modifier

Of the 42 occurrences in which the noun ***coffee*** is used attributively – ***coffee boom***, ***coffee break***, ***coffee cake / coffeecake***, ***coffee can***, ***coffee consumption***, ***coffee cream***, ***coffee creamer***, ***coffee cup***, ***coffee custard***, ***coffee essence***, ***coffee exporter***, ***coffee extract***, ***coffee filter***, ***coffee grinder***, ***coffee grounds***, ***coffee hound***, ***coffee house / coffeehouse***, ***coffee hue***, ***coffee icing***, ***coffee importer***, ***coffee jelly***, ***coffee kisses***, ***coffee klat(s)ch / kaffeeklatsch***, ***coffee maker / coffee-maker / coffeemaker***, ***coffee market***, ***coffee mill***, ***coffee plant***, ***coffee plantation***, ***coffee pot / coffeepot***, ***coffee seed***, ***coffee shop***, ***coffee solubles***, ***coffee strainer***, ***coffee substitute***, ***coffee sugar***, ***coffee table***, ***coffee tenderizer***, ***coffee tree***, ***coffee variety***, ***coffee whitener***, ***coffee wild roast*** – only 16 (38%) are defined in literature, as follows: ***coffee bean(s)*** 'The fruit of an evergreen bush, *Coffea arabica* or *C. canephora*, native to Ethiopia but now widely grown in high altitude tropical regions. The red fruit, which contains two almost hemispherical green seeds, is first fermented, the pulp removed and the seeds dried prior to roasting at or near their point of sale. After roasting they contain about 50% water soluble material including caffeine and flavours. The two main varieties are Robusta and arabica but they are often identified by their place of origin.' (Sinclair 2005), ***coffee break*** 'a short break from work during which coffee or other refreshments may be consumed' (AHDEL), ***coffee cake / coffeecake*** (AHDEL) '(*U.S.*) a plain sponge cake usually served warm with coffee' (Sinclair 2005) 'a cake or sweetened bread, often containing nuts or raisins' (AHDEL), ***coffee cream*** '(*U.S.*) pasteurized cream from cows' milk containing 18 to 30% butterfat for adding to coffee' (Sinclair 2005), ***coffee essence*** 'an aqueous extract of roasted coffee; usually about 400 g of coffee/L' (Bender & Bender 1999), ***coffee grinder*** 'a mechanically, electrically or hand-operated grinder for roasted coffee beans with an arrangement for adjusting the particle size, very fine for espresso, fine for filters, Turkish and cona, medium for cafetières and percolators and coarse for jug infusion' (Sinclair 2005), 'a machine for grinding coffee beans into powder for making coffee' (DFSN), ***coffee house / coffeehouse*** 'a restaurant where coffee and other refreshments are served, especially one where people gather for conversation, games, or musical entertainment' (AHDEL), ***coffee kisses*** 'small drop cakes or biscuits containing ground nuts, sandwiched together with coffee-flavoured butter cream' (Sinclair 2005), ***coffee klat(s)ch / kaffeeklatsch*** 'a casual social gathering for coffee and conversation' (AHDEL), ***coffee maker / coffee-maker*** (Emery) '***coffeemaker / coffee maker*** an apparatus used to brew coffee' (AHDEL), ***coffee mill*** 'a device for grinding roasted coffee beans' (AHDEL), ***coffee pot*** (DFSN) / ***coffeepot*** 'a pot for brewing or serving coffee' (AHDEL), ***coffee shop*** 'a small restaurant in which coffee and light meals are served' (AHDEL), ***coffee sugar*** 'coarse crystals, to 3 mm, of translucent, usually amber-coloured sugar but these may be mixed with crystals of other colours' (Sinclair 2005), ***coffee table / cocktail table*** 'a long, low table, often placed before a sofa' (AHDEL), and ***coffee whitener*** 'milk substitute used in tea and coffee made with glucose, fat and emulsifying salts' (Bender & Bender 1999). In the rest of the **compounds**, it is used attributively, and we can only make guesses about their meanings: ***coffee-growing region*** (AHDEL) 'a region in which they grow coffee', ***non-coffee product*** (Emery) 'a product that does not contain coffee', ***coffee-processing centre*** (AHDEL) 'a centre in which they process coffee', ***coffee-producing region*** (AHDEL) 'a region that produces coffee'. Two other **compounds** containing the word ***coffee*** are used attributively ***coffee-shop architecture*** (AHDEL) 'a cheap architecture' and ***coffee-table book*** 'an oversize book of elaborate design that may be used for display, as on a coffee table' (AHDEL). The phrase ***coffee black*** (AHDEL) is a **calque** or **loan translation** of the French ***café noir*** ('black coffee'), while ***coffee-flavoured*** (Sinclair 2005) is an adjective. (Figure 4-2)

Discussion

The meaning of the **compounds** containing the word ***coffee***, that are not mentioned by English language dictionaries or by dictionaries specialised in food as formations well-established in the English language, can only be inferred from similar phrases. This is the case of the phrases in which ***coffee*** is used as a noun, such as ***acorn coffee*** '*coffee made from acorns', ***adulterate coffee*** '*coffee that has been adulterated', ***aromatic coffee*** '*coffee that has a strong aroma', etc. The same goes for the phrases in which ***coffee*** is used as a noun modifier, such as ***coffee boom*** '*a sudden increase of the coffee price', ***coffee can*** '*a can of coffee', ***coffee consumption*** '*consumption of coffee', etc., but this does not always work for phrases such as ***coffee hound*** or ***coffee wild roast***, for instance.

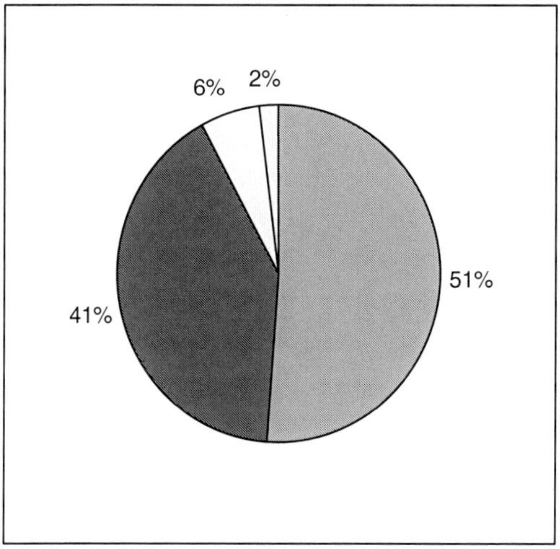

Figure 4-2. Functions of *coffee*: 51% - as a noun, 41% - as a noun modifier, 6% - used attributively, 2% - special uses

Conclusions

The hypothesis of the research that undergraduates specialising in food service should learn **compounds** containing the word ***coffee*** thoroughly (to avoid confusion) by developing meaning inference skills is, thus, confirmed, since the meanings marked with an asterisk above were suggested by our students. Similar inventories should also be done for other important semantic fields related to food service.

References

Bender, D. A. & Bender, A. E. (1999). *Benders' Dictionary of Nutrition and Food Technology*. Cambridge: Woodhead Publishing Limited.

Carstairs-McCarthy, A. (2002). *An Introduction to English Morphology. Words and Their Structure*. Edinburgh University Press.

Chalker, Sylvia & Weiner, E. (1994). *The Oxford Dictionary of English Grammar*. London - New York - Sydney - Toronto: BCA.

Dictionary of Food Science and Nutrition. (2006). London: A & C Black. (DFSN)

Emery, C. (2003). *The Encyclopedia of Country Living. An Old Fashioned Recipe Book*. Seattle: Sasquatch Books.

Petroman, Cornelia, Petroman, I. & Tolić, Snježana. (2008). Coffee: A Semantic Approach. *Journal of Linguistic Studies 1 (2008)*: 19-22.

Petroman, Cornelia, Petroman, I. & Tolić, Snježana. (2008). Café: A Semantic Approach. *Journal of Linguistic Studies 1 (2008)*: 19-22.

Sinclair, Ch. (2005). *Dictionary of Food*. London: A & C Black.

The American Heritage Dictionary of the English Language. (2008). Boston: Houghton Mifflin. (AHDEL)

SEMANTIC FIELDS: *BEEF*

ANICA PERKOVIĆ AND GEORGETA RAŢĂ

Introduction

Teaching English for Special Purposes in an agricultural university – particularly in a university where they teach not only about livestock, but also about food processing or about food services – is challenging because of the large number of specialty terms undergraduates need to acquire. One way to do it easily and properly is to learn about the entire **semantic field** of a term ('a range or system of referents that have some aspect of meaning in common', Chalker & Weiner 1994). Learning about ***beef***, for instance, also means learning about all the nouns (**compounds**, **derivatives**), adjectives, and verbs belonging to the semantic field of ***beef***: ***animals*** (*beefalo*, *steer*), ***bones*** (*aitchbone*), ***breeds*** (*Aberdeen*, *Brangus*, *Devon*), ***cuts*** (*baron*, *beefsteak*, *bottom round*), ***dishes*** (*beef bourguignon*, *beef Stroganoff*, *beef Wellington*), etc. Learning about the etymology of all of these terms broadens the cultural horizon of the undergraduates, opening a new perspective over globalization. In this paper, we present the semantic field of the term ***beef*** ('a full-grown steer, bull, ox, or cow, especially one intended for use as meat; the flesh of a slaughtered full-grown steer, bull, ox, or cow') as a model for further similar semantic fields.

Material and Method

We have inventoried all the terms related to the term ***beef*** in *The American Heritage Dictionary of the English Language* (2000). We then grouped them according to their semantic sub-fields: *animal husbandry*, *food processing*, *food service*, etc., thus making up inventories of beef-related terms specific to different subjects (animal husbandry, food processing technologies, tourism services, etc.) that undergraduates should learn as basic English vocabulary of their specialization.

Results and Discussion

The most frequently used synonym for ***beef*** 'a full-grown steer, bull, ox, or cow, especially one intended for use as meat; the flesh of a slaughtered full-grown steer, bull, ox, or cow' [< ME < OF *buef* < L *bos, bov-*] is ***red meat*** 'meat, especially beef, that is dark-coloured before being cooked'.

There is a single noun derived from ***beef*** – ***beefiness*** 'beef quality' and two **compounds** with ***beef***: one used informally – ***beefcake*** 'minimally attired men with muscular physiques, as in photographs or motion pictures' [BEEF + (CHEESE)CAKE], and ***beef tea*** '(*Cookery*) a drink made by boiling pieces of lean beef: often given to invalids to stimulate the appetite'.

Such adjectives as ***beefed-up*** 'having been made greater or stronger' and ***beefy*** 'muscular in build, brawny, substantial, filling; filled with beef' are the only adjectives derived from ***beef***, while ***au jus*** 'served with the natural juices or gravy (*roast beef au jus*)' [< F *au jus* 'with the juice'] and ***select*** 'of or relating to a lean grade of beef' are adjectives most frequently used in relation with **beef**.

A verb most frequently used in relation to ***beef*** is ***to corn*** 'to preserve (beef, for example) in brine'.

The terms making up the semantic field of the term beef can be grouped into the following semantic sub-fields (presented here in alphabetical order):

1. Anatomy: ***aitchbone*** 'the rump bone, especially of cattle'.

2. Animal husbandry:

- *animals*: ***beefalo*** (*pl.* ***beefalo*** or ***beefalo(e)s***) or ***cattalo*** 'a hybrid that results from a cross between the American buffalo, or bison, and beef cattle and is typically 8 buffalo and 8 bovine. Beefalo yields leaner beef than conventional breeds of cattle' [BEEF + (BUFF)ALO], ***steer*** 'a young ox, especially one castrated before sexual maturity and raised for beef' [< ME steer < OE *stēor*];
- *breeds:* ***Aberdeen / Black Angus*** 'a breed of black, hornless beef cattle that originated in Scotland' [After Aberdeen and Angus, former counties of Scotland.], ***Brangus*** 'a trademark used for any of a breed of beef cattle developed from a cross between the Brahman and the Aberdeen Angus', ***Devon*** 'any of a breed of reddish cattle originally developed in the English county of Devon and raised primarily for beef', ***Durham / shorthorn*** 'any of a breed of beef or dairy cattle that originated in northern England, having short, curved horns or no horns

and usually red, white, or roan in colour', ***Hereford*** 'any of a breed of beef cattle developed in England and having a reddish coat with white markings' [< *Hereford*], ***Santa Gertrudis*** 'any of a breed of large beef cattle that are highly resistant to heat and insects, developed in the United States by crossing Brahmans and shorthorns' [< the *Santa Gertrudis* section of the King Ranch in Kingsville, Texas];

- *cattle feed*: ***monensin*** 'a broad-spectrum antibiotic, $C_{36}H_{62}O_{11}$, obtained from the actinomycete *Streptomyces cinnamonensis* and used chiefly as an additive to beef cattle feed' [< NL *(cinna)monēns(is)*, species name].

3. Dishes: ***beef bourguignon*** 'braised beef cubes simmered in a seasoned red wine sauce with mushrooms, carrots, and onions' [< F *boeuf bourguignon*], ***beef Stroganoff*** 'thinly sliced beef fillet sautéed and mixed with onions, mushrooms, sour cream, and herbs, often served on a bed of noodles or rice. [< *Count Paul Stroganoff*, 19th-century Russian diplomat], ***beef tea*** '(*Cookery*) a drink made by boiling pieces of lean beef: often given to invalids to stimulate the appetite', ***beef Wellington*** 'a fillet of beef covered with paté de foie gras, encased in pastry, and baked' [< *Wellington*?], ***bitok*** 'a dish made from ground meat mixed with milk, bread, and onions to form patties that are fried and served with a sour-cream sauce' [< Russ < F *bifteck (hache)* '(ground) beef' < BEEFSTEAK], ***bollito misto*** (*pl.* ***bolliti misti***) 'a mixture of vegetables and various meats, such as chicken, veal, beef, and sausage, cooked in a broth and usually served with a mustard-fruit sauce' [< It *bollito misto* 'mixed stew'], ***baloney / bologna / boloney*** 'a seasoned smoked sausage made of mixed meats, such as beef, pork, and veal' [< *Bologna*], ***bouillon*** 'a clear, thin broth made typically by simmering beef or chicken in water with seasonings' [< F *bouillon* < OF < *boulir* 'to boil' < L *bullīre* < *bulla* 'bubble'], ***bresaola*** 'sliced salt-cured, air-dried beef that is dressed with olive oil, lemon juice, and black pepper before serving' [< It *bresaola*, diminutive of It dialectal **bresada*, past participle of *brasare* 'to braise' < F *braiser*], ***bully (beef)*** 'canned or pickled beef' [< F *bouilli* 'boiled meat'? label on canned beef < past participle of *bouillir* 'to boil' < OF Old *boilir*], ***burger*** 'a sandwich consisting of a bun, a cooked beef patty, and often other ingredients such as cheese, onion slices, lettuce, or condiments; a sandwich with a nonbeef filling' [Short for HAMBURGER], ***burrito*** (*pl.* ***burritos***) 'a flour tortilla wrapped around a filling, as of beef, beans, or cheese' [< AmSp *burrito* < Sp *burrito*, diminutive of *burro* 'burro'], ***carpaccio*** 'very thinly sliced raw beef or tuna garnished with a sauce' [< It *carpaccio* < *Vittore Carpaccio*, who favoured red pigments.],

Chateaubriand / chateaubriand 'a doublethick, tender centre cut of beef tenderloin, sometimes stuffed with seasonings before grilling' [< Vicomte *François René de Chateaubriand*], ***chipped beef*** 'dried beef smoked and sliced very thin', ***club / Delmonico steak*** 'a small, often boned steak from the front section of the short loin of beef' [< *Lorenzo Delmonico*], ***cube steak*** 'a thin slice of beef tenderized by cubing', ***(stuffed) derma / kishke*** 'beef casing stuffed with a seasoned mixture of matzo meal or flour, onion, and suet, prepared by boiling, then roasting' [< Yiddish *gederem* 'intestines' < MHG *darm* 'intestine' < OHG], ***dolma*** (*pl.* ***dolma(de)s***) 'a grape leaf stuffed and cooked with ingredients such as ground beef, minced lamb, herbs, or rice' [< Tk *dolma* 'filling'], ***filet mignon*** (*pl.* ***filets mignons***) 'a small, round, very choice cut of beef from the loin' [< F *filet mignon*], ***fil(l)et*** 'a boneless strip of meat rolled and tied, as for roasting', ***flanken*** 'a dish prepared from this cut of beef by boiling or stewing, often served with horseradish' [< Yiddish *flanken* < G *Flanken*, pl. of *Flanke* 'flank, side' < F *flanc*], ***flauta*** 'a tortilla rolled around a filling such as beef, chicken, or cheese into a flutelike shape and sometimes deep-fried' [< Sp *flauta* 'flute' < OProv *flaut*?], ***frankforter / frankfurter*** 'a smoked sausage of beef or beef and pork made in long, reddish links' [< *Frankfurt*], ***goulash*** 'a stew of beef or veal and vegetables, seasoned mainly with paprika. [< Hung *gulyas (hus)* 'herdsman's (meat), goulash' < Hung *gulya* 'herdsman'], ***hamburg(er)*** 'ground meat, usually beef, or patty of such meat; sandwich made with a patty of ground meat usually in a roll or bun' [Short for *Hamburger steak* < *Hamburg*], ***hot pot*** '*Chiefly British.* A stew of lamb or beef and potatoes cooked in a tightly covered pot', ***junk*** 'hard salt beef for consumption on board a ship', ***meat loaf / meatloaf*** 'a mounded or moulded dish, usually baked, of ground beef or a combination of various meats and other ingredients', ***mortadella*** 'an Italian sausage made of pork, beef, and cubes of pork fat, flavoured with wine and spices and smoked, steamed, or baked' [< It *mortadella*, feminine diminutive of *murtato* 'seasoned with myrtle berries'], ***moussaka*** 'a Greek dish consisting of layers of ground lamb or beef and sliced eggplant topped with a cheese sauce and baked' [< Serbo-Croatian *moussaka* < Tk *mussakka* < Arabic *musakka*], ***paillard*** 'a slice of veal, chicken, or beef that is pounded until very thin and quickly grilled, broiled, or sautéed with high heat' [Origin unknown], ***pastrami*** (*pl.* ***pastramis***) 'a highly seasoned smoked cut of beef, usually taken from the shoulder' [< Yiddish *pastrami* < Rum *pastramă*], ***pem(m)ican*** 'a food prepared by Native Americans from lean, dried strips of meat pounded into paste, mixed with fat and berries, and pressed into small cakes; a food made chiefly from beef, dried fruit, and suet, used as emergency rations'

[< Cree *pimihkaam* 'pemmican'], ***pepperoni*** (*pl.* ***pepperonis***) 'highly spiced pork and beef sausage; a slice of this type of sausage' [< It *peperoni*, pl. of *peperone* 'pimento, red pepper', augmentative of *pepe* 'pepper' < L *piper*], ***Reuben*** 'a hot sandwich consisting of corned beef, Swiss cheese, and sauerkraut usually served on rye bread' [< *Reuben*], ***runza*** '*Nebraska.* A pastry consisting of cabbage and usually pork or beef encased in yeast dough' [Origin unknown], ***Salisbury steak*** 'a patty of ground beef mixed with eggs, milk, onions, and various seasonings and broiled, fried, or baked' [< *James Henry Salisbury* (1823-1905), American physician], ***sauerbraten*** 'a pot roast of beef marinated in vinegar, water, wine, and spices before being cooked' [< G *sauer* + *Braten* 'roast meat'], ***shabu-shabu*** 'a Japanese dish consisting of a simmering pot of broth, vegetables, and noodles in which thin-sliced beef or sometimes chicken is quickly cooked at table and then dipped into a flavourful sauce' [< Jap *shabu-shabu*, imitative of bubbling water], ***shashli(c)k*** 'a dish consisting of marinated cubes of lamb or beef grilled or roasted on a spit, often with slices of eggplant, onion, and tomato; shish kebab' [< Russ *shashlyk* < Tk], ***sloppy joe / Joe*** 'a bun filled or covered with ground beef cooked in a spicy tomato sauce', ***steak*** 'a slice of meat, typically beef, usually cut thick and across the muscle grain and served broiled or fried' [< ME *steike* < ON *steik*], ***steak tartar*** or ***tartar steak*** 'raw ground beef mixed with onion, seasoning, and raw egg, eaten as an appetizer' [< E *steak* + F *tartare* 'Tartar'], ***surf and turf*** 'seafood and beefsteak served as the main course of a meal, as in a restaurant', ***wienerwurst*** 'a smoked pork or beef sausage similar to a frankfurter' [< G *Wiener* + *Wurst* 'Viennese sausage'], and ***Yorkshire pudding*** 'a popover-like quick bread served with roast beef, made by baking a batter of eggs, flour, and milk in the drippings of the beef' [< *Yorkshire*].

4. *Food processing (cuts)*: ***aitchbone*** 'the cut of beef containing the rump bone', ***baron*** 'a cut of beef consisting of a double sirloin' [< ME < OF < G?], ***beefsteak*** 'a slice of beef, such as one taken from the loin or the hindquarters, suitable for broiling or frying', ***bottom round*** 'a cut of meat, such as steak, taken from the outer section of a round of beef', ***eye*** 'a choice centre cut of meat, as of beef', ***fil(l)et*** 'a strip or compact piece of boneless meat or fish, especially the beef tenderloin', ***flanken*** 'a cut of meat taken from the short ribs of beef' [< Yiddish *flanken* < G *Flanken*, pl. of *Flanke* 'flank, side' < F *flanc*], ***hindquarter*** 'the posterior portion of a side of beef, lamb, veal, or mutton, including a hind leg and one or two ribs', ***plate*** 'a thin cut of beef from the brisket', ***porterhouse (steak)*** 'a cut of beef taken from the thick end of the short loin, having a T-bone and a sizable piece of tenderloin', ***pot roast*** 'a cut of beef that is browned and

then cooked until tender, often with vegetables, in a covered pot', ***rib roast*** 'a cut of red meat, such as beef or venison, containing the sizable piece located along the outside of the rib', ***round*** 'a cut of beef from the part of the thigh between the rump and the shank', ***round steak*** 'a lean, oval cut of beef from between the rump and shank', ***rump*** 'a cut of beef or veal from the rump' [< ME *rumpe* < Scandinavian origin], ***shin*** 'the lower foreleg in beef cattle. Used of cuts of meat' [< ME *shine* < OE *scinu*], ***sirloin*** 'a cut of meat, especially of beef, from the upper part of the loin just in front of the round' [< ME *surloine* < OF *surlonge, surloigne* < *sur* 'above' + *longe, loigne* 'loin'], ***skirt steak*** 'a boneless cut of beef from the lower part of the brisket', ***tenderloin*** 'the tenderest part of a loin of beef, pork, or similar cut of meat' [< Tenderloin], ***top round*** 'a cut of meat, such as a steak or roast, taken from the inner section of a round of beef', ***tournedos*** (*pl.* ***tournedos***) 'a fillet of beef cut from the tenderloin, often bound in bacon or suet for cooking' [< F *tourner* 'to turn' < OF], and ***undercut*** '*Chiefly British.* The tenderloin of beef; the fillet'.

5. *Food service*: ***steak house / steakhouse*** 'a restaurant that specializes in beefsteak dishes'.

6. *Others*: ***beefeater*** 'a yeoman of the British monarch's royal guard'.

7. *Terms related only formally to* ***beef***: ***Australian pine*** or ***beefwood*** or ***casuarina*** 'any of various Australian evergreen trees or shrubs of the genus *Casuarina,* having jointed stems, scale-like whorled leaves, and small fruits grouped in woody, cone-like structures; the wood of any of these plants, often used in construction' and ***beefsteak fungus*** 'an edible fungus *(Fistulina hepatica)*, growing on living tree trunks, such as oak and ash, and having a large, irregularly shaped reddish cap'.

The terms ***beefeater*** and ***hamburger*** are a good opportunity to "teach" undergraduates British and American culture by supplying information concerning their etymology and development. Thus, the following two texts can be used as a starting point for different approaches (grammar, translation, etc.): "Tourists in England who have seen the warders of the Tower of London and the Yeomen of the Guard know that these men dressed in 15th century uniforms are called *beefeaters.* Not all tourists are aware, however, that the original use of the term (recorded in 1610) was pejorative, referring to a well-fed servant. In a work published before 1628 the word was also said to have been used contemptuously by the French for an Englishman or an English soldier. The word *beefeater* has thus risen in the world, for the well-fed, well-muscled beefeaters of today (this use was first recorded in 1671) are considered by many to be a national treasure." (for ***beefeater***) and "After having eaten countless hamburgers, one may perhaps be interested in knowing more about the origins of the name. By the middle of the 19th century people in Hamburg, Germany, the busiest port in West Germany today, enjoyed pounded beefsteak in some form. Perhaps brought to America by the large numbers of Germans who migrated around that time, this sort of dish with the name *Hamburg steak* may have been mentioned on a menu as early as 1836. The first recorded use of *Hamburg steak* is found in 1884 in the *Boston Journal,* with *hamburger steak* being first recorded in a Walla Walla, Washington, newspaper in 1889. A 1902 cookbook contains a recipe for Hamburg steak that is closer to our conception, a recipe using ground beef mixed with onion and pepper. The hamburger was on its way, as was the Americanism *hamburger*." (for ***hamburger***).

Conclusions

The largest semantic sub-fields are *dishes* (46), *processing* (23), and *animal husbandry* (9) (Figure 4-3). They should be the basis for the learning of a minimum of specialised terms in our faculties of animal husbandry, food processing technologies, and tourism services. The rest of the terms more or less related to ***beef*** should be taught for their anecdotic aspect or to make undergraduates aware of the fact that they should be avoided because of lack of relevance.

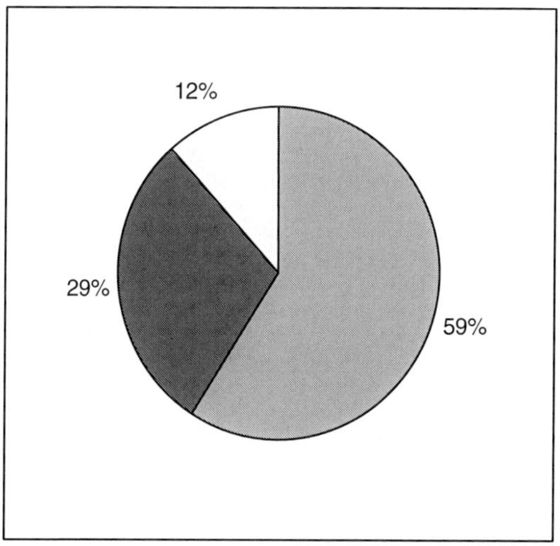

Figure 4-3. Semantic sub-fields of *beef*: 59% - dishes, 29% - processing, 12% - animal husbandry

References

Chalker, Sylvia & Weiner, E. (1994). *The Oxford Dictionary of English Grammar*. London - New York - Sydney - Toronto: BCA.

Perković, Anica & Raţă, Georgeta. (2011). Teaching Semantic Fields: Beef. *Proceedings of the 46th Croatian and 6th International Symposium on Agriculture. Opatija. Croatia*: 289-292.

The American Heritage Dictionary of the English Language. (2008). Boston: Houghton Mifflin.

Webster Encyclopedic Unabridged Dictionary of the English Language. (1996). New York: Gramercy Books.

CHAPTER FIVE

PRAGMATICS

ENGLISH IDIOMS CONTAINING NAMES OF PLANTS

OANA BOLDEA

Introduction

When teaching a new class, we usually ask our students what they expect to learn during the next semester at the English classes. The answer is always the same: "Vocabulary and grammar", meaning that they expect "to learn a lot of new words".

Little do they think of the fact that, when dealing with the vocabulary of a target language, learning new meanings of already-known words is equally important. English being a highly polysemic language, the problem of multiple meanings is not to be overlooked. We tackled the subject of **polysemy** somewhere else. In the present paper, we think of another related subject, namely **idiomatic expressions**: we cannot claim to know a foreign language without understanding **idioms**, too. For instance, it is not enough to understand the meaning of ***piece*** and ***cake*** in order to grasp the meaning of *Flying a plane is **a piece of cake** for me*.

This is why, when teaching students the basic vocabulary they are supposed to know in their field as future engineers dealing with agriculture, we find it necessary to tell them also about the **idioms** they are likely to encounter. As far as the names of plants are concerned, there is a set of idiomatic expressions commonly used in everyday life, and which can be classified according to various criteria, as shown below.

Material and Method

Unfortunately, no collection of terms was available to us with idiomatic expressions referring to plants, so the paper is based on a corpus organised with the help of some dictionaries or glossaries including what we might call "general" **idioms**, i.e. expressions containing names of animals, animal body parts, or anything else related to animals.

For the general subject of **idioms** and idiomaticity, there were some books which proved to be particularly useful. They are all mentioned in the references. Also, our ideas found support in Fengying's article (1996) on the problem of teaching **idioms** in the foreign language class.

Discussion and Conclusion

The **meaning of an idiomatic expression** given in dictionaries is that of 'phrase, construction or expression that is recognised as a unit in the usage of a given language and either differs from the usual syntactic patterns or has a meaning that differs from the literal meaning of its parts taken together' (*Webster Comprehensive Dictionary* 1995). In other words, an expression is an idiom if speakers know about its form, meaning, or use, which they wouldn't be able to figure out by simply knowing everything else about the language. Or, as Fillmore puts it, 'we don't understand what the expression means by computing its meaning from information about its form or parts: that's something we just have to know, by linguistic convention.'

Still, the aim of the present paper is not to deal with both already mentioned types of idiomatic expressions. Thus, we will not discuss here structural expressions, but only lexical ones. That is, the purpose of the paper is to draw some attention on them, because 'making use of ready-made expressions creates a sense of group solidarity, since each member of a language community can depend on the others to pick up allusions, to recognize familiar ways of thinking, and so on' (Fengying 1996).

The expressions of interest here are those in which there is not a predictable relationship between its meaning and the meanings of its parts taken separately. For instance, although we know the meanings of all the words in ***this is not my cup of tea***, the meaning of the whole expression is not that which we might predict. It actually refers to 'something the speaker does not like very much'. The words not having semantic independence, the expression cannot be translated word by word. Usually, we have to paraphrase or, in the more fortunate cases, we may even find idiomatic correspondents in Romanian. This is the case with an expression like ***to beat around the bush***, which has the Romanian equivalent *a bate şaua să priceapă iapa*.

In any case, they must be dealt with very carefully, because if not, their translation may be hilarious. Of course, this holds good the other way around: think of the effect a translation like ****for the appletree blossoms*** would have on a native English speaker: he would certainly not think of the Romanian idiom the translation started from, namely *de florile mărului* ('for nothing').

There are several possible classifications for idiomatic expressions. Thus, there are the so-called "formulaic expressions", said when performing an act. Such an idiomatic expression is ***upsy daisy***, which accompanies the act of holding a young child up in the air. One possible Romanian translation for this is *hopa sus*.

There are also the classifications according to the parts of speech making up the expression. Thus, some expressions contain a verb, like ***to bang the bush*** (translatable into R *a-i face praf pe toţi, a-i uimi*), or ***carry corn*** (R *a nu se lăsa îmbătat de succes, a nu se culca pe lauri*), or even ***tread on somebody's corn*** (R *a călca pe bătătură*). Others do not contain a verb: ***between grass and hay*** (R *între vârsta adolescenţei şi maturitate*), or ***off one's oats*** (R *indispus, fără chef, fără poftă de viaţă*).

Idiomatic expressions containing names of plants can also be classified according to the plant name they contain. If we take this criterion into consideration, very interesting meanings of the same word can occur. Thus, there are **idioms** containing ***rose***. We may find expressions such as ***life is not all roses***, ***bed of roses***, ***to lose one's roses***, or even ***under the rose***, all these having meanings which differ greatly between them. If, in the first two expressions mentioned, ***rose*** can be understood as 'joy', 'happy moment' (one possible Romanian translation being *viaţa nu e un câmp cu flori* – rendering the idiom by another idiom) and in the third we can find the equivalent *a-şi pierde bujorii din obraji* (again an idiom), the fourth expression is to be understood as 'in secret'.

Another example of the same category is represented by the expressions containing ***nut***. We may encounter something that is ***a hard nut to crack***, but we will not translate it by R *e o nucă greu de spart*, but by something more like R *e o treabă dificilă* ('it's a hard thing to do'). In the same way, we will by no means understand ***to be off one's nut*** as R *s-a dat jos de pe nucă*, or people might think we are "nuts". This may be translated as *a fi ţicnit* ('to be crazy'), or, even better, by a Romanian idiom, *a-i fila lampa* (having nothing to do with plants).

Teaching English introducing idiomatic expressions linked in a way or another to their major can be also great fun for the students. Thus, when studying the name of cereals in English, they learn also that ***to feel one's oats*** means *a fi vesel, vioi, plin de viaţa* ('to be joyful'), ***to cry barley*** can be translated by *a cere milă* ('to ask for pity'), and ***to tread on somebody's corn*** should be understood as *a călca be bătătură* ('to drive somebody mad').

There are such expressions containing the word *tree*, too. If some of them may be already known to the students, such as ***to bark up the wrong tree*** (R *a greşi adresa*), some may be totally new. Other examples are ***to***

shake the plum tree (R *a împărţi funcţii, favoruri membrilor săi* – when talking about a political party), and ***in the dry tree*** (R *în vreme de mizerie*).

References

Barbu, M. (1973). *Dicţionar de citate şi locuţiuni străine*. [A Dictionary of Foreign Quotations and Idioms]. Bucureşti: Editura Enciclopedică Română.

Boldea, Oana. (2004). Names of Plants in Food Additives Guide. *Lucrări ştiinţifice, Facultatea de Agricultură XXXVI*: 548-551.

—. (2004). Despre unele expresii idiomatice englezeşti care conţin nume de plante. [On Some English Idioms Containing Plant Names]. *Analele Universităţii din Craiova*: 91-95.

Cacciari, Cristina & Talbossi, Patrizia. (1993). *Idioms. Processing, Structure and Interpretation.* New Jersey: Lawrence Erlbaum Associates.

Chitra, F. *Idioms and Idiomaticity*. (1996). Oxford: Oxford University Press.

Fengying, M. (1996). The Ripple Effect. Word Meaning Expansion and Its Application in Teaching Vocabulary. *Forum 34 (1)*: 8-21.

Microsoft® Encarta® Encyclopedia. (2002). ©1993 – 2001 Microsoft Corporation.

Raţă, Georgeta. (1999). *English for Horticulturists.* Timişoara: Editura Mirton.

Webster Comprehensive Dictionary. (1995). Chicago: J. G. Ferguson Publishing Company.

UNIVERSAL *VS.* CULTURE-SPECIFIC IN THE DISCOURSE OF ACADEMIC AND PROFESSIONAL RHETORIC

GEORGETA RAȚĂ, IOAN PETROMAN AND SCOTT HOLLIFIELD

Introduction

"The naming of plants is a serious matter." When man discovered a new plant, he also created a new word to communicate his discovery to others. The manner in which new **common** or **vernacular** or **vulgar names of plants** are formed is not governed by other rules than acceptance and acceptability. Until communication becomes multilingual and the number of plants named becomes excessive, these names present few problems. The element of fanciful association in Romanian common plant names often drew upon comparisons with parts of the animal body, some of which still persist and strike as 'vulgar' because they have not always been 'properly' modified. In academic practice, taking into account the common plant name is rather risky, since two common names – ***barba-caprei*** 'goat's beard' and ***țâța-caprei*** 'goat's breast' (*Tragopogon major*), ***barba-lupului*** 'wolf's beard' and ***ochiul-boului*** 'ox's eye' (*Helleborus purpurescens*), ***colțul-lupului*** 'wolf's fang' and ***gura-lupului*** 'wolf's mouth' (*Bunias orientalis*), ***colțul-lupului*** 'wolf's fang' and ***gura-lupului*** 'wolf's mouth' (*Scutellaria altissima*), ***limba-broaștei*** 'toad's tongue' and ***limba-oii*** 'sheep's tongue' (*Alisma plantago*), ***limba-cerbului*** 'deer's tongue' and ***limba-vacii*** 'cow's tongue' (*Scolopendrium vulgare*), ***limba-mânzului*** 'colt's tongue' and ***limba-oii*** 'sheep's tongue' (*Plantago media*), ***ochiul-șarpelui*** 'snake's eye' and ***urechea-șoarecelui*** 'mouse's ear' (*Myosotis sylvatica*), ***țâța-oii*** 'sheep's breast' and ***țâța-vacii*** 'cow's breast' (*Primula elatior*), and ***țâța-oii*** 'sheep's breast' and ***țâța-vacii*** 'cow's breast' (*Primula veris*) – and even three common names – ***limba-broaștei*** 'toad's tongue', ***limba-oii*** 'sheep's tongue' and ***limba-șarpelui*** 'snake's tongue' (*Plantago lanceolata*) – can designate the same plant species. Even more risky is to take into account the common name, since one and the same common name can designate two – ***barba-caprei*** 'goat's beard' (*Calocera viscose* and *Tragopogon major*), ***barba-lupului*** 'wolf's beard' (*Crispis biennis* and *Helleborus purpurescens*), ***barba-ursului*** 'bear's beard' (*Equisetum* sp. and *Usnea barbata*), ***colţul-lupului*** 'wolf's fang' (*Bunias orientalis* and *Scutellaria altissima*), ***gura-lupului*** 'wolf's mouth' (*Bunias orientalis* and *Scutellaria altissima*), ***limba-boului*** 'ox's tongue' (*Anchusa officinalis* and *Plantago major*), ***limba-broaştei*** 'toad's tongue' (*Alisma plantago* and *Plantago lanceolata*), ***limba-şarpelui*** 'snake's tongue' (*Ophioglossum vulgatum* and *Plantago lanceolata*), ***ochiul-lupului*** 'wolf's eye' (*Lycopsis arvensis* and *Plantago indica*), ***spinarea-lupului*** 'wolf's back' (*Ahtynum filix femina* and *Asplenium ruta muraria*), ***ţâţa-caprei*** 'goat's breast' (*Tragopogon major* and *Tragopogon pratensis*), ***ţâţa-vacii*** 'cow's breast' (*Primula elatior* and *Primula veris*), and ***urechea iepurelui*** 'rabbit's ear' (*Bupleurum* sp. and *Stachys lanata*); four – ***limba-cucului*** 'cuckoo's tongue' (*Gentiana bulgarica*, *Gymnadenia conopsea*, *Orchis maculata* and *Dactylorchis maculata*) and ***ochiul-şarpelui*** 'snake's eye' (*Eritrichium nanum*, *Myosotis arvensis* and *Myosotis sylvatica*); and even five different plant species or one and the same plant having different scientific names at two different times – ***limba-oii*** 'sheep's tongue' (*Alisma plantago*, *Cirsium canum*, *Plantago gentianoides*, *Plantago lanceolata* and *Plantago media*) and ***ţâţa-oii / -oilor*** 'sheep's breast' (*Arnica Montana*, *Digitalis ambigua*, *Primula elatior*, *Primula officinalis* and *Primula veris*).

Material and Method

Our interest was in plant common names including **animal body parts**, the element suggesting the linguistic choice in the naming of the plants. We have identified 134 common plant names including reference to **animal body parts**, names that have the structure **N** (animal)**'s** + **N** (animal body part). In Romanian, the structure is **N** (animal body part) + **N** (animal) in the Genitive (there is no ***'s*** to mark the genitive case).

Results and Discussion

The following **animal body parts** occur in these common plant names: ***tongue*** (23): ***limba-boului*** 'ox's tongue' (*Anchusa officinalis*) '*alkanet, common bugloss*', ***limba-boului*** 'ox's tongue' (*Plantago major*) '*common plantain, greater plantain, soldier's herb*', ***limba-broaştei*** 'toad's tongue' (*Alisma plantago*) '*common water-plantain, mad-dog weed*', ***limba-broaştei*** 'toad's tongue' (*Plantago lanceolata*) '*English plantain, narrow* *leaf plantain, ribwort plantain*', ***limba-câinelui*** 'dog's tongue' (*Cynoglossum officinale*) '*hounds tongue, dog's tongue, gypsy flower, rats and mice*', ***limba-caprei*** 'goat's tongue' (*Ballota nigra*) '*black horehound*', ***limba-cerbului*** 'deer's tongue' (*Scolopendrium vulgare*), ***limba-cucului*** 'cuckoo's tongue' (*Botrychium lunaria*) '*common moonwort*', ***limba-cucului*** 'cuckoo's tongue' (*Gentiana bulgarica*), ***limba-cucului*** 'cuckoo's tongue' (*Gymnadenia conopsea*) 'fragrant orchid', ***limba-cucului*** 'cuckoo's tongue' (*Orchis maculata, Dactylorchis maculata*) '*heath spotted orchid, moorland spotted orchid*', ***limba-mânzului*** 'colt's tongue' (*Plantago media*) '*hoary plantain*', ***limba-mielu(şelu)lui*** 'lamb's tongue' (*Borago officinalis*) '*borage, starflower*', ***limba-oii*** 'sheep's tongue' (*Alisma plantago*) '*common water-plantain, mad-dog weed*', ***limba-oii*** 'sheep's tongue' (*Cirsium canum*) '*Queen Anne's thistle*', ***limba-oii*** 'sheep's tongue' (*Plantago gentianoides*), ***limba-oii*** 'sheep's tongue' (*Plantago lanceolata*) '*English plantain, narrow leaf plantain, ribwort plantain*', ***limba-oii*** 'sheep's tongue' (*Plantago media*) '*hoary plantain*', ***limba-peştelui*** 'fish's tongue' (*Limonium vulgare*) '*common sea-lavender*', ***limba-şarpelui*** 'snake's tongue' (*Ophioglossum vulgatum*) '*Southern adders tongue*', ***limba-şarpelui*** 'snake's tongue' (*Plantago lanceolata*) '*English plantain, narrow leaf plantain, ribwort plantain*', ***limba-vacii*** 'cow's tongue' (*Scolopendrium vulgare*), ***limba-vrăbiei*** 'sparrow's tongue' (*Thymelaea passerina*) '*spurge flax*'; ***breast*** (11): ***ţâţa-caprei*** 'goat's breast' (*Tragopogon major*), ***ţâţa-caprei*** 'goat's breast' (*Tragopogon pratensis*) '*Jack-go-to-bed-at-noon, meadow salsify, showy goat's-beard*', ***ţâţa-mielului*** 'lamb's breast' (*Sempervivum tectorum*) '*common houseleek, hen and chicks*', ***ţâţa-oii*** 'sheep's breast' (*Arnica montana*) '*leopard's bane, mountain arnica, mountain tobacco, wolf's bane*', ***ţâţa-oii*** 'sheep's breast' (*Digitalis ambigua*), ***ţâţa-oii*** 'sheep's breast' (*Primula elatior*) '*(true) oxlip*', ***ţâţa-oii*** 'sheep's breast' (*Primula officinalis*) '*artetyke, arthritica, buckles, cowslip, crewel, cuy lippe, cuy, drelip, fairy cups, frauenchlussel, herb Peter, key flower, key of Heaven, Lady's key, lippe, Mayflower, Our Lady's keys, paigle, palsywort, paralysio, password, peggle, Petty Mulleins, plumrocks*', ***ţâţa-oii*** 'sheep's breast' (*Primula veris*) '*artetyke, arthritica, buckles, cowslip, crewel, cuy lippe, cuy, drelip, fairy cups, frauenchlussel, herb Peter, key flower, key of Heaven, Lady's key, lippe, Mayflower, Our Lady's keys, paigle, palsywort, paralysio, password, peggle, Petty Mulleins, plumrocks*', ***ţâţa-oilor*** 'sheep's breast' (*Arnica montana*) '*leopard's bane, mountain arnica, mountain tobacco, wolf's bane*', ***ţâţa-vacii*** 'cow's breast' (*Primula elatior*) '*(true) oxlip*', ***ţâţa-vacii*** 'cow's breast' (*Primula veris*) '*artetyke, arthritica, buckles, cowslip, crewel, cuy lippe, cuy, drelip,* *fairy cups*, *frauenchlussel*, *herb Peter*, *key flower*, *key of Heaven*, *Lady's key*, *lippe*, *Mayflower*, *Our Lady's keys*, *paigle*, *palsywort*, *paralysio*, *password*, *peggle*, *Petty Mulleins*, *plumrocks*'; ***eye(e)*** (8): ***ochii-păsăruicii*** 'bird's eyes' (*Myosotis palustris*) '*true forget-me-not, water forget-me-not*', ***ochii-şoricelului*** 'mouse's eyes' (*Saxifraga adscendens*) '*ascending saxifrage*', ***ochiul-boului*** 'ox's eye' (*Helleborus purpurescens*), ***ochiul-lupului*** 'wolf's eye' (*Lycopsis arvensis*), ***ochiul-lupului*** 'wolf's eye' (*Plantago indica*), ***ochiul-şarpelui*** 'snake's eye' (*Eritrichium nanum*) '*arctic forget-me-not*', ***ochiul-şarpelui*** 'snake's eye' (*Myosotis arvensis*) '*field forget-me-not*', ***ochiul-şarpelui*** 'snake's eye' (*Myosotis sylvatica*); ***beard*** (7): ***barba-caprei*** 'goat's beard' (*Calocera viscose*), ***barba-caprei*** 'goat's beard' (*Tragopogon major*), ***barba-lupului*** 'wolf's beard' (*Crispis biennis*), ***barba-lupului*** 'wolf's beard' (*Helleborus purpurescens*), ***barba-ţapului*** 'he-goat's beard' (*Spiraea aruncus*), ***barba-ursului*** 'bear's beard' (*Equisetum* sp.) '*horsetails*', ***barba-ursului*** 'bear's beard' (*Usnea barbata*); ***ear*** (5): ***urechea iepurelui*** 'rabbit's ear' (*Bupleurum* sp.), ***urechea iepurelui*** 'rabbit's ear' (*Stachys lanata*) '*lamb's ear*', ***urechea-porcului*** 'pig's ear' (*Salvia verticillata*), ***urechea-şoarecelui*** 'mouse's ear' (*Myosotis sylvatica*), ***urechea-ursului*** 'bear's ear' (*Primula auricula*) '*auricula, bear's ear*'; ***mouth*** (4): ***gura-leului*** 'lion's mouth' (*Antirrhinum majus*) '*(common) snapdragon*', ***gura-lupului*** 'wolf's mouth' (*Bunias orientalis*), ***gura-lupului*** 'wolf's mouth' (*Scutellaria altissima*), ***gura-mâţei*** 'cat's mouth' (*Linaria vulgaris*) '*toadflax*'; ***nail*** (4): ***unghia-caprei*** 'goat's nail' (*Lactarius piperatus*) '*peppery milk-cap*', ***unghia-găii / găinii*** 'hen's nail' (*Astragalus glycyphyllos*) '*liquorice milkvetch, wild licorice, wild liquorice*', ***unghia-păsării*** 'bird's nail' (*Viola declinata*), ***unghia-ursului*** 'bear's nail' (*Carduus acanthoides*) '*spiny plumeless thistle, welted thistle*'; ***sole*** (4): ***talpa-gâştei*** 'goose's sole' (*Leonurus cardiaca*) '*lion's ear, lion's tail, motherwort, throw-wort*', ***talpa-lupului*** 'wolf's sole' (*Chaiturus marrubiastrum*), ***talpa-mâţei / -pisicii*** 'cat's sole' (*Antennaria dioica*) '*catsfoot, cudweed, mountain everlasting, stoloniferous pussytoes*', ***talpa-ursului*** 'bear's sole' (*Acanthus longifolius*); ***foot*** (3): ***piciorul-caprei*** 'goat's foot' (*Aegopodium podagraria*) '*bishop's weed, goutweed, ground-elder, herb Gerard, snow-in-the-mountain*', ***piciorul-cocoşului*** 'cock's foot' (*Ranunculus* sp.) '*buttercups, lesser celandine, spearworts, water crowfoots*', ***piciorul-viţelului*** 'calf's foot' (*Arum maculatum*) '*Adam and Eve, bobbins, cows and bulls, cuckoo-pint, devils and angels, Jack in the pulpit, lords and ladies, naked boys, starch-root, wake robin, wild arum*'; ***back*** (2): ***spinarea-lupului*** 'wolf's back' (*Ahtynum filix femina*), ***spinarea-lupului*** 'wolf's back' (*Asplenium ruta muraria*) '*wall rue*'; ***fang*** (2): ***colţul-***

lupului 'wolf's fang' (*Bunias orientalis*), ***colţul-lupului*** 'wolf's fang' (*Scutellaria altissima*); ***hair*** (2): ***părul-ciutei*** 'do's hair' (*Rhamnus cathartica*) '*buckthorn, common buckthorn, purging buckthorn*', ***părul-porcului*** 'pig's hair' (*Equisetum telmateja*)'; ***bladder*** (1): ***băşica-porcului*** 'pig's bladder' (*Lycoperdon bovista*); ***bone*** (1): ***osul-iepurelui*** 'rabbit's bone' (*Ononis spinosa*) '*(spiny) restharrow*'; ***face*** (1): ***faţa-mâţei*** 'cat's face' (*Galeopsis ladanum*); ***head*** (1): ***capul-câinelui*** 'dog's head' (*Pulsatilla pratensis*) '*small pasque flower*'; ***nose*** (1): ***nasul-curcanului*** 'turkey's nose' (*Amaranthus caudatus*) '*foxtail amaranth, love-lies-bleeding, pendant amaranth, quilete, tassel flower, velvet flower*'; ***penis*** (1): ***puţa-cocoşului*** 'cock's penis' (*Listera ovata*) '*European common twayblade*'; ***tooth*** (1): ***măseaua-ciutei*** 'do's tooth' (*Erythronium dens canis*) '*dog's tooth violet, dogtooth violet*'.

A. The analysis of these common plant names shows that they can be ranged within six main categories, as follows:

- **Same N** (animal)**'s + same N** (animal body part). Only two plant species whose names include reference to animal body parts are identical in both Romanian and English: ***limba-câinelui*** 'dog's tongue' (*Cynoglossum officinale*) '*dog's tongue*' and ***urechea-ursului*** 'bear's ear' (*Primula auricula*) '*bear's ear*'. Other two plant species names are almost similar: for the Romanian ***limba-câinelui*** 'dog's tongue' (*Cynoglossum officinale*) the English language has *HOUNDStongue*, and for ***limba-şarpelui*** 'snake's tongue' (*Ophioglossum vulgatum*) it has *Southern ADDERStongue.*
- **Same N** (animal)**'s + different N** (animal body part). Four plant species have names that include the same animal in both languages but different animal body parts: ***talpa-mâţei / -pisicii*** cat's sole' (*Antennaria dioica*) '*catsFOOT*', ***talpa-mâţei / -pisicii*** cat's sole' (*Antennaria dioica*) '*stoloniferous pussyTOES*', ***ţâţa-caprei*** 'goat's breast' (*Tragopogon pratensis*) '*showy goat's-beard*', ***ţâţa-vacii*** 'cow's breast' (*Primula elatior*) '*(true) OXlip*', and ***ţâţa-vacii*** 'cow's breast' (*Primula veris*) '*cowslip*'.
- **Different N** (animal)**'s + same N** (animal body part). One plant genus and two plant species have names belonging to this category: ***măseaua-ciutei*** 'do's tooth' (*Erythronium dens canis*) '*dog's tooth violet, dogtooth violet*', ***piciorul-cocoşului*** 'cock's foot' (*Ranunculus* sp.) '*water crowfoots*', and ***urechea iepurelui*** 'rabbit's ear' (*Stachys lanata*) '*lamb's ear*'.
- **Different N** (animal)**'s + different N** (animal body part). One plant genus and four plant species have names belonging to this category:

barba-ursului 'bear's beard' (*Equisetum* sp.) '*horsetails*', ***talpa-gâştei*** 'goose's sole' (*Leonurus cardiaca*) '*lion's ear*', ***talpa-gâştei*** 'goose's sole' (*Leonurus cardiaca*) '*lion's tail*', ***ţâţa-oii*** 'sheep's breast' (*Primula elatior*) '*(true) oxlip*', ***ţâţa-oii*** 'sheep's breast' (*Primula officinalis*) '***cowslip***', and ***ţâţa-oii*** 'sheep's breast' (*Primula veris*) '*cowslip*'.

- **Different N** (animal)**'s + no N** (animal body part). Five plant species have names that can be ranged in this category: ***gura-leului*** 'lion's mouth' (*Antirrhinum majus*) '*(common) snapdragon*', ***gura-mâţei*** 'cat's mouth' (*Linaria vulgaris*) '*toadflax*', ***limba-broaştei*** 'toad's tongue' (*Alisma plantago*) '*mad-dog weed*', ***limba-oii*** 'sheep's tongue' (*Alisma plantago*) '*mad-dog weed*', ***piciorul-viţelului*** 'calf's foot' (*Arum maculatum*) '*wake robin*', ***ţâţa-oii / -oilor*** 'sheep's breast' (*Arnica montana*) '*leopard's bane*', and ***ţâţa-oii / -oilor*** 'sheep's breast' (*Arnica montana*) '*wolf's bane*'.

B. There is a single case of correspondence between the Romanian ***limba-câinelui*** 'dog's tongue' and the English *dog's tongue*, on the one hand, and the scientific name ***Cynoglossum*** *officinale* (L *Cynoglossum* < Gk *cyno* 'dog' + Gk *glossum* 'tongue'), on the other hand; the other English common name, *hounds tongue*, comes from the belief that the plant could, if a leaf was worn in the shoe, ward off dog attacks (*Cynoglossum officinale*. Online: http://en.wikipedia.org/wiki/Cynoglossum_officinale). There are four cases of correspondence between the Romanian common name and the scientific name of the plant: R ***limba-şarpelui*** 'snake's tongue' and L ***Ophioglossum*** *vulgatum* (L *Ophioglossum* < Gk *ophio* 'snake' + Gk *glossum* 'tongue'), with partial equivalence in E *Southern adders* ***tongue***; R ***ochiul-lupului*** 'wolf's eye' and L ***Lycopsis*** *arvensis* (L *Lycopsis* < Gk *lyco* 'wolf' + Gk *opt* 'eye'), with *Plantago indica* as a synonym; R ***piciorul-caprei*** 'goat's foot' and L ***Aegopodium*** *podagraria* (L *Aegopodium* < Gk *aego* 'goat' + Gk *pod* 'foot'); R ***urechea-şoarecelui*** 'mouse's ear' and L ***Myosotis*** *sylvatica* (< Gk *my(os)* 'mouse' + Gk *otis* 'ear'), with ***ochii-păsăruicii*** 'bird's eyes' (*Myosotis palustris*) and ***ochiul-şarpelui*** 'snake's eye' (*Myosotis arvensis* and *Myosotis silvatica*). There are two cases of correspondence between the English common name and the scientific name of the plant: E ***dog's tooth*** *violet* and ***dogtooth*** *violet* and L *Erythronium* ***dens canis*** (L ***dens canis*** < L *dens* 'tooth' and L *canis* 'dog'), with R ***măseaua-ciutei*** 'do's tooth', and with the following explanation: "The cormous white rootstocks are oblong and resemble a dog's tooth in appearance; hence the name". (*Erythronium dens-canis*. Online: http://en.wikipedia.org/wiki/Erythronium_dens-canis);

E ***horsetails*** and L ***Equisetum*** (< L *equus* 'horse' and L *seta* 'bristle'), with R ***barba-ursului*** 'bear's beard' (*Equisetum* sp.) and ***părul-porcului*** 'pig's hair' (*Equisetum telmateja*)': the common name *horsetail* arose because the branched species somewhat resemble a horse's tail (*Equisetum*. Online: http://en.wikipedia.org/wiki/Equisetum). There is no correspondence between the Romanian common name and the scientific name of the plant containing both animal references: R ***băşica-porcului*** 'pig's blatter' and L ***Lycoperdon*** *bovista* (L *Lycoperdon* < Gk *lyco* 'wolf' + Gk *perdon* 'to break wind'), where *lycoperdon* means 'wolf-farts' (*Lycoperdon*. Online: http://en.wikipedia.org/wiki/Lycoperdon). There is no correspondence whatsoever between the Romanian and English common names and their scientific counterparts containing all animal references: R ***gura-leului*** 'lion's mouth', L ***Antirrhinum*** *majus* (L *Antirrhinum* < Gk *anti* 'like' + Gk *rhis* 'nose' + Gk *inus* 'pertaining to') meaning 'like a nose' and E *(common) snap**dragon***, with the explanation "[it] probably refers to the nose-like capsule in its mature state" (*Antirrhinum*. Online: http://en.wikipedia.org/wiki/Antirrhinum); R ***piciorul-cocoşului*** 'cock's foot', L ***Ranunculus*** sp.) (LL *ranunculus* 'little frog') and E '*water **crowfoots***': "This probably refers to many species being found near water, like frogs." (*Ranunculus*. Online: http://en.wikipedia.org/wiki/Ranunculus).

Acknowledgement

The paper is the result of research conducted within the project no. 178002 *Languages and cultures across space and time* funded by the Ministry of Science and Technological Development of the Republic of Serbia.

Conclusion

The only conclusions we can draw after the analysis of the common plant names containing references to animals and animal body parts in English and Romanian is that there are no detectable "universals" as well as there are no possible correlations with their scientific name either. To note, though, the huge impact of Greek in the invention of not only scientific and technical words (Carstairs-McCarthy 2002), but also of common plant names.

References

Antirrhinum. Online: http://en.wikipedia.org/wiki/Antirrhinum.

Bejan, D. (1991). *Nume româneşti de plante.* [Romanian Plant Names]. Cluj-Napoca: Dacia.

Carstairs-McCarthy, A. (2002). *An Introduction to English Morphology. Words and Their Structure*. Edinburgh: Edinburgh University Press.

Cynoglossum officinale. Online: http://en.wikipedia.org/wiki/Cynoglossum_officinale.

Equisetum. Online: http://en.wikipedia.org/wiki/Equisetum.

Erythronium dens-canis. Online: http://en.wikipedia.org/wiki/Erythronium_dens-canis.

Gledhill, D. (2008). *The Names of Plants*. Cambridge: CUP.

Lycoperdon. Online: http://en.wikipedia.org/wiki/Lycoperdon.

Ranunculus. Online: http://en.wikipedia.org/wiki/Ranunculus.

Raţă, Georgeta. (2010). Universal vs. specific în terminologia botanică: numele comune de plante. [Universal vs. Specific in Botanical English: Common Plant Names]. *Agrobuletin AGIR 5*: 35-39.

Schroeter, A. I. & Panasiuk, V. A. (1999). *Dictionary of Plant Names*. Koenigstein: Koeltz Scientific Books.

Stearn, W. T. (1985). *Botanical Latin. History, Grammar, Syntax, Terminology and Vocabulary*. Newton Abbot: David & Charles.

Creative Metaphors in Poisonous Plant Names

Alina-Andreea Dragoescu
and Petru Dragoescu

Introduction

Metaphor is a widespread linguistic device highly used in common or vernacular plant name coinage. The word **metaphor** is derived from Gk *metaphora* meaning 'transfer' (< Gk *metaphero* 'to carry over', 'to transfer' < Gk *meta* 'between' + Gk *phero* 'to bear', 'to carry') (OED). Thus, **metaphor** is largely defined as semantic change based on similarity, *i.e.* a similarity in form or function between the original concept named by a word and the target concept named by this word. Besides this lead-in on **metaphor**, introducing onomasiology would also be of import to the present analysis. Onomasiology (< Gk *onoma* 'name') is a branch of lexicology concerned with the study of words as names and with the question of how things are named (AHDEL). It is important to note that the coinage of a new designation usually aims at classifying the referent as member of a familiar concept. In the name-giving process, the speaker first selects one or two salient aspects, based on direct or indirect analogies. Our exploration of the motivations of a chosen name draws on a variety of cognitive associative relations. These relations may be: contiguity relations, similarity relations, partiality relations, contrast relations. As something which is named in terms of something else, popular plant names abound in metaphoric phrases.

Several authors describe metaphorical processes as the basic representation of denominational practice in popular botanical vocabulary (Bejan 1991). **Metaphor** in this particular context may generally be regarded as the conversion of an object's name into another object's name, on condition there be a certain connection between the two, whatever the semantic distance. In our case, a plant (A) and an object (B) are associated by means of a third element (C) serving as *tertium comparationis*, which measures the semantic distance between the former two elements. Thus, only the second transferred element remains activated, constituting the

metaphor, on account of common features identified between the two (Bejan 1991). The second element is a determinant or a specification, which is added to the generic name, the carrier of the basic meaning of the **compound**. The criterion Bejan (1991) considers as essential for the inclusion of a plant name in the category of vernacular names is its being created inside the language in an original manner, by contrast to scientific names and names borrowed, translated, or calqued from Latin or other languages. Few **metaphors** in poisonous plant names can only be understood in terms of their Latin etymology, which allows for a lesser degree of creativity and originality.

Generally regarded as implicit comparisons, **metaphors** are based on the principles of analogy, mimesis, and similarity. Nonetheless, they are more than merely elliptical comparisons, or ornamental substitutes, as they also contain cognitive value. Names in naturally occurring contexts are essentially labels for that which they designate. Yet, cognitive linguists emphasize a more profound aspect of the issue, going beyond the decorative notion of *figure of style*. In cognitive linguistics, conceptual or cognitive **metaphors** imply the understanding of an idea or concept in terms of another, which shape human perception, resulting in particular names devised for the surrounding world.

This scope of linguistics maintains that **metaphors** are foundational to our conceptual system, as they serve to facilitate the understanding of concepts through expressions that relate to more familiar and concrete notions. I. A. Richards introduced useful terminology to address this matter, such as *topic* or *tenor*, *vehicle*, *interaction*, *reference*, and *referent*. The tenor refers to the underlying idea or principal subject of the **metaphor**. The vehicle conveys the borrowed idea or what the tenor resembles. For example, in the name *five-fingered root*, the tenor is the root or whole plant, while the vehicle is the finger analogy, by transferring characteristics of a human hand to the plant's root. The tenor and vehicle display patterns of interaction in order to produce a new emergent meaning for the entire phrase or compound plant name, in our case. Thus, in the metaphorical name *dead tongue*, for example, the plant is viewed as a tongue, but the interpretation of both terms is altered. As a literal interpretation is incompatible with the context, the **metaphor** needs to be addressed and reinterpreted figuratively. Furthermore, Richards describes reference as the indicator of that realm of memory where recollections of past knowledge and contexts are located, whereas the referent represents the objects which are perceived and which generate the impression stored in the thought area (Ogden & Richards 1923). **Metaphors** play upon a similarity between the term used metaphorically (the vehicle) and the thing the name is a **metaphor** for. We conclude that, although a **metaphor** is largely a statement of similarity, the examination of similarities nonetheless uncovers the differences between the literal and the metaphorical. Thus, **metaphors** are revealed by semantic intersections, as they draw attention to the similar in the apparently dissimilar. However, Max Black (1962) explains the 'interaction' model of **metaphor** as fundamentally irreducible to a literal paraphrase. Interaction, unlike mere 'substitution' or 'comparison' of terms, emphasizes the cognitive and emotional content of **metaphor** with their added implications. As he explains in *Models and Metaphors*, a **metaphor** 'selects, emphasizes, suppresses, and organizes features of the principal subject by implying statements about it that normally apply to the subsidiary subject'. (Black 1962) Thus, the present interpretation of **metaphor** primarily aims at explaining how metaphorical meaning is created, based on a wide range of referential correspondences. This study exhibits all aspects and functions of **metaphor**: denominational, ornamental, expressive, and cognitive, as they often overlap and display complex linguistic characteristics.

Material and Method

The data considered consists of an extensive list of names attributed to poisonous plants, as recorded by botanical dictionaries. Vernacular plant names are analysed in order to reveal connotations people have attached to plants. Thus, we have drawn a corpus of metaphorical constructions, proposing onomasiological explanations with the help of etymology and pragmatic inference. The task of accounting for the designation of plants by means of particular **metaphors** is also examined from a semantic point of view, focusing on the cognitive metaphor approach.

Results

The following poisonous plant names have been analyzed: ***angel wings***, ***angel's trumpet***, ***beaver poison***, ***belladonna***, ***birthroot***, ***bitter(sweet) nightshade***, ***bittersweet***, ***black bryony***, ***black cherry***, ***black hellebore***, ***black locust***, ***black nightshade***, ***bleeding heart***, ***bloodroot***, ***bloodwort***, ***bloody fingers***, ***boneset***, ***chandelier plant***, ***Christ herb***, ***Christmas rose***, ***climbing nightshade***, ***common hyacinth***, ***cowbane***, ***crazyweed***, ***dead man's thimbles***, ***dead men's bells***, ***dead nettle***, ***dead tongue***, ***deadly nightshade***, ***devil drug***, ***devil's apple***, ***devil's cherries***, ***devil's ear***, ***devil's helmet***, ***devil's herb***, ***devil's trumpet***, ***devil's weed***, ***doll's-eyes***, ***dragon root***, ***drumsticks***, ***dumbcane***, ***Dutchman's breeches***,

elephant ear, ***fairy thimbles***, ***fairy's glove***, ***fairy-folks-fingers***, ***false hellebore***, ***five-fingered root***, ***folk's glove***, ***foxglove***, ***gloves of Our Lady***, ***heart of Jesus***, ***helleboros***, ***hemlock***, ***henbane***, ***herb paris***, ***hogweed***, ***holy rope***, ***horsebane***, ***Jamestown-weed***, ***jimson weed***, ***lambkill***, ***lark's claw***, ***larkspur***, ***lice-bane***, ***lousewort***, ***mad seeds***, ***mandrake***, ***memory root***, ***mother of millions***, ***mother of thousands***, ***naughty man's cherries***, ***nightshade***, ***one-berry***, ***palm of Christ***, ***Palma Christi***, ***pepper turnip***, ***pheasant's eye***, ***poison hemlock***, ***poison ivy***, ***poison nut***, ***poison parsley***, ***poison-berry***, ***poison-flower***, ***rattlebox***, ***rattlepods***, ***scarlet berry***, ***snakeberry***, ***stinking nightshade***, ***stinkweed***, ***tall boneset***, ***trailing bittersweet***, ***trailing nightshade***, ***true love***, ***violet bloom***, ***Virgin's glove***, ***water hemlock***, ***water lovage***, ***weed of madness***, ***weed paris***, ***white baneberry***, ***white snakeroot***, ***wild turnip***, ***witches' gloves***, ***wolfsbane***, ***woody nightshade***, and ***woollyheads***.

In all cases, **metaphors** activate specific aspects or features of the plant: physical appearance, effects, and uses. The interpretation of metaphorical names in the case of poisonous plants may be divided into two categories. On the one hand, there are plant names which instantly reveal the poisonous or even lethal nature of the plant. On the other hand, there are poisonous plant names which fail to capture this crucial piece of information about the designated object, displaying other minor onomasiological motivations. The latter are relatively fewer in number: they are usually the result of metaphoric analogies of the plant's shape or other secondary features. Most poisonous plant names, however, allude to the fact that the plant has toxic effects (e.g. ***cowbane***, ***horsebane***, ***lambkill***). Only few names are **dull** or **dead metaphors** (e.g. ***poison-berry***, ***poison-flower***, ***poison hemlock***), while most other plant names are highly creative, often displaying a large array of synonymic series: ***bittersweet nightshade*** displays a lengthy synonymic series of interrelated **metaphors** (***bittersweet***, ***bitter nightshade***, ***climbing nightshade***, ***poisonberry***, ***poisonflower***, ***scarlet berry***, ***snakeberry***, ***trailing bittersweet***, ***trailing nightshade***, ***violet bloom***, and ***woody nightshade***). The plant is poisonous, while being considered by some a herbal remedy, hence its name ***bittersweet*** incorporates both positive and negative aspects. Its other names refer to physical features like shape and colour.

Most poisonous plant names are explicit, self-descriptive and edifying in an instantly recognizable manner. ***Deadly nightshade*** has very illustrative names, such as ***devil's cherries***, ***devil's herb***, and ***naughty man's cherries***. It is also called ***belladonna*** (It 'fair lady'), because it was used by women as a beauty enhancer (OED). Another possible analogy may be drawn between the cathartic effect of beauty and the plant's paralyzing effect. The name ***naughty man's cherries*** might be an allusion, beside the shape of the fruit, to a peculiar symptom in those poisoned by ***belladonna*** manifested by frequent movements of the hands and fingers accompanied by bending forward (Grieve 1995).

Plant names containing the word 'black' are sometimes used not in reference to their colour, but on account of their toxicity, as ***black*** stands for ***poisonous***. The name of the ***black locust*** contains a **metaphor** which alludes to its feature of being invasive, as its toxic pods are easily carried long distances by the wind. Likewise, ***black bryony*** has toxic berries which are usually red, but the name carries the epithet 'black' on account of the same identification of darkness with negative connotation.

By the same token, ***nightshade*** has several self-explicatory names: ***devil's weed***, ***devil's apple***, or ***stinkweed***. It is also known as ***jimson weed***, derived from ***Jamestown-weed*** in reference to Jamestown, Virginia colony, where it was discovered by the Europeans in the 17th century, when British soldiers mistook it for an edible plant and hallucinated for eleven days (OED). Another story informs that the British soldiers were drugged with it while attempting to suppress Bacon's Rebellion and they subsequently appearing to have gone insane. In South Africa, it is known as ***mad seeds***, as intoxication typically produces delirium (WE). Another name, ***devil's trumpet***, alludes to the tubular shape of the funnel-like flowers, while also connecting the plant with evil for its poisonous effects. In the case of ***henbane***, also known as ***stinking nightshade***, the origins of the word are unclear but 'hen' probably goes back to the root ***-hen***, which originally meant death rather than referring to fowl. The former name of the plant was ***hen bell*** (as in ***death bell***), a **metaphor** which was based on an obvious analogy. However, after the meaning of 'death' was lost, no association could be established between ***hen*** and ***bell***, hence the latter element was altered to the more practical ***bane***, given its toxic effect (Liberman 2008). OED also mentions ***hennebelle*** as an OE variant of the same plant. ***Horsebane*** is also called ***dead tongue***, probably because of the paralyzing effect of this plant on the organs of speech (Grieve 1995). Similarly, ***dumbcane*** is also thus called because it is a poisonous plant which produces the immobility of the organs of speech. ***Hemlock***, or ***poison hemlock***, ***beaver poison***, and ***poison parsley***, belongs to the same family of plants as the parsley and parsnip, with which it can perilously be confused. The name ***hemlock*** is derived from the Anglo-Saxon words ***hem*** ('border, shore') and ***leac*** ('leek' or 'plant') (Grieve 1995).

Hemp has few popular names (***holy rope***), but a bulky synonymic series of slang names (***devil drug***, ***weed of madness***, ***assassin of youth***, ***Mexican ditch weed***, ***nickel bag***, ***rope***, ***weed***) illustrating its perilous effects. The plant produces an exhilarating intoxication, with hallucinations, being used as an intoxicant, hence its metaphorical names ***leaf of delusion***, ***increaser of pleasure***, ***cementer of friendship***, etc. Another intoxicant, ***herb paris***, or ***true love***, ***one berry*** (because its fruit consists of one sole berry) contains ***paris*** (< L *par*, *paris* 'equal') calqued after the scientific name on account of the regularity of its leaves. The plant has also been used as an aphrodisiac for its opium-like, though poisonous, seeds and berries, thus suggesting the deceitful name ***true love***.

As to the plants whose onomasiological motivations include references to secondary features other than toxicity, most names seem to have been given in reference to the shape of part of the plant, thus being **synecdochic metaphors**. ***Angel's trumpet*** has trumpet-shaped flowers, and the angelic **metaphor** was probably suggested by its delicacy, rather than other injurious attributes which remain clandestine. A **synonym** for ***belladonna*** is ***black cherry***, in allusion of the poisonous dark berries which are full of a dark, inky juice. Likewise, ***black nightshade*** is thus named due its masses of black berries. ***Black hellebore*** (< Gk ***helleboros***, probably meaning 'plant eaten by fawns' < Gk *ellos* / *hellos* 'fawn' + *bora* 'food of beasts') is the name given to a plant with poisonous and medicinal qualities, reputed to cure madness (OED). Another etymology claims the plant name is derived from Gk *elein* 'to injure' and *bora* 'food', clearly indicating its poisonous nature. The epithet refers to the dark coloured rootstock (Grieve 1995). It is also called ***Christ herb*** or ***Christmas rose***, because it flourishes around Christmas time and it is similar in shape to the rose. Although the name ***bloodroot*** or ***bloodwort*** seems unsuitable for this snow-white flower, it describes the bright red root containing a red sap, which seems to be 'bleeding' when cut. Some small ball-shaped flowers bear the names ***belly buttons***, ***woollyheads***, or ***drumsticks***, all alluding to their spherical shape. This was seemingly more perceptible to the eye than the plant's toxicity, as it is only toxic to cattle and other livestock.

Numerous plants were given names containing words related to parts of the body, which presented obvious analogies close at hand. For instance, ***bleeding heart*** has bisymmetric heart-shaped flowers. Its other popular name is ***Dutchman's breeches***, which is derived from the white flowers that look like breeches. Another plant called ***elephant ear***, ***heart of Jesus***, or ***angel wings*** obviously alludes to the shape of the leaves. ***Devil's ear*** or ***dragon root*** has several names which are connected to the shape or the utility of the root because this is the most used part of the plant (***birthroot***, ***memory root***, ***pepper turnip***, and ***wild turnip***). The most striking feature of ***doll's-eyes*** or ***white baneberry*** is its fruit, a poisonous white berry, whose size, shape, and black stigma scar explain its names.

The root of ***dropwort*** consists of clusters of fleshy tubers, hence its other popular name ***five-fingered root***. A poisonous plant belonging to the nightshades family, ***mandrake***, was thus called because its branched root with sinuate bifurcations resembles the human body (AHDEL). ***Mother of thousands*** or ***mother of millions*** received its name due to its impressive ability for vegetative reproduction. The name was observably suggested by the myriad plantlets growing on the fringes of its leaves, which is a more promptly perceivable feature than its toxicity in animals. It is also called ***chandelier plant***, which is an obvious analogy to the pendulous shape of its inflorescence.

Foxglove has a synonymic chain of metaphoric names: ***bloody fingers***, ***dead men's bells***, ***fairy thimbles***, ***fairy's glove***, ***folk's glove***, ***gloves of Our Lady***, ***Virgin's glove***, and ***witches' gloves***. Its common names are derived from the shape of the flowers, whose most prominent feature is the finger-like tubular aspect, upon which fingers could easily be fitted as into a glove. The association with witches, dead men, and blood is more relevant in this case than that with Our Lady, the Virgin, or fairies. The apparently zoomorphic **metaphor** ***foxglove*** seems to be a corruption of an anthropomorphic **metaphor**, originally ***folksgloves***. This is also confirmed by its other vernacular names, ***fairy-folks-fingers*** (Wales) and ***bloody fingers*** in (Scotland), which both allude to the same **metaphor** (WE). One legend accounts that the spots on the ***foxglove*** were a warning sign of the baneful juices the plant produces, hence the familiar name in Ireland ***dead man's thimbles*** (Grieve 1995).

Names containing animals or parts of animal body are usually named on account of their similarity of shape that readily presents itself to the denominator's mind. For example, ***larkspur*** or ***lark's claw*** is so called from the obvious resemblance to the bird's large hind claws. However, ***hogweed***, a name given in reference to plants which neither resemble, nor are eaten by hogs, but are sooner deemed fit only for them, as it is noxious for humans. Other names reflect the utility of the plant, such as ***lice-bane*** or ***lousewort*** indicates that its poisonous seeds are considered to have magical properties as herbal medicine or pesticide. ***Rattlepods*** or ***rattlebox*** is thus named because its dry pods emit a rattling sound, such as dice in a box. Despite their being poisonous, the name was only given on account of the shape and onomatopoeic analogy, not in allusion of the rattlesnake, as it may cross one's mind. ***White snakeroot***, whose roots remind of crawling snakes, is also known as ***tall boneset***, in reference to another plant, ***boneset***, with which it shares similar feature. The name ***boneset*** comes from the time when the plant was used to treat the flu.

Vernacular metaphors are often based upon mythological elements, fairytales, and popular legends, and their names account for imaginary events or beings. For instance, ***common hyacinth*** is derived from Gk *hyakinthos*, the name of a blue gem and a purple flower fabled to have sprouted from the blood of Hyakinthos, a youth in Greek mythology who was beloved and accidentally slain by Apollo in a jealous fight (OED). ***False hellebore***, also called ***pheasant's eye*** due to its tiny brilliant scarlet and black blossoms, is also known as ***Adonis***, a **calque** after its botanical name, derived from Adonis, the beautiful youth from whose blood it sprang, according to the Greek legends (Grieve 1995). The common name of *Ricinus communis*, ***castor oil***, probably comes from its use as a replacement for castoreum, a perfume base made from the dried perineal glands of the beaver, the Latin word for which is *castor*. Its other common name, ***palma Christi***, or the calqued ***palm of Christ***, is probably a **metaphor** alluding to the fact this oil may be used to heal wounds and cure ailments (WE).

Finally, **compound** or **loose metaphors** contain several onomasiological motivations, concurring in plant names which exhibit both the main feature (*i.e.* their being poisonous), and another element, usually relating to physical appearance. For instance, ***devil's helmet*** alludes to the evil (poisonous aspect), but also to the shape of the flower, which resembles a helmet. The plant name ***poison nut*** was named after the shape of its nut-resembling seeds, while also indicating toxicity.

Discussion

Upon analysing the examples discussed above, several types of **metaphor** have been identified:

- **absolute metaphors** or **anti-metaphors** displaying no discernible point of resemblance between the tenor and the vehicle (***gloves of Our Lady***, ***heart of Jesus***, or ***true love*** for poisonous plants);
- **compound** or **loose metaphors** containing several points of similarity (***bleeding heart*** suggests the heart shape of the root, as well as the coloured juice it contains, which resembles blood, but also its dangerous effects);
- **simple metaphors** or **tight metaphors** presenting one point of resemblance between the two elements (most examples discussed above);
- **implicit metaphors** not specifying but implying the tenor (***black locust*** refers to a poisonous tree);

- **synecdochic metaphors** sharing the attributes of synecdoche, in which a part of something stands for the whole (e.g. ***one-berry***, ***bloodroot***, etc).

Moreover, the examination of poisonous plant names has prompted us to remark that a particular salient feature is often associated with the encoded concept activated by each **metaphor**. This feature appears to be, in most cases, the poisonous nature of the plant, expressed in diverse, imaginative, and vibrant language in English. In few cases, plant names allude more unsuspectingly to less significant facts upon which metaphorical analogy is built: shape of the root (***mandrake***), shape of the berries (***doll's-eyes***), colour of foliage, root, inflorescence, or fruit (***black nightshade***, ***black bryony***, ***bloodroot***), a combination of several analogies (***bleeding heart***).

Besides referring to the shape of the plant, **metaphors** allude especially to the features typical of poisonous plants. For instance, the naming of some plants reflects the effect they have upon humans or animals (***dumbcane***, ***dead tongue***, and ***weed of madness***). Toxic properties are especially perceptible in intimidating names which call attention to the plant's ominous properties. The denominator rarely had a neutral stance towards the plant named, which is reflected by the profusion of epithets: some of the popular names for ***Atropa belladonna*** are ***deadly nightshade***, ***devil's cherries***, and ***devil's herb***, which allude to the toxic effect of the plant, as do many other plant names which connote evil-doing or evil-doers. On the other hand, ***dead nettle*** is a misleading name, as it alludes to the fact that, unlike ordinary nettles, it does not sting when touched, rather than to its poisonous effect on animals, which might have been perceived as secondary information.

Plant name **metaphors** often allude to objects within reach, generally revealing connections established in their makers' mind (***woolyheads***, ***drumsticks***). Numerous names include parts of the body, which provide various circumstances for analogies (***bloody fingers***, ***bleeding heart***, ***devil's ear***, ***doll's-eyes***, ***elephant ear***, and ***pheasant's eye***). Only few plant name **metaphors** are limited by the understanding of their etymology (***weed paris***, ***hyacinth***), while the greater majority display a high degree of originality. Their complexity illustrates popular psychology, the imagination and power of observation of their creators, and their profound knowledge of the surrounding world.

Conclusions

This research has prompted us to conclude that **metaphor** is not only a highly creative subject matter, but it is also relevant to the general study of language functioning. What is more, the paper has hopefully proved that **cognitive metaphors** are genuine facilitators of understanding, highlighting the sway of **vernacular metaphors**. Significantly, poisonous plant names illustrate a vast grasp of the surrounding world and reveal crucial information which could be of use to any language user. Most names appear to have potent onomasiological motivations, as **metaphors** convey content and reflect the way people conceptualize the world. According to reference and context dependence, metaphorical 'truth' mediates the inference of truth about plants from name **metaphors**. For that reason, the study of **denominational metaphors** is consequential, despite the fact that their creators are not scientific authorities. Plant name **metaphors** represent an accurate source of knowledge, thus being highly relevant to students of agricultural and horticultural studies.

References

Bejan, D. (1991). *Nume româneşti de plante*. [Romanian Plant Names]. Cluj-Napoca: Editura Dacia.

Black, M. (1962). *Models and Metaphors: Studies in Language and Philosophy*. Ithaca, NY: Cornell University Press.

Dragoescu, Alina-Andreea & Dragoescu, P. (2010). Creative Metaphors in Poisonous Plant Names. In Georgeta Raţă (Ed.), *Teaching Foreign Languages: Languages for Special Purposes*, Newcastle upon Tyne: Cambridge Scholars Publishing: 320-329.

Grieve, M. (1995). *A Modern Herbal*. Online: http://botanicaLcom/.

Liberman, A. & Mitchel, J. L. (2008). *An Analytic Dictionary of English Etymology: An Introduction*. Minneapolis: University of Minnesota Press.

List of Garden Plants. Online: http://en.wikipedia.org/wiki/list_of_garden_plants. (WE)

Ogden, C. K. & Richards, I. A. (1923). *The Meaning of Meaning*. New York: Harcourt, Brace & World, Inc.

Online Etymology Dictionary. Online: http://www.etymonline.com/. (OED)

The American Heritage Dictionary of the English Language. (2008). Boston: Houghton Mifflin. (AHDEL)

CHAPTER SIX

ETYMOLOGY

WORDS OF LATIN ORIGIN IN BOTANICAL ENGLISH

GEORGETA RAŢĂ AND SCOTT HOLLIFIELD

Introduction

The **purpose** of the research was to see if there are any patterns in the naming of plants based on Latin words in contemporary English, patterns that would make plant names easier to identify and/or learn.

The **hypothesis** of the research was that, given the fact that all these names were coined more or less at the dawn of the age of scientific discovery, in the 17th and 18th centuries, as new words to describe new knowledge, their naming must have relied on certain patterns. The research method consisted of analysing English plant names from an etymological point of view.

The **results** show that English plant names of Latin origin are either metonyms or English formations. Results show that Botanical English is a good example of borrowings that have become fully parts of the English vocabulary, despite the fact they retain traces of their foreign origin in their pronunciation, spelling, or inflection.

The **conclusion** is that there is a pattern in the naming of plants with Latin names: the proof – the large number of English formations (particularly **backformations** and **derivatives**), on the one hand, and the large number of plant names derived from genus names. The implications of the research and results are considerable: they can be useful to both students in natural sciences and teachers and/or researchers in the field of zoology. Additional research should focus on the words of Latin origin in the field of Botany.

The **background** information consisted in what is generally given as common sense knowledge about scientific names in general and about botanical names in particular. Apart from some lists containing English words of Latin origin posted on Internet sites, there is no exhaustive list of words of Latin origin in Botanical English.

Literature Review

The material used in the research consists of words of Latin origin specific to the English of zoology that we have picked up from the *American Heritage Dictionary of the English Language* (2008), words that we analysed from an etymological point of view (we were interested only in the changes in form and sense) using Harper's (2001), Partridge's (2006), and Glare's (1983) works. As background information of classic culture nature, we used Farrell's (2004) and Janson's (2002) works. The linguistic terminology is the one supplied by Chalker & Weiner's (1994) and Sheehan's (2000) works.

State of the Art

We have identified 199 names of plants of Latin origin in Botanical English. These common names of plants are **metonyms**, **formations** or **compounds**.

1. Metonyms are words that are used as substitutes for other words with which they are in close semantic relationships (Chalker & Weiner 1994). Metonyms of Latin origin in Botanical English (a total of 129, i.e. 64%) are *genus names* or *species names*.

1.1. Genus Names. There are 127 (98%) common names from *New Latin genus names of different origin*:

- 82 (65%) are *New Latin genus names of Greek origin*: ***acanthus*** (pl. ***acanthuses / acanthi***) 'Any of various herbs or shrubs of the genus *Acanthus*' [< NL *Acanthus*, genus name < Gk *akanthos* 'thorn plant'], ***achillea*** 'Any of several plants of the genus *Achillea*, especially *A. millefolium*' [< NL *achillēa* < L < Gk *achilleios* < *Achilles* 'plant that healed wounds' < *Achilleus, Achilles*], ***acidanthera*** 'Any of several African plants of the genus *Acidanthera*' [< NL < Gk *akis*, *akid-* 'needle'], ***aechmea*** 'Any of various bromeliads of the genus *Aechmea*' [< NL < Gk *aikhmē* 'point of a spear'], ***agapanthus*** 'A rhizomatous plant (*Agapanthus africanus*)' [< NL *Agapanthus*, genus name < Gk *agapē* 'love' + Gk *anthos* 'flower'], ***agave*** 'Any of numerous plants of the genus *Agave*' [< NL *Agave*, genus name < Gk *agauē*, feminine of *agauos* 'noble'], ***ageratum*** 'Any of various plants of the genus *Ageratum*, especially *A. houstonianum*' [< NL *Agēratum*, genus name < Gk *agēratos* 'ageless'], ***alocasia*** 'Any of various aroids of the genus *Alocasia*' [< NL, probably alteration of *Colocasia* < Gk *kolokasia* 'lotus root'], ***alyssum*** '1. An herb. (*Lobularia maritima*) 2. Any of various weeds or ornamentals of the genus *Alyssum.* 3. An herb (*Berteroa incana*)' [< NL *Alyssum*, genus name, < L *alysson* 'kind of madder' < Gk *alusson* 'a plant believed to cure rabies'], ***amanita*** 'Any of various mushrooms in the genus *Amanita*' [< NL *Amanīta*, genus name < Gk *amanitai* 'a fungus'], ***amaryllis*** '1. Any of several plants of the genus *Hippeastrum.* 2. An herb (*Amaryllis belladonna*)' [< NL *Amaryllis*, genus name < L, name of a shepherdess < Gk *Amarullis*], ***anabaena*** 'Any of various freshwater algae of the genus *Anabaena*' [< NL *Anabaena*, genus name < Gk *anabainein* 'to go up'], ***anthurium*** 'Any of various plants of the genus *Anthurium*' [< NL *Anthurium*, genus name < ANTH(O)- + Greek *oura* 'tail'], ***aspidistra*** 'Any of several plants of the genus *Aspidistra*, especially *A. elatior*' [< NL *Aspidistra*, genus name < Gk *aspis, aspid*- 'shield'], ***astilbe*** 'Any of various herbs of the genus *Astilbe*' [< NL *astilbē* < A-1 + Gk *stilbos* 'glittering'], ***calla*** '**1.** A calla lily. **2.** A plant (*Calla palustris*)' [< NL *Calla*, genus name < Gk *kallaia* 'wattle of a cock'], ***calliopsis*** 'A plant (*Coreopsis tinctoria*)' [< NL < Gk *kalli*- 'beautiful' < Gk *kallos* 'beauty' + Gk *opsis* 'appearance'], ***ceanothus*** 'Any of various shrubs or trees of the genus *Ceanothus*' [< NL *Ceanothus*, genus name << Gk *keanothos* 'corn-thistle'], ***celosia*** 'A plant (*Celosia cristata*)' [< NL *Cēlosia*, genus name < Gk *kēlos* 'dry, burnt' (from its colour)], ***chlorella*** 'Any of various algae of the genus *Chlorella*' [< NL *Chlorella*, genus name < Gk *khloros* 'green'], ***coleus*** 'Any of various herbs of the genus *Coleus*' [< NL *Coleus*, genus name < Gk *koleos* 'sheath' (from the way its filaments are joined)], ***coreopsis*** 'Any of various plants of the genus *Coreopsis*' [< NL *Coreopsis*, genus name < Gk *koris* 'bedbug' + -OPSIS], ***corydalis*** 'Any of various herbs of the genus *Corydalis*'[< NL *Corydalis*, genus name < Gk *korudallis* 'crested lark' (from the shape of the flowers)], ***crinum*** 'Any of various plants of the genus *Crinum*' [< NL *Crinum*, genus name < Gk *krinon* 'lily'], ***croton*** '**1.** Any of various plants of the genus *Croton.* **2.** A shrub (*Codiaeum variegatum*)' [< NL *Croton*, genus name < Gk *kroton* 'castor oil plant'], ***cryptomeria*** 'A tree (*Cryptomeria japonica*)' [< NL *Cryptomeria*, genus name < CRYPTO- + Gk *meros* 'part'], ***cyclamen*** 'Any of various plants of the genus *Cyclamen,* especially *C. persicum*' [< NL *Cyclamen*, genus name < L *cyclamīnos* < Gk *kuklaminos* probably < *kuklos* 'circle, wheel' (perhaps from its bulbous roots)], ***cymbidium*** 'Any of various orchids of the genus *Cymbidium*' [< NL *Cymbidium*, genus name < L *cymba* 'boat' < Gk *kumbē*], ***delphinium*** 'A plant of the genus *Delphinium*' [< NL *Delphinium*, genus name < Gk *delphinion* 'larkspur' diminutive of *delphis, delphin*- 'dolphin' (from the shape of the nectary)], ***dendrobium*** 'Any of numerous species of the orchid genus *Dendrobium*' [< NL *Dendrobium*, genus name < DENDRO- + Greek *bios* 'life'], ***derris*** 'Any of various vines of the genus *Derris*' [< NL *Derris*, genus name < Gk covering], ***dianthus*** 'A plant of the genus *Dianthus*' [< NL *Dianthus*, genus name, probably alteration of Gk *diosanthos* 'carnation' < *Dios*- 'of Zeus' + *anthos* 'flower'], ***dicentra*** 'A plant of the genus *Dicentra*' [< NL *Dicentra*, genus name < DI-1 + Gk *kentron* 'point'], ***dichondra*** 'An herb (*Dichondra micrantha*)' [< NL *Dichondra*, genus name < DI-1 + Gk *khondros* 'granule'], ***elodea*** 'Any of various herbs of the genus *Elodea*' [< NL *Elodea*, genus name < Gk *helodēs* 'marshy'], ***eremurus*** 'Any of several plants of the genus *Eremurus*' [< NL *Eremurus*, genus name < Gk *erēmos* 'solitary' + Gk *oura* 'tail'], ***eucalyptus*** 'Any of numerous trees of the genus *Eucalyptus*' [< NL *Eucalyptus*, genus name < Gk *eu*- + Gk *kaluptos* 'covered'], ***euglena*** 'Any of various organisms of the genus *Euglena*' [< NL *euglena* < Gk *eu*- + Gk *glēnē* 'eyeball'], ***galax*** 'A plant (*Galax urceolata*)' [< NL *Galax*, genus name < Gk *gala* 'milk'], ***geranium*** '1. Any of various plants of the genus *Geranium*'. 2. Any of various plants of the genus *Pelargonium*' [< NL *Geranium*, genus name < L *geranium* 'crane's bill' < Gk *geranion*, diminutive of *geranos* 'crane'], ***gypsophila*** 'Any of various plants of the genus *Gypsophila*' [< NL *Gypsophila*, genus name < Gk *gupsos* 'chalk' + Gk *philos* 'loving'], ***hemerocallis*** 'Any of several perennial Eurasian herbs of the genus *Hemerocallis*' [< NL *hemerocallis* < Gk *hēmerokalles* 'a kind of lily'], ***hydrangea*** 'Any of various shrubs of the genus *Hydrangea*' [< NL *Hydrangēa*, genus name < Gk *hudro-, hudr-* 'hydro-' + Gk *angeion* 'vessel'], ***lychnis*** 'Any of various plants of the genus *Lychnis*' [< NL *Lychnis*, genus name < L *lychnis* 'a red flower' < Gk *lukhnis* 'lamp'], ***lycopodium*** 'A plant of the genus *Lycopodium*' [< NL *Lycopodium*, genus name < Gk *lukos* 'wolf' + Gk *podion*, diminutive of *pous* 'foot'], ***mimosa*** '1. Any of various mostly tropical herbs, shrubs, and trees of the genus *Mimosa*. **2.** An Asian tree (*Albizzia julibrissin*)' [< NL *Mimosa*, genus name < L *mīmus* 'mime' (from the plant's apparent mimicry of animal reactions) < Gk *mimos*], ***myosotis*** 'Any of various plants of the genus *Myosotis*' [< NL *Myosotis*, genus name < L *myosotis* 'mouse-ear, a kind of plant' < Gk *muosotis* < *muos*, genitive of *mus* 'mouse' + *ous, ot*- 'ear'], ***oxalis*** 'Any of numerous plants of the genus *Oxalis,* having often clover-like compound leaves with three leaflets and variously collared flowers that are usually clustered in umbels. Also called *wood sorrel*' [L *oxalis* 'wood sorrel', from Gk, from *oxus* 'sour'.],

pachysandra 'Any of several plants of the genus *Pachysandra,* especially the evergreen *P. terminalis*' [< NL *Pachysandra*, genus name < Gk *pakhus* 'thick' + NL *-andrus* '-androus' (after its thick stamens)], ***pelargonium*** 'Any of various herbs and shrubs of the genus *Pelargonium*' [< NL *Pelargonium*, genus name < Gk *pelargos* 'stork' (from the resemblance of a stork's bill to its capsules)], ***polyanthus*** 'Any of a group of hybrid garden primroses' [< NL *Polyanthus* < Gk *poluanthos* 'having many flowers'], ***psyllium*** 'An annual Eurasian plant (*Plantago psyllium*)' [< NL *Psyllum* < Gk *psullion* 'small flea' (from the plant's use against fleas)], ***sphagnum*** 'Any of various pale or ashy mosses of the genus *Sphagnum*' [< NL *Sphagnus* < L *sphagnos* 'a kind of moss' < Gk *sphagnos* 'a kind of shrub'], ***spirogyra*** 'Any of various filamentous freshwater green algae of the genus *Spirogyra*' [< NL *Spīrogyra*, genus name < L *spīra* 'coil' + Gk *guros* 'ring'], ***tarragon*** 'An aromatic Eurasian herb (*Artemisia dracunculus*)' [< NL *tarchon* < MGk *tarkhon* < Ar *tarhūn*, perhaps from Gk *drakon* 'dragon, tarragon'], and ***thuja*** 'Any of several North American or eastern Asian evergreen trees or shrubs of the genus *Thuja*' [< NL *Thuja*, arborvitae genus < ML *thuia* 'cedar' < Gk].

- 26 (20%) common names come from *New Latin genus names of Latin origin*: ***abutilon*** 'Any of various plants of the genus *Abutilon*' [< NL *Abutilon*, genus name], ***arnica*** 'Any of various herbs of the genus *Arnica*' [< NL *Arnica*, genus name], ***astragalus*** (pl. ***astragali***) 'the bone of the ankle that articulates with the tibia and fibula to form the ankle joint' [< NL *astragalus* 'vertebra'], ***calathea*** 'Any of various plants of the genus *Calathea*' [< NL *Calathea*, genus name < L *calathus* 'basket'], ***calceolaria*** 'Any of various plants of the genus *Calceolaria*' [< NL *Calceolaria*, genus name < L *calceolus* 'small shoe'], ***campanula*** 'Any of various plants of the genus *Campanula*' [< NL *Campanula*, genus name < LL *campāna* 'bell'], ***capsicum*** 'Any of various plants of the genus *Capsicum,* especially *C. annuum* and *C. frutescens*'[< NL *Capsicum*, genus name, perhaps < L *capsa* 'box'], ***cereus*** 'Any of several cacti of the genus *Cereus*' [< NL *Cēreus*, genus name < L *cēreus* 'candle' (from its shape)], ***cineraria*** 'Any of several plants (*Senecio x hybridus*)' [< NL *Cineraria*, genus name < L *cinerārius* 'of ashes' (from the ash-collared down on its leaves)], ***cleome*** 'Any of various plants of the genus *Cleome*' [< NL *Cleome*, genus name], ***cotoneaster*** 'Any of various shrubs of the genus *Cotoneaster*' [< NL *Cotoneaster*, genus name < L *cotoneum* 'quince' + L *-aster* 'partially resembling'], ***cypripedium*** 'Any of various orchids of the genus *Cypripedium*' [< NL *Cypripedium*, genus name < LL

Cypris 'Venus' (< Gk *Kupris* 'Aphrodite') + NL *-pedium* (alteration of Gk *pedilon* 'sandal')], ***fraxinella*** 'A plant (*Dictamnus albus*)' [< NL, diminutive of L *fraxinus* 'ash tree'], ***fritillary*** 'Any of various plants of the genus *Fritillaria*' [< NL *Fritillāria*, genus name < *fritillus* 'dice-box'], ***hydrilla*** 'A submersed Old World Plant (*Hydrilla verticillata*)' [< NL *Hydrilla*, genus name, diminutive of L *hydra* 'hydra'], ***laburnum*** 'Any of several trees or shrubs of the genus *Laburnum*' [< NL *Laburnum*, genus name < L *laburnum* 'broadleaved bean-trefoil' perhaps of Etruscan origin], ***liquidambar*** 'Any of several deciduous trees of the genus *Liquidambar*' [< NL *Liquidambar*, genus name < L *liquidus* 'liquid' + ML *ambar* 'amber'], ***nostoc*** 'A freshwater blue-green alga of the genus *Nostoc*' [< NL *nostoc* coined by Paracelsus], ***santonica*** 'A perennial or shrubby Eurasian plant (*Artemisia maritima*)' [< NL *Santonica* < L *(herba) santonica* 'of the Santoni' (a people of Aquitania)], ***spinifex*** 'Any of various clump-forming, perennial Australian grasses, chiefly of the genus *Triodia*' [< NL *Spīnifex*, former genus name < L *spīna* 'thorn' + L *-fex*], ***stramonium*** 'A coarse, poisonous plant (*Datura stramonium*)' [< NL *Stramonium*], ***streptocarpus*** 'Any of various chiefly African plants of the genus *Streptocarpus*' [< NL *Streptocarpus*, genus name < *strepto-* + *-carpus* '-carpous'], ***tuberose*** 'A tuberous perennial Mexican herb (*Polianthes tuberosa*)' [< NL *tūberosa*, species name, 'full of lumps'], ***veronica*** 'Any of various plants of the genus *Veronica*' [< NL *Veronica*, genus name], ***volvox*** 'Any of various freshwater green algae of the genus *Volvox*' [< NL *Volvox*, genus name < L *volvere* 'to roll'], and ***zooxanthella*** 'Any of various yellow-green algae that live symbiotically within the cells of other organisms' [< NL < *zoo-* + *xanth(o)-* + *-ella* diminutive suffix].

- 19 (15%) have been coined as *New Latin genus names of different other origins*: Ambonese (the Austronesian language of Ambon, an island of eastern Indonesia in the Moluccas near Ceram) (1): ***ailanthus*** 'Any of several trees of the genus *Ailanthus*' [< NL *Ailanthus*, genus name, alteration of Ambonese *ai lanto*]; Arabic (3): ***curcuma*** 'Any of various plants of the genus *Curcuma*' [< NL *Curcuma*, genus name < Ar *kurkum* 'saffron'], ***doronicum*** 'A plant of the genus *Doronicum*' [< NL *Doronicum* < Ar *dorīnaj* < Pers *darīnak*], ***senna*** 'Any of various plants of the genus *Cassia*' [< NL *Senna* < Ar *sanā'*]; Chinook (the language of the Chinook, a Native American people formerly inhabiting the lower Columbia River valley and adjoining coastal regions of Washington and Oregon, now located in western Washington) (1): ***quamash*** 'Any of several plants of the genus

Camassia in the lily family, especially *C. quamash*' [< NL *Quamash*, species name, variant of Chinook Jargon *kamass*]; English (1): ***osmund(a)*** 'Any of several ferns of the genus *Osmunda'* [< NL *Osmunda*, genus name < ME *osmunde* 'a kind of fern' < OF *osmonde*]; French (1): ***citronella*** 'A grass (*Cymbopogon nardus*)' [< NL < F *citronnelle* 'lemon oil' diminutive of *citron* 'citron']; Hindi (1): ***datura*** 'Any of several plants of the genus *Datura*' [< NL *Datura*, genus name < Hindi *dhatûrû* < Sanskrit *dhattûrû* 'thorn-apple']; Italian (2): ***lantana*** 'Any of various aromatic, chiefly tropical shrubs of the genus *Lantana*' [< NL *Lantana*, genus name < It dialectal *lantana* 'wayfaring tree, viburnum'], ***sorghum*** 'An Old World grass (*Sorghum bicolor*)' [< NL *Sorghum*, genus name < It *sorgo* 'a tall cereal grass' probably < ML *surgum*, perhaps variant of VL **syricum* 'Syrian, from Syria']; Japanese (1): ***aucuba*** 'Any of several shrubs of the genus *Aucuba*, especially *A. japonica*' [< NL possibly < Jap *auku* 'green' + Jap *ba* 'leaved']; Malay (4): ***caladium*** 'Any of various plants of the genus *Caladium*' [< NL *Caladium*, genus name < Malay *kāladi* 'an aroid'], ***casuarina*** 'Any of various trees or shrubs of the genus *Casuarina*' [< NL *Casuarina*, genus name < Malay *kĕsuari* 'cassowary' (from the resemblance of its twigs to the drooping feathers of the cassowary)], ***nipa*** 'A large palm *(Nipa frutescens)*' [< NL *nipa* < Malay *nipah*], ***pandanus*** 'Any of numerous palm-like dioecious trees and shrubs of the genus *Pandanus*' [< NL *Pandanus*, genus name < Malay *pandan* 'screw pine']; Spanish (3): ***cohune*** 'A palm (*Orbignya cohune*)' [< NL perhaps < Am Sp Mosquito *okhun*], ***guaiacum*** 'A tree of the genus *Guaiacum*' [< NL *guaiacum* < Sp *guayacan* < Taino], ***yucca*** 'Any of various evergreen plants of the genus *Yucca*' [< NL *Iucca*, genus name < Sp *yuca* 'cassava' < Taino]; and Swedish (1): ***trillium*** 'Any of various plants of the genus *Trillium*' [< NL *Trillium*, genus name, probably < Sw *trilling* 'triplet' (from its three leaves)].

1.2. Species Names. There are only 2 (2%) common names from New Latin species names of different origin: Latin (1): ***japonica*** 'An ornamental shrub (*Chaenomeles japonica*)' [< NL *Japonica*, species name < *Japonia* 'Japan'] and Malay (1): ***bilimbi*** 'A tree (*Averrhoa bilimbi*)' [< NL *Bilimbi*, species name < Malay *bĕlimbing* < Malay *belimbing*].

2. Formations are new words formed other than by **compounding** (Chalker & Weiner 1994). There are four types of formations in our corpus of words of Latin origin belonging to Botanical English – **backformation**, **derivation**, **folk etymology** and **loan translation** – with a total of 62 occurrences (31%).

2.1. Backformations are new words formed by the removal of (real or apparent) affixes from existing words (Chalker & Weiner 1994). They are usually revealed by the fact that the date of its first sue is later than that of its apparent **derivative**. There are 10 (16%) **backformations** in our corpus, formations that seem to have lost the following 'suffixes': *-a* (2): ***diatom*** 'Any of various algae of the class Bacillariophyceae' [< NL *diatoma* < Gk *diatomos* 'cut in half'] and ***osmund(a)*** 'Any of several ferns of the genus *Osmunda'* [< NL *Osmunda*, genus name < ME *osmunde* 'a kind of fern' < OF *osmonde*]; *-aceae* (3): ***chenopod*** 'Any plant of the goosefoot family' [< NL *Chenopodiaceae*, family name < NL *Chenopodium*, type genus < Gk *khēn* 'goose' + *-podium*, neuter of *-podius* 'pod'], ***desmid*** 'Any of various algae of the family Desmidiaceae' [< NL *Desmidiaceae*, family name < *Desmidium*, type genus < Gk *desmos* 'bond'], and ***orchid*** 'A member of the orchid family' [< NL *Orchideae*, family name < L *orchis* < Gk *orkhis* 'testicle, orchid' (from the slope of its root)]; *-is* (1): ***calamite*** 'Any of various trees of the genus *Calamites*' [< NL *Calamītīs*, genus name< LGk *kalamitīs* 'reed-like' < Gk *kalamos* 'reed']; *-um* (1): ***longan*** 'An Indian evergreen tree (*Euphoria longan*)' [< NL *longanum*, specific epithet, from Chinese (Mandarin) *long yăn* < *lóng* 'dragon' + *yăn* 'eye']; *-us* (3): ***abelmosk*** 'An annual or biennial plant (*Abelmoschus moschatus*)' [< NL *abelmoschus* < Ar *Uabbalmusk* 'grain of musk'], ***amaranth*** 'Any of various annuals of the genus *Amaranthus*' [< NL *amaranthus*, genus name, alteration of L *amarantus* < Gk *amarantos* 'unfading'], and ***brome*** 'Any of various grasses of the genus *Bromus*' [< NL *Bromus*, genus name < L *bromos* 'oats' < Gk].

2.2. Derivatives are words formed from other words by a process of **derivation**, i.e. by addition of affixes such as prefixes or suffixes (Chalker & Weiner 1994). There are 48 (77%) **derivatives** from proper names of admirals, botanical illustrators, botanists, countesses, diplomats, explorers, gardeners, governors, horticulturists, librarians, missionaries, naturalists, patrons of botany, physicians, politicians, princesses, queens, scientists, and soldiers:

- 2 (4%) have been derived with the suffix *-ad*: ***bromeliad*** [< NL *Bromelia*, type genus < Olaf *Bromelius* (1639-1705), Swedish botanist] 'Any of various plants of the family *Bromeliaceae*' and ***gesneriad*** 'Any of numerous herbs or shrubs of the family Gesneriaceae' [< NL *Gesneria*, type genus < Konrad von *Gesner*];
- 46 (96%) of them have been derived with the suffix *-ia*$_2$ 'Things derived from, relating to, or belonging to: *personalia*' [L, neuter pl. of *-ius*, and Gk, neuter pl. of *-ios*.]: ***abelia*** 'Any of various shrubs of the

genus *Abelia*' [< NL < Clarke *Abel* (1780-1826), British botanist], ***allamanda*** 'Any of several shrubs of the genus *Allamanda*' [< NL *Allamanda*, genus name < Jean Nicholas Sébastian *Allamand* (1713-1787), Swiss scientist], ***alstroemeria*** 'Any of several herbs of the genus *Alstroemeria*' [< NL *Alstroemeria*, genus name < Baron Clas *Alstroemer* (1736-1794), Swedish naturalist], ***bauhinia*** 'Any of various trees, shrubs, or vines of the genus *Bauhinia*' [< NL < Jean *Bauhin* (1541-1612) and Gaspard *Bauhin* (1560-1624), Swiss botanists], ***begonia*** 'Any of various plants of the genus *Begonia*' [< NL *Begonia*, genus name < Michel *Bégon* (1638-1710), French governor in the West Indies], ***bignonia*** 'A vine (*Bignonia capreolata*)' [< NL *Bignonia*, genus name < Jean Paul *Bignon* (died 1743), French royal librarian], ***boltonia*** 'Any of several herbs of the genus *Boltonia*' [< NL *Boltonia*, genus name < James *Bolton*, 18th-c. British botanist], ***bougainvill(a)ea*** 'Any of several South American woody shrubs or vines of the genus *Bougainvillea*' [< NL *Bougainvillea*, genus name < Louis Antoine de *Bougainville*], ***bouvardia*** 'Any of several shrubs or herbs of the genus *Bouvardia*' [< NL *Bouvardia*, genus name < Charles *Bouvard* (1572-1658), French physician], ***browallia*** 'Any of various herbs of the genus *Browallia*' [< NL *Browallia* < Bishop John Browall (1707-1755), Swedish botanist], ***buddleia*** 'Any of various shrubs of the genus *Buddleja*' [< NL *Buddleia* < Adam *Buddle* (died 1715), British botanist], ***camellia*** 'Any of several shrubs or trees of the genus *Camellia*' [< NL *Camellia*, genus name < Georg Josef Kamel (1661-1706), Moravian Jesuit missionary], ***cattleya*** 'Any of various orchids of the genus *Cattleya*' [< NL *Cattleya*, genus name, after William *Cattley* (died 1832), British patron of botany], ***cinchona*** 'Any of several trees and shrubs of the genus *Cinchona*' [< NL *Cinchona*, genus name < Francisca Henriquez de Ribera (1576-1639), Countess of *Chinchon*], ***clarkia*** 'Any of various plants of the genus *Clarkia*' [< NL *Clarkia*, genus name < William *Clark*], ***clintonia*** 'Any of various herbs of the genus *Clintonia*' [< NL *Clintonia*, genus name < DeWitt *Clinton*], ***collinsia*** 'Any of various plants of the genus *Collinsia*' [< NL *Collinsia*, genus name < Zaccheus *Collins* (1764-1831), American botanist], ***columnea*** 'Any of various plants of the genus *Columnea*' [< NL *Columnea*, genus name < Fabius *Columna*, Latin name of Fabio *Colonna* (1567-1650?), Italian botanist], ***dahlia*** 'Any of several plants of the genus *Dahlia*' [< NL *Dahlia*, genus name < Anders *Dahl* (1751-1787), Swedish botanist], ***dieffenbachia*** 'Any of several plants of the genus *Dieffenbachia*' [< NL *Dieffenbachia*, genus name < Ernst *Dieffenbach* (1811-1855), German naturalist], ***echeveria*** 'Any of numerous plants of the genus *Echeveria*' [< NL *Echeveria*, genus name < Atanasio *Echeverria*, 19th-c. Mexican botanical illustrator], ***forsythia*** 'Any of several shrubs of the genus *Forsythia*' [< NL *Forsythia*, genus name < William *Forsyth* (1737-1804), Scottish horticulturist], ***franseria*** 'Any of various herbs or shrubs of the genus *Franseria*' [< NL *Franseria*, genus name < Antonio *Franseri*, 18th-c. Spanish botanist], ***freesia*** 'Any of several plants of the genus *Freesia*' [< NL *freesia* < Friedrich Heinrich Theodor *Freese* (1795-1876), German physician], ***fuchsia*** 'Any of various shrubs or trees of the genus *Fuchsia*' [< NL *Fuchsia*, genus name < Leonhard *Fuchs* (1501-1566), German botanist], ***gaillardia*** 'Any of several plants of the genus *Gaillardia*' [< NL *Gaillardia*, genus name < *Gaillard* de Marentonneau, 18th-c. French botanist], ***gardenia*** 'Any of various shrubs and trees of the genus *Gardenia,* especially *G. jasminoides*' [< NL *Gardenia*, genus name < Alexander *Garden*], ***gloxinia*** 'Any of several plants of the genus *Sinningia*' [< NL *Gloxinia* < Benjamin Peter *Gloxin*, 18th-c. German botanist], ***gmelina*** 'A tree (*Gmelina arborea*)' [< NL *Gmelina*, genus name < Georg *Gmelin* (1709-1755), German botanist], ***gunnera*** 'Any of several plants of the genus *Gennera*' [< NL *Gunnera*, genus name < Johann Ernst *Gunnerus* (1718-1773), Norwegian botanist], ***hosta*** 'Any of several eastern Asian plants of the genus *Hosta*' [< NL *Hosta*, genus name < Nicolaus Thomas *Host* (1761-1834), Austrian botanist], ***hoya*** 'Any of several evergreen climbing vines or shrubs of the genus *Hoya*' [< NL *Hoya*, genus name < Thomas *Hoy* (1750?-1822), British gardener], ***lespedeza*** 'Any of various plants of the genus *Lespedeza*' [< NL *Lespedeza*, genus name < V. M. de *Cespedez* (misread as *Léspedez*), Spanish governor of Florida], ***lobelia*** 'Any of numerous plants of the genus *Lobelia*' [< NL *Lobelia*, genus name < Matthias de *Lobel* (1538-1616), Flemish botanist and physician], ***magnolia*** 'Any of numerous evergreen or deciduous trees and shrubs of the genus *Magnolia*' [< NL *Magnolia*, genus name < Pierre *Magnol* (1638-1715), French botanist], ***monarda*** 'Any of various aromatic plants of the genus *Monarda*' [< NL *Monarda*, genus name < Nicolas *Monardes* (1493-1588), Spanish botanist], ***nicotiana*** 'Any of various flowering annual or perennial herbs of the genus *Nicotiana*' [< NL *(herba) nicotiāna* '(herb of) Nicot, nicotiana' < Jean *Nicot* (1530?-1600), French diplomat], ***paulownia*** 'Any of several Chinese deciduous trees of the genus *Paulownia*' [< NL *Paulownia*, genus name < Princess Anna *Paulovna* (1795-1865), Russian princess and queen of William II of the Netherlands], ***poinsettia*** 'A tropical American shrub (*Euphorbia* *pulcherrima*)' [< NL *Poinsettia* < Joel Roberts *Poinsett*], ***stapelia*** 'Any of various plants of the genus *Stapelia*' [< NL *Stapelia*, genus name < Jan Bode van *Stapel* (died 1636), Dutch botanist], ***stewartia*** 'Any of various deciduous trees or shrubs of the genus *Stewartia*' [< NL *Stewartia*, genus name < John *Stuart*, Third Earl of Bute (1713-1792), British politician], ***tillandsia*** 'Any of various usually epiphytic bromeliad plants of the genus *Tillandsia*' [< NL *Tillandsia*, genus name < Elias *Tillands* (1640-1693), Finno-Swedish botanist], ***weigela*** 'Any of various deciduous shrubs of the genus *Weigela* of Asia, especially *W. florida*' [< NL *Weigela*, genus name < Christian E. *Weigel* (1748-1831), German physician], ***wistaria / wisteria*** 'Any of several climbing woody vines of the genus *Wisteria*' [< NL *Wisteria*, genus name < Caspar *Wistar* (1761-1818), American physician], ***zinnia*** 'Any of various plants of the genus *Zinnia,* native to tropical America, especially *Z. elegans*' [< NL *Zinnia*, genus name < Johann Gottfried *Zinn* (1727-1759), German botanist], and ***zoysia*** 'Any of several creeping grasses of the genus *Zoysia*' [< NL *Zoysia*, genus name < Karl von *Zois* zu Laubach (1756-1800?), Austrian botanist].

2.3. Folk etymology or **popular etymology** is a popular modification of the form of a word, in order to render it apparently significant (Chalker & Weiner 1994). There is a single (2%) **folk etymology** in our corpus: ***Alyce / alyce clover*** 'An herb (*Alysicarpus vaginalis*)' [< Probably by folk etymology from NL *Alysicarpus (vaginalis)*, genus name < Gk *halusis* 'chain'].

2.4. Loan translations or **calques** are expressions adopted by one language from another in more or less literally translated forms (Chalker & Weiner 1994). There are 3 (5%) loan translations in our corpus: ***buckthorn*** 'Any of various shrubs or trees of the genus *Rhamnus*' [Translation of NL *cervī spīna* < *cervī*, genitive of *cervus* 'buck' + *spīna* 'thorn'], ***multiflora rose*** 'A climbing or sprawling shrub (*Rosa multiflora*)' [Partial translation of NL *Rosa multiflora*, species name < L *rosa* 'rose' + LL *multiflora* 'multiflorous'], and ***ponderosa pine*** 'A tall timber tree (*Pinus ponderosa*)' [Translation of NL *Pīnus ponderosa* < L *pīnus* 'pine-tree' + L *ponderosa* 'heavy'].

3. Compounds are words formed by combining two or more bases (or free morphemes) (Chalker & Weiner 1994). There is a single **compound** (1%) among the words of Latin origin in Botanical English – ***macadamia nut*** 'The round, hard-shelled nut or the edible seed of the Australian tree *Macadamia ternifolia*' [< NL *Macadamia*, genus name < John *Macadam* (1827-1865), Scottish-born Australian chemist] – made of a New Latin metonymic word derived from a chemist's name and a common noun (*nut*).

4. Other Cases. There are three other types of **word formation** in our corpus that are not inventoried as such by grammarians:

- 4 (2%) cases of **alteration** – ***cauliflower*** 'An herb (*Brassica oleracea* var. *botrytis*)' [Probably alteration (influenced by FLOWER), of NL *cauliflora* < L *caulis* 'stem' + L *flos, flor-* 'flower'], ***cycad*** [< NL *Cycas, Cycad-*, genus name < Gk *kukas*, erroneous reading of *koikas*, accusative pl. of *koix* 'a kind of palm tree'] 'Any of various plants of the division Cycadophyta', ***pteridophyte*** 'Any of various vascular plants that reproduce by means of spores rather than by seeds, including the ferns and related plants' [< NL *Pteridophyta*, former division name < Gk *pteris, pterid-* 'fern'], ***tracheophyte*** 'Any of various vascular plants, including seed plants and ferns' [< NL *Tracheophyta*, division name < *tracheo-* + Gk *phuta* 'plants'];
- 1 (1%) case of **shortening and alteration**: ***pyxie*** 'A creeping evergreen shrub (*Pyxidanthera barbulata*)' [Shortening and alteration of NL *Pyxidanthera*, genus name < Gk *puxis, puxid-* 'box' + Gk *anthera* 'pollen'];
- 1 (1%) case of **derivation from a Greek word and a Latin prefix**: ***achira*** 'A South American and West Indian plant (*Canna edulis*)' [< NL a- + Gk *kheir* 'hand'], and 1 case of two-term common name: ***lignum vitae*** 'Either of two tropical American trees (*Guaiacum officinale* or *G. sanctum*)' [< NL *lignum vītae* < L *lignum* 'wood' + L *vītae* genitive of *vīta* 'life'].

Unlike the words of Latin origin in the English of zoology (Rață, 2008), ***metonymic words*** are three times more numerous than English formations. Thus, 64% of all the common names of plants of Latin origin come from ***genus names*** (98%) – with 65% *New Latin genus names of Greek origin*, 20% *New Latin genus names of Latin origin*, and 15% *New Latin genus names of different other origins* (Malay 4 words, Arabic and Spanish 3 words each, Italian 2 words, and Ambonese, Chinook, English, French, Hindi, Japanese, Swedish 1 word each) – and from ***species names*** (2%). To note the name ***nostoc***, coined from *Nosthryl* and Nasenloch by Paracelsus (Potts 1997), and the names ***spinifex*** and ***zooxanthella*** (coined from Latin elements). To also note that ***osmund(a)*** 'Any of several ferns of the genus *Osmunda'* [< NL *Osmunda*, genus name < ME *osmunde* 'a kind of fern' < OF *osmonde*] is also a case of **backformation**. The large number of plant names that have come from genus names shows that scientific discovery in the field of botany reached, in the 17th and 18th centuries, the highest level ever.

Unlike the words of Latin origin in the English of zoology (Raţă, 2008), the share of English ***formations*** (31%) of Latin origin in Botanical English is almost two times smaller than that of *metonymic words*. This could be explained by the fact that these *formations* are still perceived by scientists as **loanwords** rather than as English words. Among English **formations**, the large number of **derivatives** (77%) confirms our hypothesis: perceived as foreignisms, Latin words have been 'adapted' to the English spelling system, with the prefix *-ia* specific to plant names (96% of the **derivatives** contain it). **Backformations** – with removal of the 'prefixes' *-aceae* and *-us* (3 times each), *-a* (2 times each), and *-is* and *-us* (1 time each) – are sensibly less (16%), while **folk etymology** and **loan translations** are, practically, negligible. Though the majority of back-formations in English are verbs (Chalker & Weiner 1994), the **backformations** in our corpus are exclusively nouns. To note that ***osmund(a)*** 'Any of several ferns of the genus *Osmunda'* [< NL *Osmunda*, genus name < ME *osmunde* 'a kind of fern' < OF *osmonde*] is also a *New Latin genus name of Latin origin*.

Compounds and ***other cases*** are insignificant.

Conclusion

Results show that botanical English contains many examples of Latinate linguistic borrowings. While these borrowings have been fully integrated into the English vocabulary, they retain significant traces of their foreign origins in pronunciation, spelling, or inflection. This study suggests support for our hypothesis that there must be a pattern in the naming of animals with Latin names is confirmed: the proof – the large number of **metonyms** (plant names of Latin origin derived from genus and species names) representing 64% of the total of words of our corpus, on the one hand, and the large number of **formations** (particularly **derivatives** and **backformations**) representing 31% of the total of words of our corpus, on the other hand. The implications of the research and results are considerable: they can be useful to both students in natural sciences and teachers and/or researchers in the field of botany.

References

Chalker, Sylvia & Weiner, E. (1994). *The Oxford Dictionary of English Grammar*. London - New York - Sydney - Toronto: BCA.

Farrell, J. (2004). *Latin Language & Latin Culture. From Ancient to Modern Times*. Cambridge: Cambridge University Press.

Glare, P. G. W. (Ed.) (1983). *Oxford Latin Dictionary*. Oxford: Oxford University Press.

Harper, D. (2001). *Online Etymology Dictionary*. Online: http://www.etymonline.com.

Janson, T. (2002). *Speak. A Short History of Languages*. Oxford: Oxford University Press.

Partridge, E. (2006). *Origins. A Short Etymological Dictionary of Modern English*. London – New York: Routledge.

Potts, M. (1997). Etymology of the Genus Name Nostoc (Cyanobacteria). *International Journal of Systematic Bacteriology 47 (2)*: 584.

Rață, Georgeta. (2008). Words of Latin Origin in the English of Zoology. *Journal of Linguistic Studies 2*: 75-86.

Shipley, J. T. (1979). *Dictionary of Word Origins*. Totowa, NJ: Littlefield, Adams & Co.

Sheehan, M. J. (2000). *Word Parts Dictionary. Standard and Reverse Listings of Prefixes, Suffixes, Roots and Combining Forms*. Jefferson (NC) & London: McFarland & Company, Inc., Publishers.

The American Heritage Dictionary of the English Language. (2008). Boston: Houghton Mifflin.

Stearn, W. T. (1983). *Botanical Latin. History, Grammar Syntax, Terminology and Vocabulary*. Newton Abbot – London – North Pomfret: David & Charles.

SEEDS AND SEEDLING

GEORGETA RAȚĂ, IONEL SAMFIRA
AND ANICA PERKOVIĆ

Introduction

Studying words from an **etymological perspective** (to see when they entered a language, from what source, and how their form and meaning have changed over time) can be of real help for any specialist in agriculture and/or related branches, that needs to understand, learn, and use them more efficiently. The words in our corpus have originated through a rather limited number of basic mechanisms: **borrowing** (i.e., the adoption of **loanwords** from other languages), and **word formation** (**derivation** and **composition / compounding**).

Material and Method

Our analysis has been made on a corpus of terms specific to the field of "seeds and seedlings" such as inventoried by Ross E. Koning (1994). This "seed vocabulary", the basic vocabulary of the students in "plant biology", contains 63 terms (nouns and adjectives) belonging to the following fields related to "seeds and seedlings":

- general vocabulary (***fruit*** and ***funiculus***);
- seed surfaces (***alate***, ***circumalate***, ***comose***, ***coronate***, ***crested***, ***umbonate*** and ***verrucous***);
- seed parts (***aril***, ***embryo***, ***endosperm***, ***hilum***, ***micropyle***, ***raphe***, and ***seed coat***);
- embryo parts (***coleoptile***, ***coleorhiza***, ***cotyledon***, ***epicotyl***, ***hypocotyls***, ***mesocotyl***, ***pumule***, ***radicle***, and ***scutellum***);
- seed types (***cotylespermous***, ***endospermous***, and ***monocotyledonous***);
- cotyledon number (***dicotyledonous*** and ***monocotyledonous***);
- cotyledon size (***anisocotylous*** and ***syntropous***);

- embryo types (***bent embryo***, ***broad embryo***, ***capitate embryo***, ***dwarf embryo***, ***folded embryo***, ***investing embryo***, ***linear embryo***, ***micro embryo***, ***peripheral embryo***, and ***rudimentary embryo***);
- cotyledon position (***accumbent cotyledon*** and ***incumbent cotyledon***);
- seedling types (***cryptocotylar***, ***epigeous***, ***hypogenous***, and ***phanerocotylar***);
- seedling parts (***adventitious root***, ***cataphyll***, ***coat***, ***collet***, ***eophyl***, ***lateral root***, ***leaf primordial***, ***leaf***, ***metaphyll***, ***primary root***, ***root apex***, ***root hair***, ***root*** and ***seed hoot apex***).

We have analysed all these terms from an **etymological point of view**, in an attempt to identify a possible pattern in the origin of these nouns and adjectives, that can make them easier to understand, learn and use.

Results and Discussion

We have inventoried 29 nouns and 34 adjectives, i.e. a total of 63 terms to "seeds and seedlings".

1. The 29 nouns have 6 different origins – Greek, Latin, Old English, Germanic, Old High German, and Scandinavian – which they share in an uneven manner. Thus, of 29 nouns: 13 have a Greek origin (46%), i.e. they are direct or indirect (via Latin, Late Latin, Middle Latin and New Latin) **loanwords** from Greek: ***cataphyll*** < *cata-* [< Gk *katā-* 'down, through, against, according to, toward, during'] + *phyll* [< Gk *phỳllon* 'leaf'] 'rudimentary scale leaf in cryptocotylar species', ***coleoptile*** [< ML *coleoptil(um)* < Gk *koleōn* 'sheath' + Gk *ptílon* 'soft feather, dawn'] 'protective first leaf sheath surrounding epicotyl in grasses', ***coleorhiza*** (pl. ***coleorhizae***) [< ML < Gk *koleōn* 'sheath'+ Gk *rhīza* 'root'] 'protective sheath around radicle in grasses', ***cotyledon*** [< L *cotyledon* 'navelwort' < Gk *kotylēdōn* 'cup-shaped cavity' < Gk *kotulē* 'cup'] 'embryonic or seed leaf', ***embryo*** [< ML *embryon-* < Gk *embryon* < Gk *embryos* 'in growing'] 'young sporophyte consisting of epicotyl, hypocotyls, radicle, and one or more cotyledons', ***endosperm*** < *endo-* < [Gk *éndon-* 'within'] + *sperm* [< ME *sperme* < LL *sperma* < Gk *spérma* 'seed'] 'food reserve formed from syngamy of central cell and sperm', ***eophyll*** < *eo-* [< Gk *eós-* 'dawn'] + *phyll* [< Gk *phýllon* 'leaf'] 'the first few green leaves developed by seedlings in transition to adult leaves', ***epicotyl*** < *epi-* 'upon, on, over, near, at, before, after'] + *cotyl(edon)* [< L *cotyledon* 'navelwort' < Gk *kotylēdōn* 'cup-shaped cavity' < *kotulē* 'cup'] 'apical end of embryo axis above cotyledon', ***hypocotyl*** < *hypo-* [< Gk *hypō-* 'under'] + *cotyl* < *cotyl(edon)* [< L *cotyledon* 'navelwort' < Gk *kotylēdōn* 'cup-shaped

cavity' < *kotulē* 'cup'] 'embryonic stem axis between scutellum and coleoptile in grasses', ***metaphyll*** < *meta-* [< Gk *meta* 'after, along with, beyond, among, behind'] + *phyll* [< Gk *phýllon* 'leaf'] 'adult leaves', ***micropyle*** < *micro-* [< Gk *mīcrós* 'small'] + *pyle* [< Gk *pylē* 'gate'] 'hole through seed coat for delivery of sperm cells', and ***raphe*** [< NL < Gk *raphē* 'seam, suture'] 'ridge on seed coat formed from fused funiculus'; 10 have a Latin origin (35%), i.e. they have been borrowed directly or indirectly (via French, Old French) from Latin: ***apex*** (pl. ***apexes / apices***) (as in *root apex*) [< L *apex* 'conical cap, tip, point, summit'] 'root meristem including terminal root cap', ***aril*** [< NL *arillus* < ML *aril(ī)* 'dried grape pips'] 'fleshy outgrowth of funiculus, raphe, or integuments (seed, coat)', ***collet*** [< F *collet* 'little collar' < *col* 'collar'< L *collum* 'neck'] 'external demarcation between hypocotyl and root', ***fruit*** [< ME < OF < L *frūct(us)* 'enjoyment, proceeds, fruit'< past participle of *frui* 'enjoy, have the use of'] 'the mature ovary wall, generally containing seeds', ***funiculus*** (pl. ***funiculi***) [< L *funiculus* 'little rope' < *fūnis* 'rope'] 'stalk by which seed is attached to placenta in fruit', ***hilum*** (pl. ***hila***) [< L *hilum* 'little thing, trifle'] 'funicular scar on seed coat', ***plumule*** [< NL < L *plūmula* 'small feather' < *pluma* 'down, feather'] 'leaves of epicotyl in seed', ***primordia*** (sg. ***primordium***) (as in *leaf primordia*) < *primordium* [< L *prīmōrdius* 'original' + *ōrdit* < *ōrdīrī* 'begin' + -us] 'young leaves having unexpanded blades', ***radicle*** [< L *rādicūla* 'little root' < *radic-* 'root'] 'based end of embryo axis, becoming root', and ***scutellum*** [< NL *scutellum*, alteration of L *scutella* 'dish, tray' as if meaning 'small shield' + L *scūtum* 'shield'] 'the cotyledon in grasses'; 3 have an Old English origin (10%), i.e. they come from Old English vie Middle English: ***leaf*** [< ME *leef*, OE *lēaf* 'leaf'] 'plant part: a flat green part that grows in various shaped from the stems or branches of a plant or tree and whose main function is photosynthesis', ***seed*** (as in *seed coat*) [< ME, OE *sēd*, *sǽd* 'seed'] 'plant part containing embryo: a plant part produced by sexual reproduction that contains the embryo and gives rise to a new individual. In flowering plants it is enclosed within the fruit', and ***shoot*** (as in *shoot apex*) [< ME *shote(n)*, OE *seceōtan* 'shoot'] 'shoot meristem with leaf primordial'; 1 has a Germanic origin (3%), i.e. it comes from Germanic *via* Old French: ***coat*** (as in *seed coat*) [< ME *cote* 'coat' < OF < Gmc] 'any thin layer that covers something'; 1 has an Old High German origin (3%), i.e. it comes from Old High German *via* Old French: ***hair*** (as in *root hair*) [< ME *heer* 'hair', OE *haer* < OF < OHG] 'epidermal cells of a young root extending outward among soil particles.'; 1 has a Scandinavian origin (3%), i.e. it comes from Scandinavian *via* Old Norse: ***root*** [< ME < ON *rōt* < Scand] 'underground base of plant: the part of a plant that has no leaves or buds and usually spreads underground, anchoring the plant and absorbing water and nutrients from the soil.’

2. The 34 adjectives have 3 different origins – Latin, Greek, and Old English – which they share in an uneven manner. Thus, of the 34 adjectives: 7 have a Latin origin (50%), i.e. they have been borrowed from Latin directly or indirectly (via Middle French, Old French): ***accumbent*** (as in *accumbent cotyledon*) [< L *accumbent*] ‘with cotyledon edges against hypocotyls’, ***adventitious*** (as in *adventitious root*) [< L *adventicius* ‘coming from outside’] ‘a root arising from stem, petiole, or leaf tissues’, ***alate(d)*** [< L *ālāt(us)* <L *ala* ‘wing’] ‘winged’, ***capitate*** (as in *capitate embryo*) [< L *capitātus* ‘having a head’< *caput* ‘head’] ‘basal head-like embryo surrounded in endosperm’, ***circumalate*** [< L *circus* ‘circle’ + L. *ālāt(us)* ‘winged’] ‘winged circumferentially’, ***comose*** [< L *comōs(us)*] ‘with a tuft of trichomes’, ***crested*** < *crest* [< ME *creste* < MF < L *crist(a)* ‘crest’] ‘with an elevated ridge or ridges; with raphe’, ***dicotyledonous*** < *dicotyledon* [< NL *dicotyledonē(s)* ‘two cotyledons’] + *-ous* [< ME < OF < L *-ōsus* ‘full of, abounding in, given to, characterized by, having, like, etc.’] ‘two-cotyledons’, ***incumbent*** (as in *incumbent cotyledon*) [< ME < L incumbent < L *incumbere* ‘lie in or on’ + L *-cumbere* ‘lie down’] ‘with cotyledon sides against hypocotyl’, ***investing*** (as in *investing embryo*) [< ME < L *invest(īre)* ‘to surround, to clothe (in)’ < L *vestis* ‘clothing’] ‘embryo with thick cotyledons as major portion with limited endosperm’, ***lateral*** (as in *lateral root*) [< L *laterāl(is)* < L *later-* ‘side’] ‘a branch root, a secondary root, arising from root tissue’, ***linear*** (as in *linear embryo*) [< L *līneāris* ‘of, belonging to lines’], ‘axial embryo several times longer than broad, with thin cotyledons’, ***monocotyledonous*** < *monocotyledon* [< NL *monocotyledon* ‘one cotyledon’] + *-ous* [< ME < OF < L *-ōsus* ‘full of, abounding in, given to, characterized by, having, like, etc.’] ‘one cotyledon’, ***primary*** (as in *primary root*) [< ME < L < *prīmāri(us)* ‘of the first rank”] ‘derived from radicle’, ***rudimentary*** (as in *rudimentary embryo*) < *rudiment* [< L *rudiment(um)* < L *rudis* “raw, rough”] + *-ary* [< L *-āris* “pertaining to, connected with”] ‘undifferentiated embryo in copious endosperm’, ***umbonate*** < *umbo* [< *umbōn* “having an umbo or projecting boss”] + *-ate* [< L *-ātus* ‘a suffix forming adjectives’] ‘with distinct projection usually from the side’, and ***verrucose / verrucous*** < *verruc(a)* [< L < *verruca* ‘wart’] + *-ous* [< ME < OF < L *-ōsus* ‘full of, abounding in, given to, characterized by, having, like, etc.’], ‘warty’; 3 have a Greek origin (38%), i.e. they have come from Greek directly or indirectly (via Latin, Old French): ***anisocotylous*** < *aniso-* [< Gk *aniso(s)* ‘unequal’] + *cotyl(edon)* [< L *cotyledon* ‘navelwort’ < Gk *kotylēdōn* ‘cup-shaped cavity’ < *kotulē* ‘cup’] + *-ous* [< ME < OF < L *-ōsus* ‘full of,

abounding in, given to, characterized by, having, like, etc.] 'unequal cotyledons', ***antitropous*** < *anti-* [< L *-antī* 'against, opposite of'] + *-tropous* [< Gk *-tropos* 'pertaining to a turn'] 'radicle pointing away from hilum', ***coronate*** < *corona* [< L *corōna* 'garland, crown' < Gk *coróne* 'crown'] + *-ate* [< L *-ātus* 'a suffix forming adjectives'] 'with a crown', ***cotylespermous*** < *cotyle* < *cotyl(edon)* [< L *cotyledon* 'navelwort' < Gk *kotylēdōn* 'cup-shaped cavity' < *kotulē* 'cup'] + *-spermous* [a word element used to form adjectives corresponding to nouns with stems ending in *-sperm*] 'food reserve in cotyledon', ***cryptocotylar*** < *crypto-* [< Gk *kryptós* 'hidden'] + *cotyl(edon)* [< L *cotyledon* 'navelwort' < Gk *kotylēdōn* 'cup-shaped cavity' < *kotulē* 'cup'] + *-ar* [< L *-ār(is)* 'pertaining to'] 'with cotyledons remaining in seed and usually below ground', ***epigeous*** < *epi-* [< Gk *epī* 'upon, on, over, near, at, before, after'] + *-genous* [a suffix of adjectives corresponding to nouns with stems in *-gen*] 'cotyledons emergent from seed and usually appearing above ground', ***hypocotylespermous*** < *hypocotyl* [< Gk] + *-spermous* [a word element used to form adjectives corresponding to nouns with stems ending in *-sperm* < Gk] 'food reserve in hypocotyl', ***hypogeal / hypogean / hypogeous*** [< L *hypogēus* < Gk *hypógeios* 'underground' < *gē* 'ground, earth'] 'cotyledons remaining in seed and usually below ground,' ***isocotylous*** < *iso-* [< Gk *ísos* 'equal'] + *cotyl(edon)* [< L *cotyledon* 'navelwort' < Gk *kotylēdōn* 'cup-shaped cavity' < *kotulē* 'cup'] + *-ous* [< ME < OF < L *-ōsus* 'full of, abounding in, given to, characterized by, having, like, etc.] 'cotyledons equal in shape and size', ***micro*** (as in *micro embryo*) [< Gk *mīkrōs* 'small'] 'axial embryo in minute seeds less than 0.2 mm long, occupying most of seed', ***peripheral*** (as in *peripheral embryo*) < *periphery* [< Gk *peripherēs* 'carrying around' < Gk *pherein* 'carry'] 'embryo pressed along side of seed coat, remainder is endosperm', ***phanerocotylar*** < *phanero-* [<Gk *phaneros* 'visible, manifest'] + *cotyl(edon)* [< L *cotyledon* 'navelwort' < Gk *kotylēdōn* 'cup-shaped cavity' < *kotulē* 'cup'] + *-ar* [< L *-ār(is)* 'pertaining to'] 'cotyledons emergent from seed and usually appearing above ground', and ***syntropous*** < *syn-* [< Gk *sýn* 'with, together'] + *-tropous* [< Gk *-tropos* 'pertaining to a turn'] 'radicle pointing toward hilum.'; 4 have an Old English origin (12%), i.e. they have come directly from Old English: ***bent*** (as in *bent embryo*) [< ME *bend(en)* < OE *bendan* 'to bind, to bend'] 'embryo with cotyledons folded back along the hypocotyls / radicle', ***broad*** (as in *broad embryo*) [< ME *brood* < OE *brād* 'broad'] 'basal globular or lenticular embryo surrounded in endosperm', ***dwarf*** (as in *dwarf embryo*) [< ME *dwarf* < OE *dweorh* 'dwarf'] 'embryo small compared to seed; seed in 0.2 to 2 mm long', and ***folded*** (as in *folded embryo*) [< ME *fold, fald* < OE *fald, falod* 'folded'] 'embryo with thin extensive cotyledons folded and pleated in various ways'.

Conclusion

The following conclusions can be drawn from the analysis above:

- the number of the nouns belonging to the field of "seeds and seedlings" (i.e., 29) is close to that of the adjectives (i.e., 34);
- the number of nouns of Greek (13 vs. 17, i.e., 46% and 50% respectively), Latin (10 vs. 13., i.e., 35% and 38% respectively), and English (3 vs. 4, i.e., 10% and 24% respectively) origin and their share are very close to those of the adjectives;
- the few nouns of other origins (Germanic, Old High German, and Scandinavian) do not belong to what is called "scientific vocabulary";
- most of the adjectives are **compounds** or **derivatives** of the nouns, which eases their understanding.

References

Koning, Ross E. (1994). "Seeds and Seedling Vocabulary". *Plant Physiology Website*. Online: http://koning.ecsu.ctstateu.edu/plant_biology/seedvocab.html.

Rață, Georgeta & Perković, Anica. (2008). Seeds and Seedling Vocabulary: An Etymological Approach. *Proceedings of the 43rd Croatian and 3rd International Symposium on Agriculture*, 18-21 February 2008, Opatija, Croatia: 720-724.

Shipley, J. T. (1979). *Dictionary of Word Origins*. Totowa, NJ: Littlefield, Adams & Co.

COMMON NAMES OF PLANT DISEASES: THE LATIN HERITAGE

GEORGETA RAŢĂ AND CAMELIA GIUCHICI

Introduction

Properly understanding and learning specialised vocabularies in agricultural academic institutions in Romania is one of the main objectives of both studying different specialised subjects and foreign languages. As the students learn the scientific name (in Latin) for the different agricultural "realities" (plants, weeds, diseases, pests, etc.), understanding the mechanisms through which they survive in the common names designating the same reality in the different languages cannot only facilitate their proper understanding and acquisition, but also their reaching that degree of autonomy in approaching references that any future specialist in agriculture longs for.

Here are the mechanisms through which 96 scientific names of plant diseases have reached, one way or another, English, French, and Romanian:

- *phonetical and graphical integration*: R ***esplanadă*** 'esplanade' < It *esplanada*;
- *phonetical integration and graphical non-integration*: R ***mass-media*** 'mass media' < E *mass media*;
- *phonetical non-integration and graphical integration*: F ***redingote*** ' riding coat' < E *riding coat*;
- *phonetical and graphical integration*: R ***sandvici / sandviş / sanviş*** 'sandwich' < E *sandwich*.

All **borrowings** produce changes in the borrowing language and they can be changed themselves, too.

These languages, as well as any other language on Earth, have a lexical structure that changes to allow the integration of the **borrowings** in the new system of correlations.

Literature Review

The lexicographical analysis of the common names of plant diseases has been done on a corpus of words from the HYPP Dictionary developed within a project coordinated by G. Wachsman (from ACTA, Paris, France) and edited by M. A. Caillaud (from INRA, Paris, France): 188 scientific names of plant diseases of which only the first part has been borrowed in English, French, and Romanian to produce common names of plant diseases.

Results and Discussion

Of the 188 Latin names of plant diseases, only 96 have been borrowed by at least one of the three languages in discussion.

1. The Latin borrowings into **English** are of the following types:

- ***phonetical and graphical integration*** (1): ***cercosporiose*** > *Cercospora cladosporioides* (anamorphic name of *Mycocentrospora cladosporioides*);
- ***phonetical integration and graphical non-integration*** (31):
  - ***Ascochyta blight*** < *Ascochyta pisi*, ***Cercospora disease*** < *Cercospora beticola*, ***Didymella stem rot*** < *Didymella lycopersici*, ***Eutypa canker*** < *Eutypa armeniacae*, ***Fusarium disease / wilt*** (7 occurrences) < *Fusarium oxysporum / roseum*, ***Mycosphaerella blight*** < *Mycosphaerella pinodes*, ***Nectria canker*** < *Nectria galligena*, ***Pestalotiopsis*** < *Pestalotiopsis menezisiana*, ***Verticillium disease / wilt*** (2 occurrences) < *Verticillium albo-atrum / dahliae*, ***Xanthomonas disease*** (2 occurrences) < *Xanthomonas campestris / cynarae / fragariae*;
  - ***Botrytis (bunch rot)*** (2 occurrences) < *Botrytis cinerea* (anamorphic name of *Botryotinia fuckeliana*), ***Cylindrosporium disease*** < *Cylindrosporium concentricum* (anamorphic name of *Pyrenopeziza brassicae*), ***Phomopsis disease*** < *Phomopsis cinerensis* (anamorphic name of *Diaporthe helianthi*), ***Septoria nodorum blotch*** < *Septoria nodorum* (anamorphic name of *Leptosphaeria nodorum*);
  - ***Ascochyta foot-rot*** < *Ascochyta pinodella* (synonym of *Phoma medicaginis*), ***Coryneum blight*** < *Coryneum beijerinckii* (synonym of *Stigmina carpophila*), ***Pinodella blight*** < *Ascochyta pinodella* (synonym of *Phoma medicaginis*).

2. The Latin borrowings into French are more numerous:

- ***phonetical and graphical integration*** (35):
    - ***alternariose*** < *Alternaria brassicae / brassicicola / dauci / helianthi* (4 occurrences), ***cercospor(i)ose*** < *Cercospora beticola* (2 occurrences), ***cladosporine*** < *Cladosporium cucumerinum*, ***eutypose*** < *Eutypa armeniacae*, ***fusariose*** < *Fusarium oxysporum* (6 occurrences), ***kabatiellose*** < *Kabatiella zeae*, ***mastigosporiose*** < *Mastigosporium rubicorum*, ***moniliose*** < *Monilina fructigena / laxa* (2 occurrences), ***pourriture phoméenne*** < *Phoma medicaginis*, ***sclérotiniose*** < *Sclerotinia minor / sclerotiorum / trifolium* (4 occurrences), ***septoriose*** < *Septoria apiicola*, ***verticilliose*** < *Verticillium albo-atrum / dahliae*;
    - ***cercosporiose*** < *Cercospora cladosporioides* (anamorphic name of *Mycocentrospora cladosporioides*), ***cylindrosporiose*** < *Cylindrosporium concentricum* (anamorphic name of *Pyrenopeziza brassicae*), ***fusariose*** < *Fusarium nivale* (anamorphic name of *Monographella nivalis*), ***septoriose*** < *Septoria nodorum* (anamorphic name of *Leprosphaeria nodorum*);
    - ***ascochytose*** < *Ascochyta pinodella* (synonym of *Phoma medicaginis*), ***cladosporiose*** < *Cladosporium fulvium* (synonym of *Fulvia fulva*).

- ***phonetical integration and graphical non-integration*** (14):
    - ***chancre à Didymella*** < *Didymella lycopersici*, ***Fusarium du pied*** < *Fusarium roseum*, ***chancre à Fusicoccum*** < *Fusicoccum amygdali*, ***chancre à Nectria*** < *Nectria galligena*, ***Pestalotiopsis*** < *Pestalotiopsis menezisiana*, ***dépérissement bactérien à Pseudomonas*** < *Pseudomonas syringae*, ***Pseudopeziza*** < *Pseudopeziza medicaginis*, ***charbon à Sphacelotheca*** < *Sphacelotheca reiliana*, ***charbon à Ustilago*** < *Ustilago maydis*;
    - ***chancre à Cytospora*** < *Cytospora cincta* (anamorphic name of *Leucostoma cinctum*), ***criblure à Corynum*** < *Coryneum beijerinckii* (synonym of *Stigmina carpophila*), ***Phomopsis*** < *Phomopsis mali* (anamorphic name of *Diaporthe eres*), ***Phomopsis*** < *Phomopsis sojae* (anamorphic name of *Diaporthe phaseolum*), ***Phomopsis du tournesol*** < *Phomopsis helianthi* (anamorphic name of *Diaporthe helianthi*).

3. The Latin borrowings into Romanian are the least:

- ***phonetical and graphical integration*** (28):
    - ***alternarioză*** < *Alternaria brassicae / brassicicola / dauci / helianthi / radicicola* (6 occurrences), ***ascochitoză*** < *Ascochyta fabae*, ***cercosporioză*** < *Cercospora beticola*, ***eutipoză*** < *Eutypa armeniacae*, ***fuzarioză*** < *Fusarium oxysporum / roseum* (9 occurrences), ***monilioză*** < *Monilina fructigena / laxa* (2 occurrences), ***septorioză*** < *Septoria apiicola*, ***verticilioză*** < *Verticillium albo-atrum / dahliae* (4 occurrences);
    - ***cercosporioză*** < *Cercospora cladosporioides* (anamorphic name of *Mycocentrospora cladosporioides*), ***septorioză*** < *Septoria nodorum* (anamorphic name of *Leptosphaeria nodorum*);
    - ***ascochitoză*** < *Ascochyta pinodella* (synonym of *Phoma medicaginis*).

The words ***cercosporioză***, ***fuzarioză***, and ***monilioză*** are, according to Romanian language dictionaries, borrowings from French. As for the words ***alternarioză***, ***eutipoză***, ***septorioză***, and ***verticilioză***, they are either **linguistic calques** after French words, or words created after Romanian words already attested in French and borrowed from the French language.

As for the **borrowings** of the phonetical integration and graphical non-integration type, it seems that any name of plant disease that does not exist in Romanian is susceptible of being borrowed as such where there is no popular synonym for it.

We need to say that any improper pronunciation of Latin words designating plant diseases is susceptible of ranging within this category: we have in mind both mispronunciations caused by the lack of knowledge in Latin pronunciation and mispronunciations caused by the patois of the speaker.

Conclusion

Table 6-1 is a synthesis of the types of integration of scientific names of plant diseases in English, French, and Romanian.

As we can see, the French language ranks first in **phonetical and graphical integration** with 29 scientific names, 4 anamorphic names and 2 synonyms, followed by the Romanian language with 25 scientific names and 2 anamorphic names. There is a single anamorphic name in English phonetically and graphically.

The English language ranks first in **phonetical integration and graphical non-integration** with 23 scientific names, 5 anamorphic names and 3 synonyms, followed from a distance by the French language with 8 scientific names, 5 anamorphic names and 1 synonym. There are no such names in Romanian if we leave aside accidental integrations caused by the lack of knowledge of Latin.

There is no phonetical non-integration and graphical integration and no phonetical and graphical integration either in any of the three languages investigated here.

Table 6-1. Types of integration of scientific names of plant diseases in English, French, and Romanian

Types of integration	Language								
	English			French			Romanian		
	Latin names	Anamorphic names	Synonyms	Latin names	Anamorphic names	Synonyms	Latin names	Anamorphic names	Synonyms
phonetical and graphical integration	-	1	-	29	4	2	25	2	-
phonetical integration and graphical non-integration	23	5	3	8	5	1	(-)	(-)	(-)
phonetical non-integration and graphical integration	-	-	-	-	-	-	-	-	-
phonetical and graphical integration	-	-	-	-	-	-	-	-	-

From the point of view of languages for special purposes (English and French for agriculturists), being a native speaker of Romanian can make learning French easier since the former has borrowed most common names of plant diseases from the latter, adapting them after the French model. Though English phonetics is more difficult for native speakers of Romanian, a good knowledge of the scientific names of plant diseases can ease both comprehension and acquisition of this terminology.

References

Giuchici, Camelia & Raţă, Georgeta. (2002). Common Names of Plant Diseases: The Latin Heritage. *Al IV-lea Simpozion Internaţional „Tinerii şi Cercetarea Multidisciplinară", 14-15 noiembrie 2002, Timişoara*. Timişoara: Editura Sudura. 837-842.

HYPP. Version 1.0 copyright © 1994

Robert, P. (1996). *Dictionnaire alphabétique & analogique de la langue française*. [An Alphabetical & Analogical Dictionary of the French Language]. Paris: S.N.L.

Shipley, J. T. (1979). *Dictionary of Word Origins*. Totowa, NJ: Littlefield, Adams & Co.

The Random House Dictionary of the English Language. (1968). New York: Random House.

SPICES

ASTRID-SIMONE GROSZLER

Introduction

In this paper we try to bring an **etymological perspective** on *spice names*. By etymological perspective we mean an etymological approach in the sense of the *Explanatory Dictionary of the Romanian Language* (1998) where we find out that *etymology* means 'establishing of a word's origin by explaining its formal and semantic evolution'. We try to establish whether or not there is a connection between the origin of the word and its meaning. We also look at the Romanian and German name of the plant and try to establish a connection with the English term.

Aside from the etymological aspect of the origin of the words, we take a look at the orthographic and phonetic form of the analysed plant names.

Material and Method

We have chosen thirteen English *spice names*. First we have identified the origin of the English common name and, in most cases, that of the scientific name using specific dictionaries; then we have looked for Romanian and German corresponding names of the spices and tried to establish whether there is a connection between their names in the three languages.

Results

The *spice names* we analysed are the following:

Allspice (*Pimenta dioica*) takes its name from its aroma. It smells like a combination of spices, especially cinnamon, cloves, ginger and nutmeg. In much of the world, allspice is called *pimento* because the Spaniards mistook the fruit for *black pepper*, which they called *pimienta.* Another English name for this herb is ***Jamaican pepper***, which relates to the Spanish name, but also *pimento* (< L *pimento*). The German name for ***allspice*** is *Allgewürz*, which is **calque** of the English name ***allspice*** and

Jamaicapfeffer again, an exact translation of the English term, but also *Piment*, again relating to the Spanish origin. In Romanian, there is *piper de Jamaica,* the corresponding common name of *Jamaican pepper*. We could not find an occurrence of *pimento* or of any derived form of the word. If we are to compare the orthographic and the phonetic forms of the herb, we notice that there is an identity between these two forms (for the English variant) only in the overtaken synonym, namely the Spanish *pimienta.* The other forms (***allspice***, ***Jamaican pepper***) undergo the rules of pronunciation present in the English language. As for the Romanian and German languages, since they are more phonetic than English (especially Romanian), the phonetic changes are minor, and again there is also a complete identity in the Spanish overtaken name, *Piment,* in German.

When we analyzed the spice ***anise*** (< L *Pimpinella anisum*), the MWDT revealed as its origin the Middle English *anis* (< AF < L *anisum* < Greek *annēson, anison*). In turn, the Greek *anison* and the Latin *anisum* are derived from the Arabic name *anysum*, since ***anise*** is native to the eastern Mediterranean region, the Levant, and Egypt. As for its meaning, ***anise*** is a Eurasian annual herb of the carrot family having carminative and aromatic seeds. When we look for the German name of the herb, there is *Anis*, which is an almost perfect match to the English word, but also *Kuchengewürz*, which means 'spice for cake', thus leading to the conclusion that ***anise*** is a spice used in cake recipes. The Romanian name for the herb is *anason*, derived from the same Greek *annison*. When analysing the orthographic and phonetic occurences of the words, there are phonetic changes in the English term, since *anise* is pronounced [æni:s]. As for the German term *Anis* and for the Romanian one *anason,* there are no phonetical changes.

Cardamom (< L *Elettaria cardamomum*) is native to the East, originating in the forests of the Western Ghats in southern India, where it grows wild. Today it also grows in Sri Lanka, Guatemala, Indo China and Tanzania. The ancient Egyptians chewed ***cardamom*** seeds as a tooth cleaner; the Greeks and Romans used it as a perfume. Vikings came upon the spice about one thousand years ago, in Constantinople, and introduced it into Scandinavia. In the MWDT, we find its origin in the Latin *cardamomum* (which is also the second half of the spice's scientific name) from the Greek *kardamōmon,* blend of *kardamon* 'peppergrass' and *amōmon*. The German name of the spice is *Kardamom* – the same as the English name, and the Romanian name for it is *cardamom verde* – again, the same as the English name, but with the specification of the colour, *verde* 'green'. Thus we can infer that all three names derive from the Greek *kardamōmon*. If we analyse the terms from a phonetical point of view, we notice that for the English term there are some differences between the orthographic and the phonetic form, since the pronunciation of ***cardamom*** is [kærdəməm], while for the Romanian and German term there are no differences between the two forms.

Ginger (< L *Zingiber officinale*) is native to India and China. It takes its name from the Sanskrit word *stringa-vera,* which means 'with a body like a horn' (according to the online *Encyclopaedia of Spices*). The MWDT records as origin Middle English, alteration of Old English *gingifer* (< ML *gingiber* < L *zingiber* < Gk *zingiberi*). Regarding the meaning, there is '1. a thickened pungent aromatic rhizome that is used as a spice and sometimes medicinally. 2. the spice usually prepared by drying and grinding ginger.' The origin of the word thus relates to its shape and not necessarily its meaning. In German, there is *Ingwer*, which is clearly derived from the Latin and Greek name, and in Romanian there is *ghimber* or *ghimbir*, word originating from the Hungarian *gyömber*, which, again, clearly derives from the Latin and Greek words. With ***ginger***, phonetical changes can be observed in all three languages. The English term ***ginger*** is pronounced [dʒindʒə], while the German *Ingwer* is pronounced [iŋvə] and the Romanian *ghimbir* [gimbir].

Lavender (< L *Lavandula stoechas*) originates from the Middle English *lavendre* (< AF < Med L *lavandula*). It is native of the Mediterranean and became widely distributed throughout southern Europe, but nowadays it is a worldwide known herb. Widely used by the ancient Egyptians, Greeks, Romans and Arabs for mummification, medicinal purposes and scented baths and soaps, its botanical name comes from the Latin *lavare* which means 'to wash'. It was and is still used as a refresher for laundry. As in the case of ***ginger***, when we analyze the German name *Lavendel* and the Romanian *lavandă,* we notice that they originate from the Latin *lavare* as well. We can notice a resemblence in the phonetic forms of the terms in the different languages as well, although the English pronunciation has undergone more changes than the other two. ***Lavender*** is read [lævəndə], the German *Lavendel* [lavendəl] and the Romanian *lavandă* [lavandə].

Lemon grass (< L *Cymbopogon ciatrus*) is a tall tropical grass. The fresh stalks and leaves have a clean lemon-like odour because they contain an essential oil, which is also present in lemon peel. The English name originates from the Middle English *lymon* (< MF *limon* < ML *limon-, limo* < Ar *laymūn, līmūn* < Pers *līmū, līmun*). In this case, the herb gets its name form the resemblance in odour with the lemon fruit, and also because it contains the same essential oil as the fruit. There are three German corresponding names: *Zitronengras* (a perfect translation of the English name), *Citronella* (derived from the scientific name), and *Lemongras* (an almost perfect transfer of the English name). As for the Romanian corresponding name, there is *iarba lămâioasă* (an almost perfect translation of the English name). When we take a look at the English term and its equivalents in German and Romanian, we find out that the English and German terms are almost perfect equivalents in writing and pronouncing, with ***lemon grass*** [lemən gra:s] in English and *Lemongras* [lemo:ngra:s] in German. As for the Romanian term, the letters *ă* and *â* are the orthographic signs for the phonetic transcripts of [ə] and [î], sounds inexistent in English. Thus, there are only minor changes in the Romanian phonetic form [jarba təmîjoasə].

Licorice (< L *Glycyrrhiza glabra*) gets its scientific name from the Greek words *glyks* 'sweet' and *rhiza* 'root'. It is the sweet tasting rhizomes (underground stems) and roots that are used as flavourings. The MWDT records points its origin in the Middle English *licorice* (< AF *licoris* < LL *liquiritia,* alteration of L *glycyrrhiza* < Gk *glykyrrhiza* < *glykys* 'sweet' and *rhiza* 'root'). The German corresponding name is *Süßholz* 'sweet wood', relating to the Greek origin not formally, but semantically. In Romanian, there is *lemn dulce*, an exact translation not of the English, but of the German term. From a phonetic point of view, we can notice that, like in the cases of the herbs discussed above, the pronunciation of ***licorice*** has brought changes compared to the orthographic form [likəris]. The German term is pronounced [zy:sholts], thus including a few changes, while the Romanian term pronounced [lemn dultʃe] presents almost no difference between the orthographic and phonetic form.

According to the online *Encyclopaedia of Spices*, ***melegueta pepper*** (< L *Aframomum melegueta*) is native to tropical West Africa and grows mainly in Ghana. The spice is practically unknown in modern Western cuisine, although it was used in Europe in the Middle Ages and the Renaissance. It was a flavouring for the old wine 'Hippocras' and is still used for the production of beer, wine and spirits, and for the flavouring of vinegar. It was known under the name of ***Grains of paradise***. Given its use in preparing alcoholic beverages, which may induce a euphoric state if consumed in large quantities, we can understand where it got this name from. We notice that the first half of the name observes the Latin *melegueta*, while the other half refers to the shape, the appearance of the seeds of the herb, the part that is used. The German and Romanian names are actually **calques** of the English name, thus bearing the same etymology: *Meleguetapfeffer* or *Paradieskörner* in German, and *piper de melengueta* or *grăuntele paradisului* in Romanian, respectively. When we take a look at the phonetic forms of the terms in the three discussed languages, we see that, in the case of ***melegueta pepper***, almost no changes occur, regardless the language analysed: English [melegweta pepə], German [melegwetapfefə] and Romanian [piper de melegweta].

Mint (< L *Mentha* spp.) comes from the Greek legend of the nymph Minthe, who caught the eye of Hades. Hades' wife, the jealous Persephone, attacked Minthe and was in the process of trampling her to death when Hades turned her into an herb (which remained sacred to him). It is a very well known and used herb. The nowadays English term originates from the Middle English *minte* (< OE < L *mentha, menta* < Gk *minthē*). In ***mint***, there is complete identity between the orthographic and phonetic ([mint]) forms. There are some minor changes in the German phonetic form [mintse] and in the Romanian one [mentə].

Nutmeg (< L *Myristica fragrans*) is the seed kernel inside the fruit of the nutmeg tree, and also the ground seed used as a spice. It takes its name from its flavour, which is nutty, warm and slightly sweet. It originates from Middle English *notemigge, notemuge* < OPr *noz muscada* < *noz* 'nut' (< L *nuc-, nux*). Another name for it is *muscat*. In German, there is an interesting combination of the two English terms: *Muskatnuss*, *Muskat* from *muscat* and *Nuss* meaning 'nut'. The Romanian name is *nucşoară*, which could be literally translated by 'small nut', which derives from the Latin *nuc-, nux*. If we look at the phonetic forms, we notice that the English term displays a change, the pronunciation being [natmeg]. In German, there is a complete identity of the orthographic and phonetic forms [muskatnus]. The Romanian pronunciation displays one small change [nucʃoarə].

Sage (< L *Salvia officinalis*) originates from the Middle English (< AF < VL **sapius* < L *sapere* 'to taste, have good taste, be wise'). The term is related to the Old Saxon *ansebbian* 'to perceive'. Ancient populations, including the Aras, associated it with immortality. The genus name derives from the Latin for 'salvation', *salvere*. The plant was used to counterattack snakebite. If we take a look at the German and Romanian names, *Salbei* and *salvie*, respectively, we notice that both derive from the Latin name *Salvia*. When we analyse the names of ***sage*** from a phonetic point of view, we notice that the Romanian term [salvie] is the only one not undergoing major phonetic changes. The English ***sage*** [seidʒ], and the German *Salbei* [zalbai] present greater phonetic changes.

Savoury (< L *Satureja Hortensis* or *Satureja Montana*) is a reputed aphrodisiac. The genus Latin name, *Satureja*, is attributed to the Roman writer Pliny and is a **derivative** of the word for *satyr* 'the half-man, half-goat with the insatiable sexual appetite'. Legends say that satyrs lived in meadows of ***savoury***, thus implying that it was the herb that made them passionate. The English term originates from the Middle English (< AF *savur* < L *sapor,* from *sapere* 'to taste'). The English word *savour* means: '1. to enjoy and appreciate (food or drink) slowly. 2. to enjoy (a pleasure) for as long as possible', thus relating to the reputation of the plant. Regarding the German name, *Pfefferkraut* 'pepper herb', and the Romanian name *cimbru* originating from the Greek *thymbra*, or Latin *cimbrus*. We can see that there is no resemblance whatsoever between the three terms belonging to the three different languages. With ***savoury***, the English and Romanian terms are the ones undergoing phonetic changes, reading [seivəri] and [tʃimbru], respectively. The German term is almost unchanged [pfefəkraut].

Wasabi (< L *Wasabia japonica*), also called ***Japanese horseradish***, is a pale green root grown in cold mountain streams under some of the most closely guarded growing practices in agriculture. The MWDT only mentions that it is of Japanese origin. The Japanese *wasabi* actually refers to the ***mountain hollyhock*** (a perennial plant of the Malvacea family), as the plant's leaves resemble those of a member of the Malvacea family, in addition to its ability to grow on shady hillsides. The name ***Japanese horseradish*** refers to the land of origin and to the flavour of the plant, resembling the taste of horseradish, while in fact it is a mustard. When we look for the German and Romanian names of the herb, there is the same ***wasabi***, but also *Japanischer Kren* in German and *hrean japonez* in Romanian, both names representing calques of the English ***Japanese horseradish***. From a phonetic point of view, ***wasabi*** stays the same, namely [wasabi] in all three languages discussed.

Discussion

In the previous section we have presented thirteen *spice names*. The analyses of their origins revealed that most spices derive their English name from the Latin or Greek denomination of the herb. Eight names (***allspice***, ***anise***, ***cardamom***, ***ginger***, ***lavender***, ***melegueta pepper***, ***mint***, ***pimento***, ***sage***) are (almost) perfect transfers from Latin or Greek and one (***licorice***) is an alliteration of the Greek name, which, in turn, represents the origin of the Latin denomination. One spice name, ***wasabi***, preserves its Japanese origin. ***Lemon grass*** and ***nutmeg*** take their name from their shape and flavour and have no connection whatsoever with the Latin or Greek origin of the scientific name. ***Savoury*** has a special etymology, since it takes its name from its use, rather than its shape or meaning, as in most cases.

Also, three of the thirteen herb names we have analysed have a second or even a third English name. As already mentioned above, ***pimento*** or ***allspice*** (a name originating from its flavour) is also called ***Jamaican pepper*** (after the shape of the seeds, the part of the spice used in cooking); ***melegueta pepper*** is also called ***grains of paradise*** (the name originating from the use of the spice for preparing alcoholic beverages); and ***wasabi*** is also called ***Japanese horseradish*** (taking its name after its flavour). Another spice taking its name after its shape is ***ginger*** (the Indian name *stringa-vera* meaning 'with a body like a horn').

When analysing the German and Romanian counterparts of the English *spice names*, we found that, of the thirteen analysed spices, four have literal translations of the English name both in German and Romanian (i.e. ***allspice***, ***lemon grass***, ***melegueta pepper*** and ***wasabi - Japanese horseradish***) and one has partial translations (***nutmeg***); in six cases, the German and the Romanian name have the same Latin origin as the English name (i.e. ***anise***, ***cardamom***, ***ginger***, ***lavender***, ***mint***, ***sage***). For ***licorice***, the German and the Romanian names, though having the same meaning (G *Süssholz* and R *lemn dulce* meaning 'sweet wood'), have no connection whatsoever with the English name. ***Savoury*** is the only herb where there is no resemblance whatsoever between all three terms belonging to the three different languages. From a phonetic point of view, in most cases, the English term was the one undergoing changes, while the Romanian term was the one to display almost no change.

Conclusions

All world languages (English, German, and Romanian included) have, at various times in their histor, been enlarged by **loanwords** from other languages. Since the 17th and 18th centuries, English settlers and, later, German travellers and/or scientists encountered and accepted, during their stays and/or travels, many of the native inhabitants' names for their flora and fauna, and also many of the names of their tribal customs and beliefs. Among them, new types of dishes, new types of cooking, and the use of spices whose presence continue to make cookbooks and/or menus more attractive. Although we have analysed only thirteen *spice names*, we could notice that most of them take their English name after the Latin scientific name, which, in turn, originated from the Greek corresponding word. For almost half of the analysed herbs, the German and Romanian names are derived from the same Latin or Greek root as the English ones. This is understandable, since Ancient Greeks were great medicine men and the Europeans inherited a great part of their medical and herbal knowledge from them. Another conclusion that can be inferred is that many spice herbs (but also medical herbs) received a common name (see ***allspice***, ***Jamaican pepper***, ***grains of paradise***, ***Japanese horseradish***). Few people were educated in older times, so it was only natural that they named the herbs after the names of well-known concepts. This tendency can be noticed in German and Romanian as well, not just in English (for ***allspice***, for instance, there is the German *Allgewürz*, which is an exact translation of the English term; the same goes for ***grains of paradise*** and ***Japanese horseradish***, which display literal translations in both German and Romanian).

References

Dicționarul explicativ al limbii române. [The Explanatory Dictionary of the Romanian Language]. (1998). București: E.U.E. (DEX)

Encyclopaedia of Spices. Online: http://www.theepicentre.com/Spices/spiceref.html.

Groszler, Astrid-Simone. (2008). Spices: An Etymological Approach. *Journal of Linguistic Studies 1 (2)*: 29-34.

Guiraud, P. (1964). *L'étymologie.* [Etymology]. Paris: PUF.

Hristea, T. (1968). *Probleme de etimologie. Studii. Articole. Note.* [Etymology issues. Studies. Articles. Notes]. București: Editura Științifică.

Király, F. (1988). *Etimologia. Etimologii.* [Etymology. Etymologies]. Timișoara: Universitatea din Timișoara.

Leo Deutsch-Englisches Wörterbuch. [Leo German-English Dictionary]. Online: http://dict.leo.org.

Shipley, J. T. (1979). *Dictionary of Word Origins*. Totowa, NJ: Littlefield, Adams & Co.

Webster's Universal Dictionary & Thesaurus (2003). New Lanark: Geddes & Grosset. (WUDT)

Sala, F. (1999). *Introducere în etimologia limbii române*. [Introduction to Romanian Etymology]. București: Editura Univers Enciclopedic.

LAND MEASUREMENT UNITS

GEORGETA RAȚĂ AND FLORIN SALA

Introduction

We do not know who uses the greatest variety of units of measurement in the world. Caught in a slow-moving transition from customary to metric units, many peoples employ a fascinating and sometimes frustrating mixture of units in talking about the same thing: thus, they measure the length of a race in meters, but the length of the long jump event in feet and inches.

Furthermore, customary units do not always form a consistent system. Reflecting their diverse roots in a wide variety of cultures, land measurement units are often confusing and contradictory. But from a linguistic point of view, there are a few patterns common to many of the land measurement units used around the world.

Material and Method

In our analysis of land measurement units we have used Russ Rowlett's *Dictionary of Units of Measurement* (2005), a dictionary that has attracted users from all over the world. As its author mentions, many users were kind enough to point out errors, while others suggested additions and improvements. This paper ranks us among this latter category.

We have inventoried a number of 79 names of units of land area used in the past and/or nowadays, as follows: ***acre*** [before 1000; < ME *aker* < OE *acer*; cf. L *ager* 'field', Gk *agrós* 'field'] (English-speaking countries), ***alqueire*** (Portugal and Brazil), ***anna*** (South Asia), ***arpent*** [1570-1580; < MF < L *arepennis* 'half-acre' < Gaulish] (French North America), ***bigha*** (South Asia), ***bonnier*** (Belgium), ***bovate*** [< L *bovis* 'cow', 'ox'] (syn. ***oxgang***) (England], ***bunder*** (Holland), ***caballeria*** (Spain), ***cadastral hold*** (syn. ***kataszteri hold***) (Hungary), ***carreau*** (Haiti), ***carucate*** (syn. ***hide***) [1375-1425; late ME < ML < *carrūcāta* 'plough, plough team' (L travelling carriage, with the sense 'wheeled plough' in Gaul] (England), ***cawney / cawny*** [? < Tamil *cawney*] (India), ***colp / colpa / collop*** (BE) [<

Ir *colpa* 'a unit of livestock equal to one cow or horse or to 6 sheep'] (Ireland), ***commercial acre*** (U.S.A.), ***cong*** (Vietnam), ***cotka / sotka*** (Russia), ***cuerda*** (syn. ***Spanish acre***) (Puerto Rico), ***decare / dekare*** (worldwide), ***dessiatina*** [1790-1800; < Russ *desyatína* 'tithe, tenth'] (Russia), ***dhur*** (South Asia), ***djerib / jerib*** (Middle East, south-western Asia), ***donum / dunum*** [? < Turk *donum*] (Middle East, the Balkans, Israel, Palestine, Croatia, former Yugoslavia), ***fall*** (worldwide), ***fanega / fanegada*** [1495-1505; < Sp < Ar < *faniqah* 'big bag', 'large sack'] (Spain, Latin America), ***fardel*** (syn. ***farthingdale***$_2$) [1375-1425; late ME < AF, OF < OPr < OE, equivalent to *fard(a)* 'bundle' (<< Ar *fardah* 'load' + *el*] (England), ***farthingdale***$_1$ ['a fourth part'] (syn. ***rood***) (England), ***farthingdale***$_2$ ['a fourth part'] (syn. ***fardel***) (England), ***feddan*** [< Ar *faddān* 'yoke of oxen'] (Egypt, Middle East, North Africa), ***ground*** (India), ***hacienda*** [1710-1720; < Sp < L *facienda* 'things to be done or made'] (Mexico, south-western U.S.A.), ***hectare / hektare*** [1800-1810; < F] (worldwide), ***hide*** [before 900; ME *hid(e)*, *hig(i)d* 'portion of land', 'family'] (syn. ***carucate***) (England), ***hold*** (Hungary), ***homestead*** [before 1000] (U.S.A.), ***Hungarian acre*** (syn. ***Magyar hold***) (Hungary), ***Irish acre*** (Ireland), ***jitro*** [< L *jugerum* 'yoke'] (the Czech Republic), ***joch*** (pl. ***joche***) [< G *joch* 'yoke' < L *jugerum* 'yoke'] (German speaking countries), ***journal*** (< F *jour* 'day' < L *jugerum* 'yoke') (France), ***juchart / juchert*** [< L *jugerum* 'yoke'] (syn. ***tagwerk***) (Germany and German-speaking Switzerland), ***jutro*** (pl. ***jutra***) [< Cr *jutro* 'morning'] (Croatia), ***kanal*** (Pakistan), ***kappland*** (Sweden), ***kataszteri hold*** (syn. ***cadastral hold***) (Hungary), ***katta / kattha*** (South Asia), ***labor*** [< Sp *labor* 'work'] (Latin American countries), ***lanac*** [< Slav *lanac* 'chain'] (former Yugoslavia), ***legua*** (syn. ***sitio***) (Spain), ***Magyar hold*** (syn. ***Hungarian acre***) (Hungary), ***mål*** [< N *mål* 'measure'] (Norway), ***manzana*** [< Sp *manzana* 'apple' < Sp *manzanar* 'orchard'] (Central America), ***mou / mu*** (China), ***ngarn*** (Thailand), ***nook*** [1250-1300; ME *nok*] (northern England), ***olk*** (Iraq), ***oxgang*** (syn. ***bovate***) (England), ***rai*** [< Thai *rai* 'field'] (Thailand), ***rood*** [before 900; ME; OE *rōd* 'pole, crucifix] (syn. ***farthingdale***$_1$) (England), ***ropani*** (Nepal), ***sao*** (Vietnam), ***sitio*** (syn. ***legua***) (Mexico, Texas), ***soendre*** (Bhutan), ***sotka / cotka*** (Russia), ***Spanish acre*** (syn. ***cuerda***) (Puerto Rico), ***stang*** (Wales), ***stremma*** (Greece), ***tan*** (Japan), ***tarea*** [< Sp *tarea* 'job', 'task'] (the Dominican Republic, Central America), ***tarefa*** [< Port *tarefa* 'job', 'task'] (Brazil), ***tønde land*** [< D *tønde* 'barrel'] (Denmark), ***tunnland*** [< Sw *tunn* 'barrel'], ***vergee*** [< F *verger* 'orchard'] (the Channel Islands), ***virgate*** [1645-1655; < ML *virgāta (terrae)* 'pertaining to a rod'] (syn. ***yard of land***, ***yardland***)

(England), ***yard of land*** (syn. ***virgate***) (England), and ***yardland*** (syn. ***virgate***) (England).

Results and Discussion

A few patterns can be discerned among the **land measurement units**.

1. Thus, the names of some of the **land measurement units** are related to different types of 'amount':

- the 'share' of an 'amount' of land to be worked (4): ***dessiatina*** [1790-1800; < Russ *desyatína* 'tithe, tenth'] (Russia), ***farthingdale***$_1$ ['a fourth part'] (England), ***farthingdale***$_2$ ['a fourth part'] (England), *hig(i)d* 'portion of land', 'family'] (England);
- the 'amount' of land that could be ploughed in 1 day by 1 yoke of oxen (4): ***jitro*** [< L *jugerum* 'yoke'] (the Czech Republic); ***joch*** (pl. ***joche***) [< G *joch* 'yoke' < L *jugerum* 'yoke'] (German speaking countries); ***juchart / juchert*** [< L *jugerum* 'yoke'] (southern Germany and German-speaking Switzerland); ***tagwerk*** [< G *tagwerk* 'day's work'] (southern Germany and German-speaking Switzerland);
- the 'amount' of land to be worked (3): ***acre*** [before 1000; < ME *aker* < OE *acer*; cf. L *ager* 'field', Gk *agrós* 'field'] (English-speaking countries); ***arpent*** [1570-1580; < MF < L *arepennis* 'half-acre' < Gaulish] (French North America), ***rai*** [< Thai *rai* 'field'] (Thailand);
- the 'amount' of land that could be planted with a 'measure' of seed (3): ***fanega(da)*** [1495-1505; < Sp < Ar < *faniqah* 'big bag', 'large sack'] (Spain, Latin America), with a *fanega* (about 55.50 litres) of seed; ***tønde land*** [< D *tønde* 'barrel'] (Denmark), with a *tønde* (about 139 liters) of seed; ***tunnland*** [< Sw *tunn* 'barrel'], with a *tunn* (about 139 liters) of seed;
- the 'amount' of land that could be worked with a certain number of beasts of burden (3): ***bovate*** [< L *bovis* 'cow', 'ox'] (England], with a cow / ox; ***feddan*** [< Ar *faddān* 'yoke of oxen'] (Egypt, Middle East, North Africa), with a yoke of oxen; ***oxgang*** (England), with a cow / ox;
- the 'job' to be done (3): ***hacienda*** [1710-1720; < Sp < L *facienda* 'things to be done or made'] (Mexico, south-western U.S.A.), ***tarea*** [< Sp *tarea* 'job', 'task'] (the Dominican Republic, Central America); ***tarefa*** [< Port *tarefa* 'job', 'task'] (Brazil);
- the 'amount' of land that could be ploughed in 1 day (1): ***journal*** [< F *jour* 'day'] (France);

- the ‘amount’ of land that could be ploughed in 1 day by 1 team of oxen (1): ***langdo*** (Bhutan);
- the ‘amount’ of land that could be ploughed in 1 morning (1): ***jutro*** (pl. ***jutra***) [< Cr *jutro* ‘morning’] (Croatia);
- the ‘amount’ of land that could be ploughed in 1 morning by 1 yoke of oxen (1): ***morgen*** [< G *morgen* ‘morning’] (Northern Europe);
- the ‘amount’ of land that could be cultivated by a single farmer / ploughman (1): ***labor*** [< Sp *labor* ‘work’] (Latin American countries);
- the ‘amount’ of land that could be cultivated by a single farmer / ploughman + a team of 8 oxen (1): ***hide*** [before 900; ME *hid(e)*, *hig(i)d* ‘portion of land’, ‘family] (England);
- the ‘amount’ of pastureland supporting a certain unit of livestock (1): ***colp(a) / collop*** [< Ir *colpa* ‘a unit of livestock equal to one cow or horse or to 6 sheep’] (Ireland), a *colpa* (1 cow / horse or to 6 sheep).

2. A few **land measurement units** are related to types of ‘instruments’ used in working the land:

- the ‘instrument’ the land is marked with (2): ***rood*** [before 900; ME; OE *rōd* ‘pole, crucifix] (England), ***virgate*** [1645-1655; < ML *virgāta (terrae)* ‘pertaining to a rod’] (syn. ***yard of land***, ***yardland***) (England);
- the ‘instrument’ the land is worked with (1): ***carucate*** [1375-1425; late ME < ML < *carrūcāta* ‘plough, plough team’ < L travelling carriage, with the sense ‘wheeled plough’ in Gaul] (England).

Conclusions

Of the 79 names of **land measurement units**, 27 (34%) refer to different kinds of ‘amount’ related to the working of land (land, number of beasts, work) and only 3 (4%) refer to types of ‘instruments’ used in land working (Figure 6-1). For the rest of the terms (62%), Rowlett’s *Dictionary of Units of Measurement* (2005) does not supply the etymology. Nevertheless, it is interesting to note that 1/3 of the terms designating **land measurement units** refer to ‘amounts’ of:

- land to be worked;
- time units to work the land;
- number of beasts to do the job,

i.e. the things that have ever counted most in agriculturists’ lives.

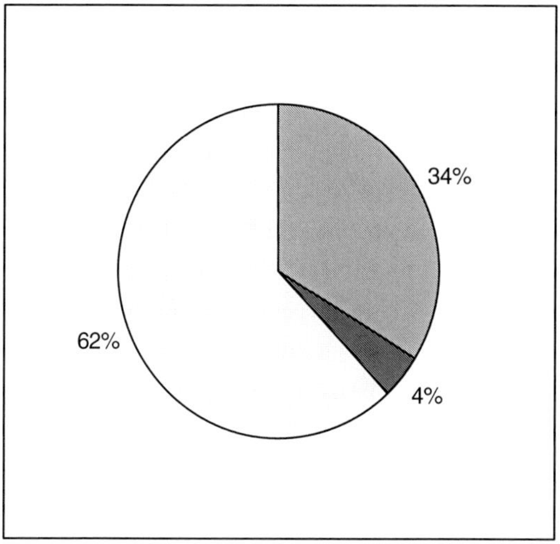

Figure 6-1. Share of land measurement units after 'meaning': 34% 'amount', 4% 'instrument', 62% unknown meaning

References

Benson, M., Benson, Evelyn & Ilson, R. (1990). *The Combinatory Dictionary of English*. Amsterdam – Philadelphia: John Benjamins Publishing Company.

Britannica Online Encyclopaedia. Online: http://www.britannica.com/EBchecked/topic/259196/hectare.

Chalker, Sylvia & Weiner, E. (1994). *The Oxford Dictionary of English Grammar*. London - New York - Sydney - Toronto: BCA.

Dicţionarul explicativ al limbii române. [An Explanatory Dictionary of the Romanian Language]. (1998). Bucureşti: Editura Univers Enciclopedic.

English Units. Online: http://en.wikipedia.org/wiki/English_units_of_measurement.

Finnish Units of Measurement. Online: http://en.wikipedia.org/wiki/Finnish_units_of_measurement.

German Units of Measurement. Online: http://en.wikipedia.org/wiki/German_units_of_measurement.

Glossary of Land Records Related Words. Online: http://etawah.nic.in/bhulekh/land_glossary.htm.

Leviţchi, L. & Bantaş, A. (1995). *Dicţionar englez-român.* [English-Romanian Dictionary]. Bucureşti: Ed. Teora.

Leviţchi, L. (Ed.). (1974). *Dicţionar englez-român.* [English-Romanian Dictionary]. Bucureşti: Ed. Academiei R.S.R

Leviţchi, L. (Ed.). (2004). *Dicţionar englez-român.* [English-Romanian Dictionary]. Bucureşti: Ed. Univers Enciclopedic.

Longman Dictionary of English Language and Culture (2003). London: Longman.

New Standard Encyclopedia. (1995). Chicago: Standard Educational Corporation.

Obsolete Russian Units of Measurement. Online: http://en.wikipedia.org/wiki/Obsolete_Russian_units_of_measurement.

Raţă, Georgeta & Sala, F. (2006). On Land Measurement Units in Romania and Hungary (A Linguistic Approach). *5th International Scientific Days of Land Management in the Great Hungarian Plain, October 26-27, 2006, Mezötúr, Hungary*: 1-5.

Romanian Units of Measurement. Online: http://en.wikipedia.org/wiki/Romanian_units_of_measurement

Rowlett, R. (2005). *Dictionary of Units of Measurement.* Online: http://www.unc.edu/~rowlett/units/index.html.

Shipley, J. T. (1979). *Dictionary of Word Origins.* Totowa, NJ: Littlefield, Adams & Co.

Spanish Customary Units. Online: http://en.wikipedia.org/wiki/Spanish_units_of_measurement.

Swedish Units of Measurement. Online: http://en.wikipedia.org/wiki/Swedish_units_of_measurement.

The New Encyclopaedia Britannica. (1992). Chicago: Encyclopaedia Britannica, Inc.

The Random House Dictionary of the English Language. (1968). New York: Random House.

Units of Measurement in France. Online: http://en.wikipedia.org/wiki/French_units_of_measurement.

Units & Systems of Units. Online: http://www.sizes.com/units/index.htm.

Webster Comprehensive Dictionary. (1995). Chicago: J. G. Ferguson Publishing Company. (WCD)

Webster Encyclopedic Unabridged Dictionary of the English Language. (1996). New York: Gramercy Books.

Webster's Universal Dictionary & Thesaurus (2003). New Lanark: Geddes & Grosset.

WORDS OF LATIN ORIGIN IN THE ENGLISH OF ZOOLOGY

GEORGETA RAȚĂ

Introduction

The purpose of the research was to see if there are any patterns in the naming of animals based on Latin words in contemporary English, patterns that would make animal names easier to identify and/or learn.

The hypothesis of the research was that, given the fact that all these names were coined more or less at the dawn of the age of scientific discovery, in the 17th and 18th centuries, as new words to describe new knowledge, their naming must have relied on certain patterns.

The background information consisted in what is generally given as common sense knowledge about scientific names in general and about zoological names in particular.

Apart from some lists containing English words of Latin origin posted on Internet sites, there is no exhaustive list of words of Latin origin in the **English of zoology**.

Material and Method

The material used in the research consisted of words of Latin origin specific to the **English of zoology** that we have picked up from the *American Heritage Dictionary of the English Language* (2008), words that we analysed from an etymological point of view (we were interested only in the changes in form and meaning) using Harper's *Online Etymology Dictionary* (2001), Partridge's *Origins. A Short Etymological Dictionary of Modern English* (2006), and Glare's *Oxford Latin Dictionary* (1983). As background information of classic culture nature, we used Farrell's *Latin Language & Latin Culture. From Ancient to Modern Times* (2004), Janson's *Speak. A Short History of Languages* (2002). The linguistic terminology is the one supplied by Chalker & Weiner's *Oxford Dictionary of English Grammar* (1994) and by Sheehan's *Word Parts Dictionary.*

Standard and Reverse Listings of Prefixes, Suffixes, Roots and Combining Forms (2000).

Results

Common names of animals are either *metonyms* ('words that are used as substitutes for other words with which they are in close semantic relationships' – Chalker & Weiner 1994) or *English formations* ('new words formed other than by **compounding**' – Chalker & Weiner 1994).

Metonyms

Metonyms of Latin origin in the **English of zoology** (a total of 96, i.e. 27%) come from *genus names*, *species names*, and *common names*.

Genus Names. There are 83 (87%) common names from New Latin genus names of Greek origin: ***acarus*** (pl. ***acari***) 'a mite, especially one of the genus *Acarus*' [< NL *Acarus*], ***actinia*** (pl. ***actiniae / actinians***) 'a sea anemone or a related animal' [< NL *Actinia*], ***aedes*** (pl. ***aedes***) 'a mosquito of the genus *Aëdes*' [< NL *Aëdes*], ***aepyornis*** 'a genus of extinct, large, flightless birds native to Madagascar' [< NL *Aepyornis*], ***alternaria*** 'any of various fungi in the genus *Alternaria*' [< NL *Alternaria*], ***am(o)eba*** (pl. ***amoebas / amoebae***) 'a protozoan of the genus *Amoeba* or related genera' [< NL *Amoeba*], ***aphid*** 'any of various small, soft-bodied insects of the family Aphididae' [< NL *Aphis, Aphid-*], ***aphis*** (pl. ***aphides***) 'an aphid, especially one of the genus *Aphis*' [< NL *Aphis*], ***apteryx*** 'any of several flightless birds of the genus *Apteryx*' [< NL *Apteryx*], ***archaeopteryx*** 'an extinct primitive bird (genus *Archaeopteryx*)' [< NL *Archaeopteryx*], ***aspergillus*** (pl. ***aspergilli)*** 'any of various fungi of the genus *Aspergillus*' [< NL *Aspergillus*], ***betta*** 'any of various species of small, brightly coloured, long-finned freshwater fishes of the genus *Betta*' [< NL *Betta*], ***bilharzia*** 'any of several chiefly tropical trematode worms of the genus *Schistosoma*' [< NL *Bilharzia*, genus name, after Theodor *Bilharz* (1825-1862), German physician], ***botrytis*** 'any of various fungi of the genus *Botrytis*; noble rot' [< NL *Botrytis*], ***brontosaurus*** 'a very large herbivorous dinosaur of the genus Brontosaurus' [< NL *Brontosaurus*], ***ceratodus*** (pl. ***ceratoduses***) 'any of various extinct lungfishes of the genus *Ceratodus*; 'any of several Australian food fishes' [< NL *Ceratodus*], ***chimaera*** 'a deep-sea cartilaginous fish of the family Chimaeridae; an organism consisting of two or more tissues of different genetic composition, produced as a result of mutation, grafting, or the mixture of cell populations from different zygotes; an organism produced by genetic engineering, in which DNA from distinct parent species is combined to produce an individual with a double chromosome complement' [< NL *Chimaera*], ***coccidium*** (pl. ***coccidia***) 'any of various parasitic protozoans belonging to the order Coccidia' [< NL *Coccidium*], ***colobus (monkey)*** 'any of various large African monkeys of the genus *Colobus*' [< NL *Colobus*], ***danio*** (pl. ***danios***) 'any of various small, often brightly coloured freshwater fishes of the genera *Danio* and *Brachydanio*' [< NL *Danio*], ***daphnia*** (pl. ***daphnia)*** 'any of various water fleas of the genus *Daphnia*' [< NL *Daphnia*], ***dasyure*** 'any of various often carnivorous marsupials of the family Dasyuridae ' [< NL *Dasyurus*, genus name], ***dentalium*** (pl. ***dentalia*** / ***dentaliums***) 'any of various tooth shells of the genus *Dentalium*' [< NL *Dentālium*], ***diplodocus*** 'a very large herbivorous dinosaur of the genus *Diplodocus*' [< NL *Diplodocus*], ***drosophila*** 'any of various small fruit flies of the genus *Drosophila*' [< NL *Drosophila*], ***dugong*** 'a herbivorous marine mammal (*Dugong dugon*)' [< NL *Dugong*], ***echinococcus*** (pl. ***echinococci***) 'any of several parasitic tapeworms of the genus *Echinococcus*' [< NL *Echinococcus*], ***filaria*** (pl. ***filariae***) 'any of various slender, threadlike nematode worms of the superfamily Filarioidea' [< NL *Fīlāria*], ***fusarium*** (pl. ***fusaria***) 'any of various pathogenic fungi of the genus *Fusarium*' [< NL *Fusarium*], ***galago*** (pl. ***galagos***) 'any of several small, nocturnal African primates of the genera *Galago* and *Euoticus*' [< NL *Galago*], ***gambusia*** 'any of various small livebearers of the genus *Gambusia*' [< NL *Gambusia*], ***gerbil*** 'any of various small, mouse like rodents of the genus *Gerbillus* and related genera; the Mongolian gerbil (*Meriones unguiculatus*)' [F *gerbille* < NL *Gerbillus*], ***hydra*** (pl. ***hydras*** / ***hydrae***) 'any of several small freshwater polyps of the genus *Hydra* and related genera' [< NL *Hydra*], ***iguanodon*** 'any of various large dinosaurs of the genus *Iguanodon*' [< NL *Iguanodon*], ***lemur*** 'any of several small arboreal, mostly nocturnal primates chiefly of the family Lemuridae' [< NL *Lemur*], ***lygus (bug)*** 'any of various North American bugs of the genus *Lygus*' [< NL *Lygus*], ***mastodon*** 'any of several very large, extinct proboscidian mammals of the genus *Mammut*' [< NL *Mastodon*], ***murex*** (pl. ***murices*** / ***murexes***) 'any of various marine gastropods of the genus *Murex*' [< NL *Mūrex*], ***mysid*** 'any of various small, shrimp like, chiefly marine crustaceans of the order Mysidacea' [< NL *Mysis, Mysid-*], ***noctiluca*** 'any of various bioluminescent dinoflagellates of the genus *Noctiluca*' [< NL *Noctilūca*], ***notornis*** (pl. ***notornis***) 'any of several flightless New Zealand birds, now rare, of the genus *Notornis*' [< NL *Notornis*], ***obelia*** 'any of various colonial marine hydroids of the genus *Obelia*' [< NL *Obelia*], ***octopus*** (pl. ***octopuses*** / ***octopi***) 'any of numerous carnivorous marine molluscs of the genus *Octopus* or related genera' [< NL *Octopus*], ***panchax*** 'any of various small, brightly coloured Old World tropical fishes of the genus *Aplocheilus* and related genera' [< NL *Panchax*, genus name], ***paramecium*** (pl. ***paramecia*** / ***parameciums***) 'any of various freshwater ciliate protozoans of the genus *Paramecium*' [< NL *Paramēcium*], ***peripatus*** 'any of numerous wormlike carnivorous animals of the phylum Onychophora' [< NL *Peripatus*], ***phalanger*** 'any of various small arboreal marsupials of the family Phalangeridae' [< NL *Phalanger*], ***phylloxera*** (pl. ***phylloxerae***) 'any of several small insects of the genus *Phylloxera*' [< NL *Phylloxera*], ***pithecanthropus*** 'an extinct primate postulated from bones found in Java in 1891 and originally designated *Pithecanthropus erectus*' [< NL *Pithecanthropus*], ***plasmodium*** (pl. ***plasmodia***) 'a multinucleate mass of cytoplasm formed by the aggregation of a number of amoeboid cells, as that characteristic of the vegetative phase of the slime moulds; a protozoan of the genus *Plasmodium*' [< NL *Plasmodium*], ***plesiosaurus*** (pl. ***plesiosaurs*** / ***plesiosauri***) 'a large extinct marine reptile' [< NL *Plesiosaurus*], ***psylla*** 'any of various jumping plant lice of the family Psyllidae' [< NL *Psylla*], ***pyrrhuloxia*** 'a large, crested finch (*Pyrrhuloxia sinuata*)' [< NL *Pyrrhuloxia*], ***quelea*** 'an African weaverbird of the genus *Quelea*' [< NL *Quelea*], ***rasbora*** 'any of various tropical fishes of the genus *Rasbora*' [< NL *Rasbora*], ***rhea*** 'any of several flightless South American birds of the genus *Rhea*' [< NL *Rhea*], ***rhizopus*** 'any of various rot-causing fungi of the genus *Rhizopus*' [< NL *Rhizopus*], ***saccharomyces*** (pl. ***saccharomyces***) 'any of several single-celled yeasts belonging to the genus *Saccharomyces*' [< NL *Saccharomycēs*], ***salpa*** 'any of various free-swimming chordates of the genus *Salpa*' [< NL *Salpa*], ***simulium*** 'any of a genus (*Simulium*) of black flies' [< NL *Simulium*], ***sinanthropus*** 'an early member of an extinct species of human beings, considered a subspecies of *Homo erectus*' [< NL *Sinanthropus*, genus name], ***smilodon*** 'any of the large sabre-toothed tigers of the genus *Smilodon*' [< NL *smilodon*], ***solenodon*** 'a large ratline insectivorous mammal of the family Solenodontidae' [< NL *Solēnodon*], ***squilla*** *(pl.* ***squillas*** / ***squillae***) 'any of various burrowing predatory marine crustaceans of the order Stomatopoda' [< NL *Squilla*], ***stegodon*** 'any of various extinct elephant like mammals of the genus *Stegodon* and related genera' [< NL *Stegodon*], ***stegosaurus*** 'any of several herbivorous dinosaurs of the suborder *Stegosauria*' [< NL *Stegosaurus*], ***streptomyces*** (pl. ***streptomyces***) 'any of various actinomycetes of the genus *Streptomyces*' [< NL *Streptomyces*], ***syrphus (fly)*** 'any of numerous flies of the family Syrphidae' [< NL *Syrphus*], ***tachina (fly)*** any of several bristly, usually greyish dipterous flies of the family Tachinidae' [< NL *Tachina*], ***teredo*** (pl. ***teredos***) 'a shipworm of the genus *Teredo*' [< NL *Terēdo*], ***tilapia*** 'any of various cichlid fishes of the genus *Tilapia*' [< NL *Tilapia*], ***tortrix*** 'a moth of the family Tortricidae; a tortricid' [< NL *Tortrix*], ***torula*** (pl. ***torulae / torulas***) 'any of a group of fungi similar to the yeasts but lacking asci' [< NL *Torula*], ***treponema*** (pl. ***treponemata*** / ***treponemas***) 'any of a group of spirochetes of the genus *Treponema*' [< NL *Treponema*], ***triceratops*** 'a herbivorous dinosaur of the genus *Triceratops*' [< NL *Triceratops*], ***trichina*** (pl. ***trichinae*** / ***trichinas***) 'a small, slender parasitic nematode worm (*Trichinella spiralis*)' [< NL *Trichina*], ***trichomonad*** 'any of various flagellate protozoans of the genus *Trichomonas*' [< NL *Trichomonas, Trichomonad-*], ***tubifex*** (pl. ***tubifex*** / ***tubifexes***) 'any of various small, slender, reddish freshwater worms of the genus *Tubifex*' [< NL *Tubifex*], ***tyrannosaurus*** 'a large carnivorous dinosaur' [< NL *Tyrannosaurus*], ***uncinaria*** 'any of numerous small parasitic nematode worms of the family Ancylostomatidae' [< NL *Uncīnria*], ***vorticella*** (pl. ***vorticellae*** / ***vorticellas***) 'any of various ciliate protozoans of the genus *Vorticella'* [< NL *Vorticella*], ***zamia*** 'any of various chiefly tropical American cycads of the genus *Zamia*' [< NL *Zamia*], and ***zyzzyva*** 'any of various tropical American weevils of the genus *Zyzzyva*' [< NL *Zyzzyva*].

Species Names. Only 3 (3%) common names come from species names: ***cecropia (moth)*** 'a large North American silkworm moth (*Hyalophora cecropia*)' [< NL *cecropia*, species name], ***luna (moth)*** 'a large, pale-green North American moth (*Actias luna*)' [< NL *lūna*, species name], and ***vivax*** 'the protozoan (*Plasmodium vivax*); malaria caused by this protozoan' [< NL *vīvāx*, species name].

Common Names. Ten (10%) Latin common names designate different animal species in contemporary English: ***astragalus*** (pl. ***astragali***) 'the bone of the ankle that articulates with the tibia and fibula to form the ankle joint' [< NL *astragalus* 'vertebra'], ***cuscus*** 'any of several nocturnal marsupials of the genus *Phalanger*' [< NL *cuscus*, probably from a New Guinean word], ***eohippus*** 'a small, herbivorous, extinct mammal of the genus *Hyracotherium* (or *Eohippus*)' [< NL *eohippus* 'small horse'], ***glochidium*** (pl. ***glochidia***) 'the parasitic larva of certain freshwater mussels of the family Unionidae' [< NL *glochidium* 'a special larval stage'], ***gorilla*** 'the largest of the anthropoid apes (*Gorilla gorilla*)' [< NL *Gorilla* < Gk *Gorillai* 'a tribe of hairy women, perhaps of African origin'], ***Homo sapiens*** 'the modern species of human beings' [< NL *Homo sapiēns*: *Homo*, genus name + L *sapiēns*, present participle of *sapere* 'to be wise'], ***ichthyosaur(us)*** (pl. ***ichthyosaurs*** / ***ichthyosauri***) 'any of various extinct fishlike marine reptiles of the order Ichthyosauria' [< NL *ichthyosaurus*], ***motmot*** 'any of several tropical American birds of the family Momotidae' [< NL *motmot*, probably of imitative origin], ***platypus*** (pl. ***platypuses***) 'a semi aquatic egg-laying mammal (*Ornithorhynchus anatinus*)' [< NL *platypus* 'flat-footed'], and ***skua*** 'any of several large predatory sea birds of the genus *Catharacta*; any of several Arctic and Boreal sea birds of the genus *Stercorarius*' [< NL *skua*, alteration of Faroese **skūvur* < ON *skūfr* 'tassel, sea gull'].

English Formations

There are four types of formations in our corpus of words of Latin origin belonging to the **English of zoology**:

- **backformations** (new words formed by the removal of real or apparent affixes from existing words);
- **derivatives** (words formed from other words by a process of **derivation**, i.e. by addition of affixes such as prefixes or suffixes – Chalker & Weiner 1994);
- **compounds** (words formed by combining two or more bases or free morphemes – Chalker & Weiner 1994);
- **blend / portmanteau words** or **lexical / word blends** (words formed by the merging of parts of two other linguistic elements – Chalker & Weiner 1994).

Backformations. There are 152 (60%) **backformations** in our corpus: ***acarid*** 'an arachnid of the order Acarina' [< NL *Acaridae*, family name], ***amphipod*** 'a small crustacean of the order Amphipoda' [< NL *Amphipoda*, order name], ***annelid*** 'any of various worms or wormlike animals of the phylum Annelida' [< NL *Annelida*, phylum name], ***arachnid*** 'any of various arthropods of the class Arachnida' [< NL *Arachnida*, class name], ***arachnoid*** 'any of various arthropods of the class Arachnida' [< NL *arachnoidēs* < Gk *arakhnoeidēs* 'cobweb like'], ***arthropod*** 'any of numerous invertebrate animals of the phylum Arthropoda' [< NL *Arthropoda*, phylum name], ***artiodactyl*** 'any of various hoofed mammals of the order Artiodactyla' [< NL *Artiodactyla*, order name], ***bombycid*** 'a moth of the family Bombycidae' [< NL *Bombycidae*, family name], ***bovid*** 'a member of the family Bovidae' [< NL *Bovidae*, family name], ***brachiopod*** 'any of various marine invertebrates of the phylum Brachiopoda' [< NL *Brāchiopoda*, phylum name], ***braconid*** 'any of several ichneumon flies of the family Braconidae' [< NL *Braconidae*, family name], ***branchiopod*** 'any of various aquatic crustaceans of the subclass Branchiopoda' [< NL

Branchiopoda, subclass name], ***buprestid*** 'any of various beetles of the family Buprestidae' [< NL *Būprēstidae*, family name], ***canid*** 'any of various widely distributed carnivorous mammals of the family Canidae' [< NL *Canidae*, family name], ***carabid*** 'any of a large family (Carabidae) of chiefly black beetles' [< NL *Cārabidae*, family name], ***carangid*** 'any of a large family (Carangidae) of marine food and game fishes' [< NL *Carangidae*, family name], ***cephalochordate*** 'any of various primitive chordate animals of the subphylum Cephalochordata' [< NL *Cephalochordata*, subphylum name], ***cephalopod*** 'any of various marine molluscs of the class Cephalopoda' [< NL *Cephalopoda*, class name], ***cestode*** 'any of various parasitic flatworms of the class Cestoda' [< NL *Cestoda*, class name], ***chaetognath*** 'any of various marine worms of the phylum Chaetognatha' [< NL *Chaetognatha*, phylum name], ***chalicothere*** 'any of various extinct ungulate mammals of the Eocene to Pleistocene epochs' [< NL *Chalicotherium*, genus name], ***characid*** / ***characin*** 'any of numerous small, colourful tropical freshwater fish of the family Characidae' [< NL *Characinidae*, family name], ***chilopod*** 'any of various arthropods of the class Chilopoda' [< NL *Chīlopoda*, class name], ***chiropter*** 'a mammal that is a member of the order Chiroptera' [< NL *Chiroptera*, order name], ***chordate*** 'any of numerous animals belonging to the phylum Chordata' [< NL *Chordāta*, phylum name], ***cirriped(e)*** 'any of various crustaceans of the subclass Cirripedia' [< NL *Cirripedia*], ***coccid*** 'any of various insects of the superfamily Coccoidea' [< NL *Coccidae*, family name], ***coelacanth*** 'any of various mostly extinct fishes of the order Coelacanthiformes' [< NL *Coelacanthus*, genus name], ***colubrid*** 'any of numerous, widely distributed, chiefly nonvenomous snakes of the family Colubridae' [< NL *Colubridae*, family name], ***comatulid*** / ***comatula*** (pl. ***comatulids*** / ***comatulae***) 'any of various marine invertebrates of the class Crinoidea' [< NL *Comātulidae*, family name], ***copepod*** 'any of numerous minute marine and freshwater crustaceans of the subclass Copepoda' [< NL *Copepoda*, order name], ***creodont*** 'any of various extinct carnivorous mammals of the suborder Creodonta' [< NL *Creodonta*, suborder name], ***cricetid*** 'any of various small rodents of the family Cricetidae' [< NL *Cricetidae*, family name], ***crinoid*** 'any of various echinoderms of the class Crinoidea' [< NL *Crinoidea*, class name], ***ctenophore*** 'any of various marine animals of the phylum Ctenophora' [< NL *Ctenophora*, phylum name], ***cyclostome*** 'any of various primitive eel like vertebrates of the class Agnatha' [< NL *Cyclostomī* and *Cyclostomata*, class names], ***cyprinid*** 'any of numerous often small freshwater fishes of the family Cyprinidae' [< NL *Cyprīnidae*, family name], ***cyprinoid*** 'a cyprinoid fish' [< NL *Cyprīnoidēa*, suborder name], ***decapod*** 'a crustacean of the order Decapoda; a cephalopod mollusc' [< NL *Decapoda*, order name], ***dinothere*** 'any of various extinct elephant like mammals of the genus *Dinotherium*' [< NL *Dinotherium*, genus name], ***elapid*** 'any of several venomous snakes of the family Elapidae' [< NL *Elapidae*, family name], ***elasmobranch*** 'any of numerous fishes of the class Chondrichthyes' [< NL *Elasmobranchii*, subclass name], ***elaterid*** 'any of numerous beetles of the family Elateridae' [< NL *Elatēridae*, family name], ***ephemerid*** 'an insect of the order Ephemeroptera' [< NL *Ephemeridae*, order name], ***eurypterid*** 'any of various large, segmented aquatic arthropods of the order Eurypterida' [< NL *Eurypterida*, order name], ***ganoid*** 'a ganoid fish' [< NL *Ganoidei*, subclass name], ***gastropod*** 'any of various molluscs of the class Gastropoda' [< NL *Gastropoda*, class name], ***gastrotrich*** 'any of various minute aquatic animals of the phylum Gastrotricha' [< NL *Gastrotricha*, phylum name], ***geometrid*** 'any of various moths of the family Geometridae' [< NL *Geometridae*, family name], ***grackle*** 'any of several American blackbirds of the family Icteridae; 'any of several Asian mynas of the genus *Gracula*' [< NL *Grācula*, genus name], ***hadrosaur*** 'any of various amphibious dinosaurs of the genus *Anatosaurus* and related genera' [< NL *Hadrosaurus*, genus name], ***hexapod*** 'a six-legged arthropod of the class Insecta (formerly Hexapoda); an insect' [< NL *Hexapoda*, class name], ***hominid*** 'a primate of the family Hominidae' [< NL *Hominidae*, family name], ***hominoid*** 'a member of the Hominoidea' [< NL *Hominoidea*, superfamily name], ***ichthyosaur*** (pl. ***ichthyosaurs*** / ***ichthyosauri***) 'any of various extinct fishlike marine reptiles of the order Ichthyosauria' [< NL *ichthyosaurus*], ***isopod*** 'any of numerous crustaceans of the order Isopoda' [< NL *Isopoda*, order name], ***lagomorph*** 'any of various plant-eating mammals having fully furred feet and two pairs of upper incisors and belonging to the order Lagomorpha' [< NL *Lagomorpha*, order name], ***lamellibranch*** 'any of the bivalve molluscs of the class Lamellibranchia' [< NL *Lāmellibranchia*, class name], ***mantid*** 'any of various predatory insects of the family Mantidae' [< NL *Mantidae*, family name], ***megalosaur*** A gigantic carnivorous dinosaur of the genus *Megalosaurus*' [< NL *Megalosaurus*, genus name], ***megapode*** 'any of various large-footed, ground-dwelling birds of the family Megapodiidae' [< *Megapodius*, genus name], ***megathere*** 'a large extinct ground sloth of the family Megatheriidae' [< NL *Megatherium*, type genus], ***meloid*** 'any of various soft-bodied beetles of the family Meloidae' [< NL *Meloidae*, family name], ***monotreme*** 'a member of the Monotremata' [< NL *Monotremata*, order name], ***mosasaur*** 'any of various very large extinct aquatic lizards of the genus *Mosasaurus*' [< NL *Mosasaurus*, genus name], ***muscid*** 'a fly of the family Muscidae' [< NL *Muscidae*, family name], ***myriapod*** / ***myriopod*** 'any of several arthropods' [< NL *Myriapoda*, class name], ***mysticete*** 'any of several usually large whales of the suborder Mysticeti' [< NL *mysticētus* < Gk *mustikītos*, alteration of *(ho) mus to kītos* '(the) whale (called) the mouse'], ***nautiloid*** 'a mollusc of the subclass Nautiloidea' [< NL *Nautiloidea*, subclass name], ***nematode*** 'any of several worms of the phylum Nematoda' [< NL *Nēmatoda*, phylum name], ***noctuid*** 'any of numerous, usually dull-coloured night-flying moths of the family Noctuidae'[< NL *Noctuidae*, family name], ***nudibranch*** 'any of various highly colourful marine gastropods of the suborder Nudibranchia' [< NL *Nudibranchia*, order name], ***nummulite*** 'a large, coin-shaped, fossil foraminifer of the genus *Nummulites*' [< NL *Nummulītēs*, genus name], ***nymphalid*** 'any of various medium to large butterflies of the family Nymphalidae' [< NL *Nymphālidae*, family name], ***octopod*** 'any of various cephalopod molluscs of the order Octopoda' [< NL *Octopoda*, order name], ***odonate*** 'any of the large predacious winged insects of the order Odonata' [< NL *Odonata*, order name], ***oligoch(a)ete*** 'any of various annelid worms of the class Oligochaeta' [< NL *Oligochaeta*, class name], ***ophiuroid*** 'any of various marine organisms of the class Ophiuroidea' [< NL *Ophiūroidea*, class name], ***opisthobranch*** (pl. ***opisthobranchs***) 'any of various marine gastropod molluscs of the subclass or order Opisthobranchia' [< NL *Opisthobranchia*, order name], ***oscine*** 'a bird of the suborder Oscines' [< NL *Oscinīs*, suborder name], ***ostracod*** 'any of various minute, chiefly freshwater crustaceans of the subclass Ostracoda' [< NL *Ostracoda*, subclass name], ***pelecypod*** 'any of the bivalve molluscs of the class Lamellibranchia' [< NL *Pelecypoda*, class name], ***pentastome*** 'any of numerous tongue-shaped, soft-bodied, colourless invertebrates of the phylum Pentastoma' [< NL *Pentastomum*, genus name], ***percoid*** 'a fish belonging to the suborder Percoidea' [< NL *Percoidea*, suborder name], ***perissodactyl*** 'a hoofed mammal of the order Perissodactyla' [< NL *Perissodactyla*, order name], ***phasmid*** 'any of various insects of the order Phasmida' [< NL *Phasmida*, order name], ***phoronid*** 'any of the small, wormlike marine animals of the phylum Phoronida' [< NL *Phoronida*, phylum name], ***pinniped*** 'a mammal of the suborder Pinnipedia' [< NL *Pinnipedia*, order name], ***platyrrhine*** 'a platyrrhine monkey' [< NL *Platyrrhina*, group name], ***platy₂*** (pl. ***platys*** / ***platies***) 'any of several small freshwater live bearing fishes of the genus *Xiphophorus*' [NL *Platypoecilus*, genus name], ***platyhelminth*** 'any of various parasitic and nonparasitic worms of the phylum Platyhelminthes' [< NL *Platyhelminthes*, phylum name], ***plectognath*** 'any of various tropical marine fishes of the order Tetraodontiformes or Plectognathi' [<

NL *Plectognathī*, order name], ***plesiosaur*** [< NL *Plesiosaurus*, genus name], ***pogonophore*** 'any of various wormlike marine invertebrates of the phylum Pogonophora' [< NL *Pogonophora*, phylum name], ***polych(a)ete*** 'any of various annelid worms of the class Polychaeta' [< NL *Polychaeta*, class name], ***pongid*** 'an anthropoid ape of the family Pongidae' [< NL *Pongidae*, family name], ***pratincole*** 'any of several Old World shore birds of the genus *Glareola*' [< NL *prātincola* < L *prātum* 'meadow' + L *incola* 'inhabitant'], ***primate*** 'a mammal of the order Primates' [< NL *Prīmātes*, order name], ***psocid*** 'any of various small, soft-bodied, sometimes winged insects of the order Psocoptera' [< NL *Psocidae*, family name], ***pterodactyl*** 'any of various small, mostly tailless, extinct flying reptiles of the order Pterosauria' [< NL *Pterodactylus*, genus name], ***pteropod*** 'any of various small marine gastropod molluscs of the subclass Opisthobranchia' [< NL *Pteropoda*, order name], ***pterosaur*** 'any of various extinct flying reptiles of the order Pterosauria' [< NL *Pterosauria*, order name], ***pycnogonid*** 'any of various marine arthropods of the class Pycnogonida' [< NL *Pycnogonidae*, family name], ***pyralid*** 'any of numerous small or medium-sized moths of the diverse, widely distributed family Pyralidae' [< NL *Pyralidae*, family name], ***reduviid*** 'any of various predatory bugs of the family Reduviidae' [< NL *Reduviidae*, family name], ***rhizopod*** 'a protozoan of the phylum Rhizopoda' [< NL *Rhizopoda*, phylum name], ***rodent*** 'any of various mammals of the order Rodentia' [< NL *Rodentia*, order name], ***rotifer*** 'any of various minute multicellular aquatic organisms of the phylum Rotifera' [< NL *Rotifera*, phylum name], ***salp*** 'any of various free-swimming chordates of the genus *Salpa*' [< NL *Salpa*, genus name], ***saturniid*** 'any of various often large and colourful moths of the family Saturniidae' [< NL *Sāturniidae*, family name], ***satyrid*** 'a butterfly of the family Satyridae' [< NL *Satyridae*, family name], ***sauropod*** 'any of various large semi aquatic dinosaurs of the suborder Sauropoda' [< NL *Sauropoda*, suborder name], ***scarabaeid*** 'any of the numerous stout-bodied, lamellicorn beetles of the family Scarabaeidae' [< NL *Scarabaeidae*, family name], ***schistosome*** 'any of several chiefly tropical trematode worms of the genus *Schistosoma*' [< NL *Schistosoma*, genus name], ***schizopod*** 'any of various shrimp like crustaceans of the orders Euphausiacea and Mysidacea' [< NL *Schizopoda*, order name], ***sciurid*** 'a sciurid rodent' [< NL *Sciūridae*, family name], ***scolopendrid*** 'any of numerous centipedes of the family Scolopendridae' [< NL *Scolopendridae*, family name], ***scombroid*** 'a scombroid fish' [< NL *Scombroidei*, suborder name], ***scorpaenid*** [< NL *Scorpaenoidei*, suborder name], ***scorpaenoid*** 'a scorpaenoid fish' [< NL *Scorpaenoidei*, suborder name], ***serotine*** 'any of a widely distributed genus (*Eptesicus*) of usually small brown bats' [< NL *sērotinus*, species name], ***serranid*** 'any of various fishes of the family Serranidae' [< NL *Serranidae*, family name], ***silurid*** 'a silurid fish' [< NL *Silūridae*, family name], ***siphonophore*** 'any of various transparent, often subtly coloured marine hydrozoans of the order Siphonophora'[< NL *Siphonophora*, order name], ***sparid*** 'a member of the Sparidae' [< NL *Sparidae*, family name], ***sphingid*** 'a moth of the family Sphingidae; a hawk moth' [< NL *Sphingidae*, family name], ***staphylinid*** 'any of numerous beetles of the family Staphylinidae' [< NL *Staphylinidae*, family name], ***stomatopod*** 'any of various marine crustaceans of the order Stomatopoda' [< NL *Stomatopoda*, order name], ***strongly(e)*** 'any of various nematode worms of the family Strongylidae' [< NL *Strongylus*, genus name], ***syrphid*** 'any of numerous flies of the family Syrphidae' [< NL *Syrphidae*, family name], ***tabanid*** 'any of various bloodsucking dipterous flies of the family Tabanidae' [< NL *Tabānidae*, family name], ***tanager*** 'any of various small New World passerine birds of the family Thraupidae' [< NL *tanagra*], ***teiid*** 'a lizard of the family Teiidae' [< NL *Teiidae*, family name], ***teleost*** 'a teleost fish' [< NL *Teleostei*, group name], ***tenebrionid*** 'a beetle of the family Tenebrionidae' [< NL *Tenebrionidae*, family name], ***testudinate*** 'a turtle or tortoise' [< NL *Testūdināta*, order name], ***tetra*** 'any of numerous small, colourful tropical freshwater fish of the family Characidae' [NL *Tetragonopterini*, group name], ***tetrabranchiate*** 'a tetrabranchiate cephalopod' [< NL *Tetrabranchiāta*, order name], ***therapsid*** 'any of various reptiles of the order Therapsida' [< NL *Thērapsida*, order name], ***theropod*** 'any of various carnivorous dinosaurs of the suborder Theropoda' [< NL *Theropoda*, suborder name], ***thylacine*** 'a large wolf like carnivorous marsupial (*Thylacinus cynocephalus*)' [< NL *Thylacinus*, genus name], ***titanosaur*** 'any of various plant-eating, amphibious sauropod dinosaurs of the genus *Titanosaurus'* [< NL *Titanosaurus*, genus name], ***titanothere*** 'any of various extinct herbivorous hoofed mammals of the genus *Brontotherium* and related genera' [< NL *Titanotherium*, genus name], ***tortricid*** 'any of various small, thick-bodied moths of the family Tortricidae' [< NL *Tortricidae*, family name], ***trematode*** 'any of numerous flatworms of the class Trematoda' [< NL *Trematoda*, class name], ***trilobite*** 'any of numerous extinct marine arthropods of the class Trilobita' [< NL *Trilobītēs*, class name], ***trypanosome*** 'any of various parasitic flagellate protozoans of the genus *Trypanosoma'* [< NL *Trypanosoma*, genus name], ***tyrannosaur*** 'a large carnivorous dinosaur' [< NL *Tyrannosaurus*, genus name], ***urodele*** 'any of various amphibians of the order Caudata' [< NL *Ūrodēla*, order name], ***vespertilionid*** 'any of various widely distributed insect-eating bats of the family

Vespertilionidae' [< NL *Vespertilionidae*, family name], ***vespid*** 'any of various widely distributed social insects of the family Vespidae' [< NL *Vespidae*, family name], and ***viverrid*** 'a member of the Viverridae' [< NL *Vīverridae*, family name].

Derivatives. There are 101 (40%) suffixed **derivatives** of Latin origin in the **English of zoology**, and some of the affixes (*-al*, *-an*, *-ate*, etc.) seem to have specialised to designate 'animals':

- *-al* 'of, relating to, or characterized by' (1): ***brachyural*** 'a member of the Brachyura' [< NL *Brachyura*, suborder name + *-al*];
- *-an* 'of, relating to, or resembling' (75): ***acanthocephalan*** 'any of various worms of the phylum Acanthocephala' [< NL *Acanthocephala* + *-an*], ***acanthopterygian*** 'any of a large group of fishes of the super order Acanthopterygii' [< NL *Acanthopterygii* + *-an*], ***actinian*** (pl. ***actiniae*** / ***actinians***) 'a sea anemone or a related animal' [< NL *Actinia* + *-an*], ***amphibian*** 'a cold-blooded, smooth-skinned vertebrate of the class Amphibia; an animal capable of living both on land and in water' [< NL *Amphibia* + *-an*], ***annelidan*** 'any of various worms or wormlike animals of the phylum Annelida' [< NL *Annelida* + *-an*], ***anuran*** 'an amphibian of the order Salientia (formerly Anura or Batrachia)' [< NL *Anura*, order name + *-an*], ***aphidian*** [< NL *Aphis, Aphid-* + *-an*], ***arachnidan*** [< NL *Arachnida* + *-an*], ***ascidian*** 'any of various sedentary marine animals of the class Ascidiacea' [< NL *Ascidia* + -*an*], ***baluchithere*** 'a very large, extinct rhinoceros like mammal of the genus *Baluchitherium*' [< *Baluchi(stan)* + *-there* 'an extinct mammal'], ***brachyuran*** 'a member of the Brachyura' [< NL *Brachyura*, suborder name + *-an*], ***bryozoan*** 'any of various small aquatic animals of the phylum Bryozoa' [< NL *Bryozoa*, phylum name + *-an*], ***cephalopodan*** [< NL *Cephalopoda* + *-an*], ***cetacean*** 'any of various aquatic, chiefly marine mammals of the order Cetacea' [< NL *Cētācea* + *-an*], ***chelonian*** 'a reptile of the order Chelonia' [< NL *Chelonia* + *-an*], ***chiropteran*** 'a mammal that is a member of the order Chiroptera' [< NL *Chiroptera* + *-an*], ***cladoceran*** 'any of various small, mostly freshwater crustaceans of the order Cladocera' [< NL *Cladocera* + -*an*], ***coleopteran*** / ***coleopteron*** 'any of numerous insects of the order Coleoptera' [< NL *Coleoptera* + *-an*], ***crossopterygian*** 'a member of the Crossopterygii, a group of bony fishes ' [< NL *Crossopterygia* + -*an*], ***crustacean*** 'any of various predominantly aquatic arthropods of the class Crustacea' [< NL *Crustācea* + *-an*], ***dermapteran*** 'any of various insects of the order Dermaptera' [< NL *Dermaptera* + *-an*], ***dibranchiate*** 'a member of the order Dibranchiata' [< NL

Dibranchiata + *-ate*], ***dipnoan*** 'any of various fishes of the group Dipnoi' [< NL *Dipnoi* + *-an*], ***entomostracan*** 'any of various small crustaceans formerly constituting the subclass Entomostraca' [< NL *Entomostraca* + *-an*], ***foraminifer(an)*** (pl. ***foraminifera*** / ***foraminifers***) 'any of the chiefly marine protozoans of the order Foraminifera' [< NL *Forāminifera* + *-an*], ***heliozoan*** 'any of various aquatic protozoans of the order Heliozoa' [< NL *Heliozoa* + *-an*], ***hydrozoan*** 'any of numerous coelenterates of the class Hydrozoa' [< NL *Hydrozoa* + *-an*], ***hymenopter·an*** / ***hymenopteron*** 'an insect of the order Hymenoptera' [< NL *Hymenoptera* + *-an*], ***lepidopteran*** 'an insect belonging to the large order Lepidoptera' [< NL *Lepidoptera* + *-an*], ***mastigophoran*** 'any of various protozoans of the class Mastigophora' [< NL *Mastigophora* + *-an*], ***mecopteran*** 'any of various carnivorous insects of the order Mecoptera' [< NL *Mecoptera* + *-an*], ***megalosaurian*** [< E *megalosaur* + *-an*], ***metazoan*** 'a multicellular animal of the subkingdom Metazoa' [< NL *Metazoa* + *-an*], ***moneran*** 'a member of the kingdom Monera (or Prokaryotae)' [< NL *Monēra* + *-an*], ***monogenean*** 'any of various trematodes of the order Monogenea' [< NL *Monogenea* + *-an*], ***nemertean*** / ***nemertine*** 'any of several velvety, usually brightly coloured worms of the phylum Nemertina (or Nemertea)' [< NL *Nēmertēs* + *-an*], ***neuropteran*** 'a carnivorous insect of the order Neuroptera' [< NL *Neuroptera* + *-an*], ***nudibranchian*** [< NL *Nudibranchia* + *-an*], ***onychophoran*** 'any of numerous wormlike carnivorous animals of the phylum Onychophora' [< NL *Onychophora* + *-an*], ***ophidian*** 'a member of the suborder Ophidia or Serpentes; a snake' [< NL *Ophidia* + *-an*], ***ornithischian*** 'a dinosaur of the order Ornithischia' [< NL *Ornithischia* + *-an*], ***orthopteran*** / ***orthopteron*** 'an insect of the order Orthoptera' [< NL *Orthoptera* + *-an*], ***percoidean*** 'a fish belonging to the suborder Percoidea' [< NL *Percoidea* + *-an*], ***phylloxeran*** [< NL *Phylloxera* + *-an*], ***planarian*** 'any of various small, chiefly freshwater turbellarian flatworms of the order Tricladida' [< NL *Plānāria* + *-an*], ***platyrrhinian*** 'a platyrrhine monkey' [< NL *Platyrrhina*, group name], ***plecopteran*** 'any of numerous weak-flying insects of the order Plecoptera' [< NL *Plecoptera* + *-an*], ***pogonophoran*** 'any of various wormlike marine invertebrates of the phylum Pogonophora' [< NL *Pogonophora* + *-an*], ***polyzoan*** 'any of various small aquatic animals of the phylum Bryozoa' [< NL *Polyzoa* + *-an*], ***poriferan*** 'any of various members of the phylum Porifera'[< NL *Porifera* + *-an*], ***proboscidean / proboscidian*** 'a mammal of the order Proboscidea' [< NL *Proboscidea* + *-an*], ***prosimian*** 'a primate of the suborder

Prosimii' [< NL *Prosīmiī* + *-an*], ***protozoan*** / ***protozoon*** (pl. ***protozoa(ns)*** / ***protozoons***) 'any of a large group of single-celled, usually microscopic, eukaryotic organisms' [< NL *Protozoa* + *-an*], ***pteropodan*** [< NL *Pteropoda* + *-an*], ***radiolarian*** 'any of various marine protozoans of the order Radiolaria' [< NL *Radiolāria* + *-an*], ***rhizocephalan*** 'any of various small aquatic crustaceans of the order Rhizocephala' [< NL *Rhizocephala* + *-an*], ***rhizopodan*** [< NL *Rhizopoda* + *-an*], ***rhynchocephalian*** 'a rhynchocephalian reptile' [< NL *Rhynchocephalia* + *-an*], ***salientian*** 'an amphibian of the order Salientia' [< NL *Salientia* + *-an*], ***sarcodinian*** 'a protozoan belonging to the super class Sarcodina' [< NL *Sarcodīna* + *-an*], ***saurian*** 'any of various reptiles of the suborder Sauria' [< New L *Sauria* + *-an*], ***saurischian*** 'a dinosaur of the order Saurischia' [< NL *Saurischia* + *-an*], ***scyphozoan*** 'any of various marine coelenterates of the class Scyphozoa' [< NL *Scyphozoa* + *-an*], ***selachian*** 'a member of this order' [Probably < NL *Selachiī* + *-an*], ***sertularian*** 'any of various colonial hydroids of the genus *Sertularia*' [< NL *Sertularia* + *-an*], ***sirenian*** 'any of several large, cylindrical, herbivorous marine mammals of the order Sirenia' [< NL *Sīrēnia* + *-an*], ***sporozoan*** 'any of numerous parasitic protozoans of the class Sporozoa' [< NL *Sporozoa* + *-an*], ***suctorian*** 'a protozoan of the class Suctoria' [< NL *Suctoria* + *-an*], ***teleostean*** 'a teleost fish' [< NL *Teleostei* + *-an*], ***testacean*** 'any of various rhizopods of the order Testacea' [< NL *Testācea* + *-an*], ***theropodan*** [< NL *Theropoda* + *-an*], ***thysanuran*** 'a wingless insect of the order Thysanura' [< NL *Thysanura* + *-an*], ***trichopteran*** 'an insect of the order Trichoptera' [< NL *Trichoptera* + *-an*], ***turbellarian*** 'any of various free-living chiefly aquatic ciliate flatworms of the class Turbellaria' [< NL *Turbellāria* + *-an*], and ***xiphosuran*** 'an arthropod of the order Xiphosura' [< NL *Xiphosūra* + *-an*];

- *-ate* 'characterized by' (3): ***coelenterate*** 'any of various invertebrate animals of the phylum Coelenterata' [< NL *Coelenterata* + *-ate*], ***dinoflagellate*** 'any of numerous minute, chiefly marine protozoans of the order Dinoflagellata' [< NL *Dinoflagellata* + *-ate*], ***nudibranchiate*** [< NL *Nudibranchia* + *-ate*];
- *-iasis* 'a pathological condition characterized or produced by' (1): ***histomoniasis*** 'an infectious disease of turkeys and some wildfowl that is caused by a protozoan (*Histomonas meleagridis*)' [< NL *Histomonas*];
- *-ine* 'of or relating to' (5): ***australopithecine*** 'any of several extinct humanlike primates of the genus *Australopithecus*' [< NL

Australopithēcus + *-ine*], ***dryopithecine*** 'an extinct ape of the genus *Dryopithecus*' [< NL *Dryopithīcus* + *-ine*], ***gregarine*** 'any of various sporozoan protozoans of the order Gregarinida' [< NL *Gregarīna* + -*ine*], ***rusine (antler)*** 'an antler with three tines, as that of the sambar' [< NL *Rusa* + *-ine*], and ***viverrine*** 'a member of the Viverridae' [< NL *vīverrīnus* + *-ine*];

- *-ite* 'resident of' (1): ***termite*** 'any of numerous pale-coloured, usually soft-bodied social insects of the order Isoptera' [< NL *Termes* + *-ite*];
- *-(o)id* 'resembling; having the appearance of; related to' (10): ***acanthocephalid*** 'any of various worms of the phylum Acanthocephala' [< NL *Acanthocephal* + *-id*], ***chalcid*** 'any of various minute wasps of the superfamily Chalcidoidea' [< NL *Chalcis* + *-id*], ***chrysomelid*** 'any of various beetles of the family Chrysomelidae' [< NL *Chrysomela* + *-id*], ***cichlid*** 'any of various tropical and subtropical freshwater fishes of the family Cichlidae' [< NL *Cichla* + *-id*], ***clupeid*** 'any of various widely distributed soft-finned fishes of the family Clupeidae' [< NL *Clupeidae* + *-id*], ***gad(o)id*** 'a fish of the family Gadidae' [< NL *Gadus* + *-oid*], ***psyllid*** 'any of various jumping plant lice of the family Psyllidae' [< NL *Psylla* + *-id*], ***pyralidid*** 'any of numerous small or medium-sized moths of the diverse, widely distributed family Pyralidae' [< NL *Pyralidae* + *-id*], ***sciaenid*** [< NL *Sciaena* + *-id*], and ***sciaenoid*** 'a sciaenoid fish' [< NL *Sciaena* + *-id*];
- *-ous* 'possessing; full of; characterized by' (1): ***brachyurous*** 'a member of the Brachyura' [< NL *Brachyura*, suborder name + *-ous*].

It is not very clear what the ending *-y* means, since none of the meanings supplied by language dictionaries fits the following **derivatives** (4 occurrences): ***cavy*** (pl. ***cavies***) 'any of various tailless South American rodents of the family Caviidae; 'any of various similar or related rodents' [< NL *Cavia*, genus name + *-y*], ***platyrrhiny*** [< NL *Platyrrhina* + *-y*], ***saury*** (pl. ***sauries***) 'any of several offshore marine fishes of the family Scomberesocidae' [< NL *saurus* + *-y*], and ***tody*** (pl. ***todies***) 'any of various small birds of the family Todidae' [Probably < F *todier* + *-y*].

Compounds. There is a single **compound** (0%) among the words of Latin origin in the **English of zoology**: ***tilefish*** (pl. ***tilefish / tilefishes***) 'a reddish-blue percoid marine food fish (*Lopholatilus chamaeleonticeps*)' [*Tile-* (short for < NL *Lopholatilus*, genus name) + FISH].

Portmanteau words. The single (0%) portmanteau word in our corpus is ***merganser*** 'any of various fish-eating diving ducks of the genus *Mergus* or related genera' [< NL < L *mergus* 'diver' + L *ānser* 'goose'] (Figure 6-2).

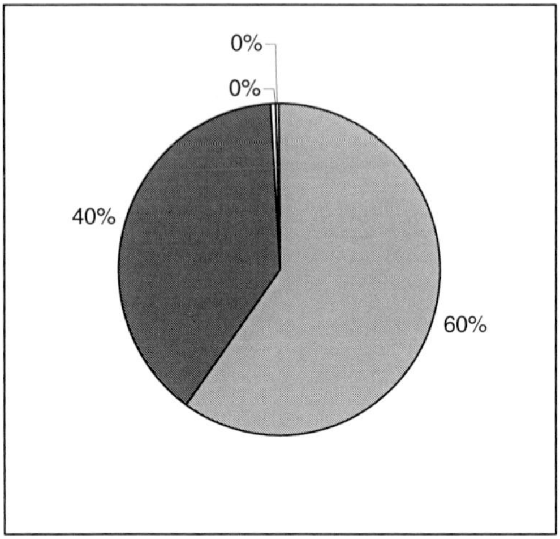

Figure 6-2. English formations in the **English of Zoology**: 60% backformations, 40% derivatives, 0% compounds, 0% portmanteaus

Discussion

It is interesting to note that the share of **English formations** (73%) of Latin origin in the **English of zoology** is almost three times larger than that of **metonymic words** (27%). This could be explained by the fact that, on the one hand, they are old formations, and in time even scientists began to perceive them more and more as English words rather than as **loanwords**. The large number of **backformations** (60%) confirms this hypothesis: perceived as foreignisms, Latin words were slightly 'adapted' to the English spelling system. The large number of **derivatives** (40%) also supports the hypothesis, since **derivation** occurs only within the system, never outside it. Moreover, certain affixes, such as *-an*, are specific to animal names (75% of the **derivatives** contain it). **Compounds** and **portmanteau words** are, practically, negligible.

As for **metonymic words**, they come from *genus names* (87%), from *Latin common names* (10%), and from *species names* (3%). The large share of animal names that have come from genus names shows the fact that scientific discovery in the field of zoology reached, in the 17th and 18th centuries, the highest level ever.

Conclusions

Results show that the **English of zoology** contains good examples of Latin borrowings that have become fully parts of the English vocabulary, despite the fact that they retain traces of their foreign origin in their pronunciation, spelling, or inflection (see the plural forms of the nouns supplied above).

Our hypothesis that there must be a pattern in the naming of animals with Latin names is confirmed: the proof – the large number of English formations (particularly **backformations** and **derivatives**) representing 73% of the total of words of our corpus, on the one hand, and the large number of animal names derived from genus names (87% of the total metonymic names).

The implications of the research and results are considerable: they can be useful to both students in natural sciences and teachers and/or researchers in the field of zoology.

Additional research should focus on the words of Latin origin in the field of Botany.

References

Chalker, Sylvia & Weiner, E. (1994). *The Oxford Dictionary of English Grammar*. London - New York - Sydney - Toronto: BCA.

Farrell, J. (2004) *Latin Language & Latin Culture. From Ancient to Modern Times*. Cambridge: Cambridge University Press.

Glare, P. G. W. (Ed.). (1983). *Oxford Latin Dictionary*. Oxford: Oxford University Press.

Harper, D. (2001). *Online Etymology Dictionary*. Online: http://www.etymonline.com.

Janson, T. (2002). *Speak. A Short History of Languages*. Oxford: Oxford University Press.

Latin Dictionary Headword Search Result. Online: http://www.perseus.tufts.edu/hopper/resolveform?redirect=true&lang=Latin.

List of Latin Words with English Derivatives. Online: http://en.wikipedia.org/wiki/List_of_Latin_words_with_English_derivatives.

Online Etymology Dictionary. Online: http://www.etymonline.com/.

Partridge, E. (2006). *Origins. A Short Etymological Dictionary of Modern English*. London – New York: Routledge.

Rata, Georgeta. (2008). Words of Latin Origin in the English of Zoology. *Journal of Linguistic Studies 2 (2008)*: 75-86.

Sheehan, M. J. (2000). *Word Parts Dictionary. Standard and Reverse Listings of Prefixes, Suffixes, Roots and Combining Forms*. Jefferson (NC) & London: McFarland & Company, Inc., Publishers.

Shipley, J. T. (1979). *Dictionary of Word Origins*. Totowa, NJ: Littlefield, Adams & Co.

The American Heritage Dictionary of the English Language. (2008). Boston: Houghton Mifflin.

Word Study Tool / Word Analysis. Online: http://latinlexicon.org/word_study_tool.php.

CHAPTER SEVEN

A COMPARATIVE APPROACH

AGRICULTURAL TERMINOLOGY IN MULTICULTURAL COMMUNICATION

ANDREEA VARGA

Terminology as an integrating part of a language constitutes a living organism which is constantly liable to alterations and enrichment. From this vantage point, it is a tool endorsing information, research, science and scientific culturem and, nonetheless, it is a means of higher qualification for an individual in his profession. Perforce, we cannot eschew the paramount importance of learning terminology in the context of the European Union standardization of scientific language, but also against the background of a European Union nation, such as Romania, in which technical language becomes a prerequisite for the employment in international multilevel enterprises or for any other foreign investment. Thus, terminology becomes a *sui generis* mechanism encompassed harmoniously in multicultural communication. We refer below to the domain of agriculture *stricto sensu* and to the difficulties emerging in the process of learning this particular scientific language.

Terminology as defined by Daniel Gouadec (1990) refers to a scientific discipline whose main focus is the system of specialized linguistic designations, their content, the nature of the interrelations between the constituents of this system, the relation between the concept on the one hand and its representation on the other, the tenets tackling the specialized representations accordingly, and their applicability in the language and professional activities. The linguistic unity for terminology is the term denoting the representation of a concept. Some specialists construe terminology as a subsystem of the lexicon pertaining to a language, eliciting the similar features of the terminological units and the lexical units. However, the technical language presupposes some idiosyncratic traits such as being accurate, univocal, unambiguous (the relationship between the representation and the concept is reflexive, each concept has only one pendant term; albeit it does not comprise the whole meaning of the concept it undoubtedly aspires to), interdependent (the terms in a scientific field are interconnected, related to each other). Nonetheless, other proponents disinter terminology as a science *per se*, as science about science: “Une science est bien d’abord une terminologie” (A science is first of all a terminology) (Bernard Quemada, in Gouadec 1990), stemming in linguistics (the representations are tantamount to the signifier while the value of the concept is equal to the signified) and ontology, the study of concepts and systems of concepts. However, if we are to minify terminologies, they can be defined as specialized vocabularies.

The coherence of the representation systems which lay the foundation of terminologies is determined by the particularity of the field (conceptual field and usage of the representations), of the object or of a process, of the referent or of a category of referents, and by the particularity of a given corpus (the terminology of a text or of a corpus of documents). However, the terminology of a given corpus is immutably circumscribed, while coherence is engendered by applying the corpus. Circumscription is subdued to subjectivity: since the conceptual fields entwine, *a fortiori*, some boundaries have to be imposed regarding what is considered to pertain to that lexical network.

The concept of standardization is corollary to the need of imposing clear boundaries within special semantic fields and communicating at the same level on an international scale. The concept is ambivalent on the basis of the fact that it comprises technical standards, also known as specification standards, and terminology standards. Technical standards expound a process of concurring upon the technical specifications and other criteria employed as rules, guidelines or definitions of features to ensure that materials, products, services, and processes are interlocking and interoperable. From the vantage point of the international communication level, the foregoing goals are achieved *via* consensus agreements between national delegations and other institutions which need to approve of specifications, criteria applied in the classification, manufacture and supply of materials, testing and analysis, terminology and provision of services. Regarding terminological applicability, standardization establishes which technical terms will be employed in a standard and underscores the criteria by means of which the term needs to be understood. This enables standardizing groups to use the necessary tools to draft terminologically and conceptually pertinent standards. What is more, terminology standards differentiate themselves from technical standards in so far as they do not provide specifications for a product, service or process: they designate the terms along with the definitions to facilitate the understanding of the key concepts in a given standard. ISO/IEC, one of the main organizations demarcating standardization, purports the standard as being “a document, established by consensus and approved by a recognized body, that provides, for common and repeated use, rules, guidelines or characteristics for activities or their results, aimed at the achievement of the optimum degree of order in a given context." (*Standardisation*. Online: http://en.wikipedia.org/wiki/Standardization)

Other concepts subsuming under the umbrella of standardization in terminology are harmonization and officialization. The former advances a process of aligning terms and definitions between languages or between variants within a language carried into effect by means of recommendations since this process is divested of the power entailed by a standard. Officialization, on the other hand, represents a process performed by an authoritative body which selects a term or a name and invests it with an official status.

Furthermore, the concept of databank is of paramount importance in the context of standardization and on the background of a technology evolving at such a rapid pace at present. If the first systems of terms date back in the 18th century, when biologist Carl von Linné wrote his work *Systema Naturae*, expounding his binominal system positing animals and plants, and continued with Antoine Laurent de Lavoisier in the field of chemistry, terminology has developed by leaps and bounds due to the impetus of the 20th century when the first efforts of technical language standardization were at an inceptive stage. Currently, there are informatized and structured systems of data collection sorting out, using and rendering the terminological databases available. The banks of data comprise terminological catalogues in which all of the enlisted terms can be easily identified in valid contexts due to the fact that the source which employed them is mentioned.

At a national level, terminology touchstones are immutably espoused in the national language(s), thus the standards become monolingual. However, multilingual standards, which are often indispensable, are adopted by each country and are carried into practice by means of equivalence, definitions in the national language or simply by borrowing the term which will subsequently be assimilated by the source language. To further proceed with our analysis, we expound the techniques of equivalence and borrowings applied to the field of agriculture.

Equivalence, defined under a larger scope, constitutes the result of a translation process revolving around the identity relation between two terms pertaining to two different languages and possessing almost the same connotation and denotation. Reduced to a more narrow meaning, equivalence is a translation technique which presupposes rendering a term or set phrase in the source language into a different term in the target language, but preserving the correspondence with the same reality. Equivalences are established between texts integrating in a certain culture, in a particular communication context: they are perceived as consequences of the interaction between the translator and the text. Several types of equivalence can be distinguished:

- cognitive (engendered by the semantics of the text and the knowledge brought forth by the translator, based on real or unreal experiences);
- emotional (intuitive);
- denotative (conveying the information advanced by the source language text);
- connotative (preserving the style, the language register, the sociolect, the geographical outreaching of the expressions);
- standard or normative (according to the genre of the text for translation, for example, a textbook, an advertisement, etc.);
- pragmatic or dynamic (adapted to the knowledge of the reader in order to be comprehended).

An eloquent example of equivalence is the translation of the term ***agribusiness*** from English into Romanian. The equivalent in the target language is, *de facto*, a definition of the term 'the practice of agriculture and the entire activities pertaining to agriculture (the manufacturing and distribution of farm equipment and supplies and the processing, storing, shipping, and marketing of agricultural products)'. Although the **combining form *agri-*** preserved from the Latin word *agricola* which further on roots in *ager* 'field' can be identified in Romanian as well (e.g., ***agricol***) or in French (e.g., ***agricole***), it cannot be translated in combination with the word ***business***. Consequently, the equivalence is rendered by means of defining the term. A similar case is that of ***county / agricultural agent***, which does not entail a correspondent in Romanian and is defined as 'government representative whose main concern is to counsel the farmers in matters of agriculture'. A term which has an equivalent both in Romanian and in French is ***fruit farming*** having the correspondents ***pomiculture*** in French and ***pomicultură*** in Romanian. In this case, however, the English language did not inherit the Latin *pomum* 'fruit', but the Latin *fructus* meaning 'something that proffers sweetness, satisfaction, pleasure'. A most conspicuous example is that of the term designating a spice, ***rosemary***, which has the equivalent of ***romarin*** in French and ***rozmarin*** in Romanian. The particularity resides in the fact that although the terms elicit the same object and resemble phonetically, they posses different meanings. The English term is preserved as a name ***Rose-Mary***, while the French and the Romanian equivalents stem in the

Latin *rosmarinus* 'dew of the sea', and constitute a concoction of the adjectives *ros* 'dew' and *maris* 'sea'.

Borrowings are also frequent in the field of agriculture. Except for the Greek and Latin terms inherited by the aforementioned languages, there are terms designating animal breeds or species which have been preserved in their inceptive form. One such example is ***chinchilla***, the name of an animal raised for its fur, original from South America. The term is Spanish and was traced back to a dialect pertaining to an Indian Peru tribe; however, it was borrowed as such in all languages. Another such example is the ***Holstein*** breed of cattle having its denomination in the German duchy bearing the same name.

From the foregoing demonstration, it is foreseeable that learning terminology is not a facile process and the multifarious problems emerging need to be analyzed and solutions espoused. Students might encounter difficulties in decoding the written texts at least in an inchoate stage. They are liable to produce errors due to the abusive extension of lexical structures and semantic sphere of the terms in the source language. Consequently, students might be liable to invent inexistent terms or they might just have a silent reaction due to lack of courage and insecurity. Thus, teaching technical language should yield a gradual process of familiarization with the fundamental structures. When introducing the new vocabulary, students should be provided with the keywords which impede the global understanding of the text. After having understood the keywords, the rest of the vocabulary can be inferred from the context. Thus, the student should be able to produce descriptive sentences regarding the acquired technical issues in his field, construct short dialogues on a given subject or write short summaries of the studied texts, the ability to reconstruct the informational message of a technical text by means of identifying some essential guidelines. The subsequent stage in teaching terminology constitutes an evolution towards more technical, specialized texts. Consequently, students can be taught terminology through methods of semantization which can be direct comprising extralinguistic methods – images – and linguistic methods – synonyms, antonyms, formal structure. The indirect methods make reference to the technique of translation. Thus, students should be able to comprehend the entire text and thus provide equivalents from the source language to the target language. Any future specialist should easily deal with technical documents in foreign languages. The final purpose of learning terminology is to acquire oral comprehension skills of the technical information, to write reports, to deliver speeches, to build international relations of cooperation. However, the student should ferret out that terminology falls within a corpus of words, that it is an entire mechanism which should be understood as such and which should be used as an instrument in his field.

On balance, employing terminology as a means of communication at a multicultural level and its interlocking operating concepts of standardization and bank of data has become a prerequisite in the field of technology *per se*. The institutions working in this sense, improving the standards and keeping pace with the language alterations and technological novelties are a paragon for communication as a live and permanent process.

References

Gouadec, D. (1990). *Constitution des données*. [Constitution of Data]. Paris: Afnor. Online: http://www.gouadec.net/publications.html.

Lungu Badea, G. (2003). *Dictionary of Terms Employed in the Theory, Practice and Didactics of Translation*. Timişoara: Ed. Orizonturi.

Raţă, Georgeta. (1999). *Curs practic de limba engleză – profilul agricol (anul al II-lea de studiu)*. [A Practical Course of English – Agriculture (2nd year)]. Timişoara: Editura Mirton.

Standardisation. Online: http://en.wikipedia.org/wiki/Standardization.

Varga, Andreea. (2007). Terminology in the Context of Multicultural Communication. *Proceedings of Lingua Summit 2007, Trenčin, Slovakia*: 4 p.

Vintilă-Rădulescu, I. (1999). *Terminology and Its Current Problems*. Bucureşti: Editura Academiei Române.

ROMANIAN AGRICULTURAL TERMS AND THEIR ENGLISH EQUIVALENTS

OANA BOLDEA

Introduction

When dealing with the terminology of a certain field, it is sometimes not an easy task to translate the terms accurately, whether the translation is from Romanian into English or the other way around. The task becomes ever more difficult when the terms are not formed by one word only, but when we have to work with phrases, complex terms. These need our closer attention, or else we run the risk of not only giving the wrong equivalents, but also of coming up with ridiculous ones.

However, giving the perfect equivalent is not a matter of guessing; if we look closely at the way the phrases are made up, in general, we can notice that there are some rules governing their formation. By noticing these rules and by taking them into consideration, we can supply the perfect equivalent in the language we need to convey the message.

It is the purpose of this paper to identify the possible rules which govern phrase formation in Romanian as well as in English. We believe that emphasizing them is helpful for the linguists, but also for the Romanian specialist in agriculture who, at one point, may find himself/herself in the position of having to convey a message in English with no dictionary at hand at the time.

We start from the premise that Romanian uses more words to name the same thing than English does. We base our assumption on the fact that Romanian is a synthetic language, while English is an analytic one. Hence, Romanian makes use of inflections and prepositions, while English prefers juxtaposition. It remains to see whether the analysis proves our assumption to be right or wrong.

Material and Method

For the purpose of our research, we chose a corpus found in *Fodder Crops and Agriculture* (Raţă *et al.* 2002), a compendium of agricultural technical terms in Romanian, English and French. As French is not of interest for the present study, all the terms in this language have been left aside. We only worked with English and Romanian terms.

More precisely, we had 1,900 terms in view for the study, grouped in the chapter "general vocabulary". Thus, we have not taken into consideration the terminology in the chapter dedicated to vocabulary having to do with botany, nor have we dealt with the names of plants occurring in the third chapter of the book. Our interest lay exclusively on the second chapter, already mentioned, tackling so-called "general" terminology.

The method used is that of comparative lexical analysis. Also, we used the quantitative method to establish the frequency with which a certain construction occurs. And then, on the basis of word count, we identified the general rule or trend in phrase formation both in Romanian and in English.

Some books were used for the theoretical background of the paper. Thus, Crystal's *Linguistics* (1990) proved extremely useful. The same can be said about Matthew's *Morphology* (1993). However, books by Romanian researchers were equally useful, for instance, Ciobanu's *Adaptation of the English Element in Romanian* (1997).

Results and Discussion

After analyzing the 1,900 entries in the corpus, we found 751 terms made up of more than one word. One cannot but notice that out of these, 349 phrases contain a noun and an adjective in Romanian. The interesting part is that many of them have English equivalents in which no adjective is found. For example:

- R *argilă aluvionară* = E ***silt loam***;
- R *comunitate vegetală* = E ***plant community***;
- R *defoliere estivală* = E ***summer grazing***;
- R *diminuare cromatică* E ***colour reduction***.

Thus, in 173 cases, English displays a clear preference for "Noun + Noun" constructions, as opposed to Romanian, which favours the "Noun +

Adjective" construction. This means that, in almost half the cases, English will use the "Noun + Noun" construction.

Going further with our analysis, we find out that, out of the same number of phrases (751), 415 of the Romanian ones contain a preposition. The prepositions we found were *cu* 'with', *de* 'of', *din* 'from', *în* 'in', *la* 'to', *pe* 'on', 'per', *pentru* 'for', and *prin* 'through'. Generally speaking, the phrases are made according to the same rule, namely "Noun + Preposition + Noun" (expressions such as *infestare* ***cu*** *buruieni* 'weed infestation', *îngrăşământ* ***de*** *primăvară* 'spring fertiliser', *irigare* ***prin*** *inundare* 'flooding irrigation', *uscat* ***la*** *aer* 'air-dried', *zonă* ***de*** *păşunat* 'grazing area', etc.).

Still, there are some cases (all involving the preposition *în* 'in') where the phrases are made up either by "Adjective + Preposition + Noun" or "Participle + Preposition + Noun", as in *bogat* ***în*** *frunze* 'leaf rich'. However, these cases are extremely rare (only 3 occurrences in 415 phrases, i.e. less than 1%). That is why we can conclude that the general feature is "Noun + Preposition + Noun", in what the Romanian phrases containing prepositions are concerned.

Table 7-1 presents the findings regarding the phrases containing prepositions in Romanian and English, comparatively.

Table 7-1. Prepositions and their occurrence in agricultural terms

Romanian preposition	Number of Romanian phrases containing the preposition	Number of phrases containing a preposition in the English equivalent
cu 'with'	27	-
de 'of'	311	3
din 'from'	5	-
în 'in'	34	-
la 'to'	18	-
pe 'on', 'per'	5	1
pentru 'for'	5	2
prin 'through'	10	-

From the table, we notice that the preposition *de* 'of' occurs in most Romanian phrases, namely in 75% of the cases. It is, by far, the most often used preposition, as it results from the analysis. The other two prepositions occurring in a fair share of phrases are *în* 'in' and *cu* 'with', but they are far behind the abovementioned preposition, occurring in only 8.19% and 6.50% of the cases, respectively. The fourth most frequent preposition is *la* 'to' (occurring in 4.33% of the phrases), followed by *prin* 'through' (2.40%). The other prepositions used in Romanian phrases concerning agriculture are *din* 'from', *pe* 'on', 'per' and *pentru* 'for', each occurring only 5 times, i.e. in 1.20% of the cases (Figure 7-1).

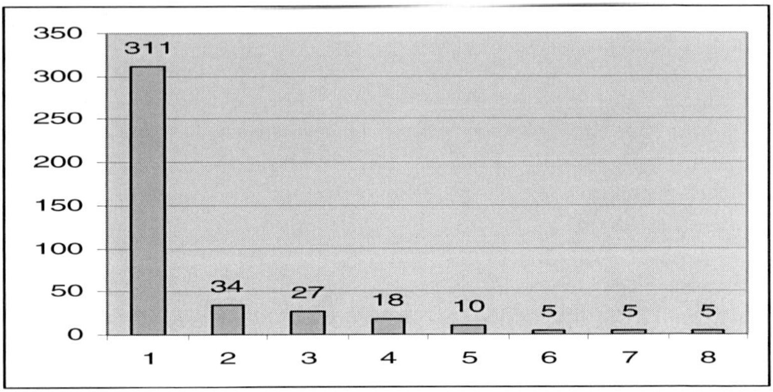

Figure 7-1. Prepositions and their occurrence in agricultural terms: 311 *de*, 34 *în*, 27 *cu*, 18 *la*, 10 *prin*, 5 *din*, 5 *pe*, 5 *pentru*

If we look at the third column of Table 7-1, namely that which contains data referring to the occurrence of prepositions in the English equivalents of the Romanian phrases discussed above, we can see that we have only 6 phrases containing prepositions, i.e. a little over 1% of the Romanian phrases. The English prepositions found here are *of* (occurring 3 times, in equivalents for Romanian terms containing preposition *de*), *for* (which occurs 2 times in equivalents for Romanian terms containing preposition *pentru*), and *per* (occurring only 1 time, in an equivalent for a Romanian term containing preposition *pe*).

As only 1% of the Romanian terms containing prepositions have equivalents similarly formed in English, one is entitled to say that making up phrases by means of prepositions is not the general rule for phrase formation in English. Instead, we can notice that all English phrases (except those very few already discussed) are of the "Noun + Noun" type. Thus, Romanian phrases have the following English equivalents:

- R *câmp **de** ameliorare* = E ***improvement field***;
- R *floră **de** pajişte* = E ***meadow flora***;
- R *indice **de** nutriţie*) = E ***nutrition index***;
- R *izolare **în** hibridare* = E ***hybridation isolation***;

- R *rație* ***de*** *furaj* = E ***feed intake***;
- R *reacție* ***la*** *fertilizare* = E ***fertilizer response***;
- R *ritm* ***de*** *producție* = E ***production rate***;
- R *vaci* ***de*** *lapte* = E ***dairy cows***;
- R *zonă* ***de*** *restituire* = E ***restitution area***.

Another aspect of our study was related to the Genitives in Romanian, i.e. we wanted to find the equivalents for those Romanian phrases that contain genitive forms. We were interested in seeing whether the English terms also make use of the genitive or not. The findings are inciting: of the 144 Romanian terms using genitive forms, only 3 have English equivalents which use a genitive also, be it the "'s genitive" or the "of genitive". Thus:

- R *greutatea lăstarului* = E ***tiller's weight***;
- R *sănătatea animalului rumegător* = E ***ruminant's health***;
- R *umiditatea solului* = E ***soil's moisture***.

Given the fact that only 2% of the Romanian phrases containing genitives have English equivalents formed in the same way, we can say that, as a rule, the English expressions avoid using the genitive, as opposed to Romanian ones.

Conclusions

The results of the research confirmed our hypothesis. We started from the initial idea that English uses fewer words to express a concept than Romanian does. Thus, we analyzed a corpus of 1,900 agricultural terms in order to get an idea about how they are made up, according to what rules, and, at the same time, we wanted to establish whether the English equivalents for these terms were formed according to the same rules.

Much as we had predicted, the English phrases proved to use a lot less prepositions and genitive forms than the Romanian ones. The structure is less periphrastic: the preferred structure seems to be the "Noun + Noun", avoiding genitives and prepositions as much as possible.

References

Boldea, Oana. (2005). On some Romanian Agricultural Terms and Their English Equivalents. *Scientific Papers. Faculty of Agriculture, XXXVII*: 677-681.

Ciobanu, Georgeta. (1997). *Adaptation of the English Element in Romanian*. Timişoara: Editura Mirton.

Crystal, D. (1990). *Linguistics*. London: Penguin Books.

Leech, G. (1996). *An A – Z of English Grammar & Usage*. London: Longman.

Matthews, P. H. (1993). *Morphology*. Cambridge: Cambridge University Press

Raţă, Georgeta, Moisuc, Al. & Samfira, I. (2002). *Fodder Crops and Agriculture*. Timişoara: Editura Orizonturi Universitare.

Raţă, Georgeta, Samfira, I., Diana-Andreea Boc-Sânmărghiţan & Butnariu, Monica. (2012). *Compendium of Forage Technical Terms in English, French and Romanian*. Newcastle upon Tyne: Cambridge Scholars Publishing.

Swan, M. (1995). *Practical English Usage*. Oxford: Oxford University Press.

Webster Comprehensive Dictionary. (1995). Chicago: J. G. Ferguson Publishing Company.

ENGLISH LOANWORDS IN THE ROMANIAN LAND IMPROVEMENT VOCABULARY

GEORGETA RAȚĂ AND ANICA PERKOVIĆ

Introduction

One of the most outstanding features of Modern Romanian is its large and varied vocabulary. Romanian has borrowed many words from many other languages and made them its own. In the modern period, a large number of new affixes have been taken from French, Latin, Greek, and a few others. Beginning with the middle of the 19th century, and still going apace, is the fertile coinage of scientific, technical, scholarly, commercial, and political terms that have swelled the Romanian vocabulary to such proportions that even the largest unabridged dictionaries can select and define only a small portion of the total number of words that actually exist.

The Romanian language has taken over a large number of words from English. These **borrowings** or **loanwords** have, nevertheless, become a permanent part of Romanian language. Most of them have been modified to bring them into line with the phonological rules of Romanian, and as such they can help a non-native English speaking Romanian or they can, on the contrary, confuse them. This is also the case of the **land improvement vocabulary**.

Material and Method

Our corpus of words was selected from Pricop & Marcu's *Dictionary of Land Improvement* (1987-1989). Of the 4,801 entries in the dictionary, only 30 are **loanwords** of English origin, a rather small number if we compare it with French **loanwords** (378) or German **loanwords** (80).

Results and Discussion

Loanwords can attain different degrees of assimilation into the language. Linguists (Chalker & Weiner 1994) have identified four types of assimilation and, consequently, four groups of **loanwords**: *totally* *assimilated words*, *words retaining traces of their origin*, *words assimilated in their form but semantically tied to a foreign context*, and *words which have not yet achieved general currency*. All the English **loanwords** in our corpus must have fallen, initially, into the last category, but they now belong to the following groups:

1. **Totally assimilated words**, i.e. *words assimilated to the native word-stock and that are phonetically and orthographically integrated*: R ***badland*** < E *badlands* 'a barren area characterized by numerous ridges, peaks, and mesas cut by erosion' (W); R ***buldozer*** < E *bulldozer* '1. A powerful, tractor-driven machine equipped with a heavy steel blade: used for clearing wooded areas and moving soil in road construction' (W); R ***caterpillar*** < E *Caterpillar* 'a tractor whose driving wheels gear with self-laid, wide metal belts whereby its weight is distributed over a large area, permitting the tractor to move over soft or rough terrain; a trade name' (W); R ***dispecer*** < E *dispatcher* '2. One who directs the movements of trains, trucks, etc., and maintains records of such movement' (W); R ***macroelement*** < E no mention in any of the two English language dictionaries mentioned in References; R ***mulci*** < E *mulch* 'any loose material, as straw, placed about the stalks or plants to protect their roots' (W); R ***prefiltrare*** < E *pre-filtering*, for which the *Explanatory Dictionary of the Romanian Language* mentions "after English"; R ***screper*** < E *scraper* '2. A horse-drawn or motor-driven apparatus having a large metal scoop or scoops, for scraping up, transporting, and dumping dirt' (W); R ***stoc*** '1. water supply. 2. materials stored' (DIF) < E *stock* '2. A quantity of something accumulated, as for future use' (RHDEL) or '10. A quantity of something acquired or kept for future use; 14. Raw material' (W); R ***stres*** < E *stress* '3. Mech. A force or system of forces which tends to produce deformation in a body on which it acts' (W); R ***termogramă*** < E, F, or G, but without any mention in the English language dictionaries cites in References. To note that the trade name *Caterpillar* has become a common name in Romanian. Other **loanwords** are not mentioned in the Romanian explanatory dictionary, which makes it impossible for us to trace their origins. Thus, R ***bermă*** might come from either the E *berm(e)* or from the F *berme*. As for R ***conveyer***, it presumably comes from the E *conveyer*, but the latter does not occur as such in Romanian language dictionaries or in English language dictionaries, where only E *conveyor* is mentioned. Some of these **loanwords** have been assimilated so well that they have produced **derivatives** in Romanian. Though not mentioned by the Romanian language dictionary mentioned in References, they do exist and they are used in land improvement: R *dispecerizare* < R ***dispecer***, R *dispecerat* < R ***dispecer***, and R *mulcire* < R ***mulci***.

2. **Words retaining traces of their origin**, i.e. *words fully part of the Romanian vocabulary, but that retain traces of their foreign origin in their pronunciation, spelling, or inflection*: R ***bullgrader*** 'the trade name of a soil-moving bulldozer' < E *Bullgrader* is not mentioned by Romanian or English language dictionaries; R ***bushel*** < E *bushel* '1. A measure of capacity; four pecks, 35.238 l or 2150.42 cubic inches' (W); R ***Darcy*** < E *Darcy* 'a rock permeability measure unit equal to 0.987 μm^2' is not mentioned by Romanian or English language dictionaries; R ***dash-pot*** 'a hydraulic damper used to soften a mechanism movements' < E *dash-pot* is not mentioned by Romanian or English language dictionaries; R **data processing** < E *data processing* 'processing of information, esp. the handling of information by electronic machines in accordance with strictly defined systems of procedure' (RHDEL) or 'the operation of digital or analogue computers' (W); R ***dry farming*** < E *dry farming* 'in an arid or dry country, the raising of crops without irrigation, mainly by saving the moisture of the soil and by raising drought-resisting crops' (W); R ***hardpan*** < E *hardpan* '1. A layer of firm detritus under soft soil' (W); R ***jet*** < E *jet* '1. That which spurts out from a narrow orifice; a gushing flow' (W); R ***know-how*** < E *know-how* ' Colloq. Knowledge of how to perform a complicated operation or procedure; technical skill' (W); R ***marketing*** < E *marketing* 'the total of activities by which transfer of title or possession of goods from seller to buyer is effected, including advertising, shipping, storing, and selling' (RHDEL); R ***plotter*** < E *plotter* '3. A contrivance, as for plotting coordinates' (W); R ***reset*** < E *reset* 'the act of resetting, or that which is reset; specifically, a resetting of type' (W); R ***vortex*** < E *vortex* '1. A mass of rotating or whirling fluid, especially when sucked spirally toward a centre; a whirlpool; an eddy' (W). To note that the trade name *Bullgrader* has become a common name in Romanian.

3. **Words assimilated in their form but semantically tied to a foreign context**: R ***airlift*** 'a device for raising water, the functioning of which is based on specific weight difference between water and air-water emulsion'(DIF) < E *airlift* '1. The operation of transporting foodstuffs and other commodities […] by airplane […]; 3. The load carried by such a transport method' (W); R ***display*** 'a device (screen) for displaying alphanumerical and graphical information' (DIF) < E *display* '1. The act of spreading out, unfolding, exhibiting, or bringing to the view or to the mind. 2. Ostentatious show. 3. Printing. A style of type calculated to attract attention. 4. The matter so displayed'; R ***mull*** 'intermediary humus; in most cases, a humus characterised by an intimate mixture between organic matter and mineral soil, and having a slightly acid or neutral reaction' (DIF) < E *mull*[3] '1. A thin, soft, cotton, rayon, or silk dress

goods. 2. A variety of soft, thin muslin used as a base for medicated ointments, as mulla' (W).

Conclusions

The richness of the Romanian vocabulary is in large part due to **borrowing** from many other languages of the world, sometimes in such a way as to allow fine denotative, connotative, or stylistic distinctions between semantically related or nearly synonymous words to grow up. We have seen that 11 of the **loanwords** (36%) are *totally assimilated words* (they are phonetically and orthographically integrated), while other 14 **loanwords** (47%) are *words retaining traces of their origin* (in pronunciation, spelling, or inflection), and other 3 **loanwords** (10%) are *words assimilated in their form but semantically tied to a foreign context.* Two **loanwords** (7%) are untraceable and, therefore, impossible to range within one of the three categories above. Despite the small number of **loanwords** from English and their rather restricted use, the high percentage of words totally assimilated and the even higher percentage of words fully part of the Romanian vocabulary show the extraordinary ability of the Romanian language to turn **loanwords** into permanent words.

References

Chalker, Sylvia & Weiner, E. (1994). *The Oxford Dictionary of English Grammar*. London - New York - Sydney - Toronto: BCA.

Dicţionarul explicativ al limbii române. [An Explanatory Dictionary of the Romanian Language]. (1998). Bucureşti: Editura UE.

Pricop, G. & Marcu, C. (1987-1989). *Dicţionar de îmbunătăţiri funciare*. [A Dictionary of Land Improvement]. Vol. I-II. Bucureşti: Editura Ceres. (DIF)

Raţă, Georgeta & Perković, Anica. (2007). On English Loanwords in the Romanian of Land Improvement. *Scientific Papers. Faculty of Agriculture XXXIX (2)*: 675-678.

The Random House Dictionary of the English Language. (1968). New York: Random House. (RHDEL)

Webster Comprehensive Dictionary. (1995). Chicago: J. G. Ferguson Publishing Company. (W)

AGRICULTURAL ENTOMOLOGY IN ENGLISH AND ROMANIAN: *COLOUR* AND/OR *COLORATION*?

GEORGETA RAŢĂ
AND ELENA-MIRELA SAMFIRA

Introduction

Teaching **English for Special Purposes** in a university centred upon nature and its uses (and abuses) arises a series of questions engendered mainly by the fact that sometimes the language we teach/learn or translate from/into has not always coined the right terms or phrases for the right items or situations. This seems also to be the case for noun phrases containing ***colour*** and ***coloration***. The problem is that, on the one hand, these words are, at least, partially interchangeable, i.e. they are defined by the same features, and, on the other hand, Romanian, for instance, has one word to render ***colour*** (R *culoare*) and two words to render ***coloration*** (R *colorare* and *coloraţie*). Therefore, there are two sources of error in the approach of the fields of insects and insect behaviour.

Material and Method

The method we used is the comparative one. We inventoried all the noun phrases containing ***colour*** and ***coloration*** and looked up for their definitions. We then compared the number of occurrences in both types of noun phrases and checked for the proper Romanian equivalents.

Results and Discussion

In Gordth & Hendrick's *Dictionary of Entomology* (2001), a large number of noun phrases designating insect colours include either ***colour*** or ***coloration***. From the very beginning, Gordth defines ***coloration as*** '1. The capacity or quality of colour. 2. In insects, an arrangement of colours by pigments or structural modification of the integument', completely ignoring to define ***colour***.

Thus, 7 of the 26 noun phrases (26.92%) contain the word ***colour***: ***cuticular colour*** 'colour contained mostly in the epidermis, including permanent browns, blacks and yellow'; ***diffraction colour*** 'iridescent colouring due to the presence of a diffraction grating'; ***interference colour*** 'one of the three forms of structural (physical) colour'; ***pigmentary colour*** 'any of several colours that are derived from substances of a definite chemical composition'; ***scattering colour*** 'structural colour that is produced from the scattering of light from a surface, such as irregular and diffuse light reflected from ground glass, or the scattering of light within the tegument'; ***structural colour*** 'physical colour produced by surface structure (setae, scales, sculpture) that reflect or break up light (by refraction or diffraction) into spectral colours or combinations of them'; and ***subhypodermal colour*** 'any colour that is contained in the fat-body and haemolymph of an insect'.

In these noun phrases, ***colour*** is defined in two ways (Table 7-2).

Table 7-2. Definitions of *colour*

Noun phrases with *colour*	***Colour* defined as**	
	'colour'	**'colouring'**
cuticular colour	+	
diffraction colour		+
interference colour	+	
pigmentary colour	+	
scattering colour	+	
structural colour	+	
subhypodermal colour	+	

As for ***coloration***, it has 18 occurrences: ***advancing coloration*** 'any colour (e.g., yellow) or combination of colours that appears nearer the observer than other viewed in the same plane'; ***alluring coloration*** 'any colour, colour pattern or combination of colours that are attractive to some species of insects (prey) and displayed (used) by predaceous species as part of their feeding strategy'; ***anticryptic coloration*** 'any combination of colours on the body that expose, reveal or draw attention to an organism'; ***apetetic coloration*** 'combinations of colour or colour patterns that resemble some part of the environment or the appearance of another species'; ***aposematic coloration*** 'warning coloration'; ***combination coloration*** 'colours arising from a combination of pigmentary and structural colours'; ***cryptic***

coloration 'coloration patterns of animals that make their outline form, shape or body patterns less apparent, and thereby conferring a degree of protection from predation'; ***directive coloration*** 'directive marks of colours that tend to divert the attention of any enemy or predator from more vital parts of the organism'; ***disruptive coloration*** 'colour patterns of an animal that obscure the outline of an animal and thereby confer protection from predators'; ***epigamic coloration*** 'combination of colours or colour patterns displayed by animals during the attraction of mates or courtship'; ***episematic coloration*** 'recognition colours in the theory of mimicry'; ***procryptic coloration*** 'combination of colours or patterns of colour on the body or appendage that serve in protective resemblances and for concealment as a protection against predators'; ***protective coloration*** 'any pattern or arrangement of colours in an animal that enables it to escape detection by resembling or blending into its surroundings'; ***pseudepisematic coloration*** 'colours displayed in a body that are critical in aggressive mimicry and alluring coloration'; ***pseudoaposematic coloration*** 'combination of colours or colour patterns that lead to protective mimicry'; ***seasonal coloration*** 'a change or difference in aspect ad in colour in two or more succeeding broods in the same species in the same or in different seasons'; ***sematic coloration*** 'warning and signalling colours in insects'; and ***warning coloration*** 'conspicuous colours or patterns of colour which are frequently associated with qualities which render their possessor unpalatable, offensive or dangerous to predators'.

In these noun phrases, ***coloration*** is defined in 6 different ways, except for ***alluring coloration*** (defined in 3 different ways), and ***apetetic coloration***, ***aposematic coloration***, ***procryptic coloration***, ***protective coloration***, and ***pseudoaposematic coloration*** (defined in 2 different ways) (Table 7-3).

Conclusions

Noun phrases containing ***colour*** seem to be better coined since in 7 of the 8 occurrences it is defined as 'colour'. Noun phrases containing ***coloration*** are not so well coined as the former ones, since only 1 of the 18 occurrences defines coloration as 'coloration'. It is interesting to note, though, that some noun phrases with ***colour*** and some noun phrases with ***coloration*** are defined as 'colours' or as 'colour patterns'.

Table 7-3. Definitions of *coloration*

Noun phrase with *coloration*	***Coloration* defined as**					
	change / difference in colour	**(marks of) colour(s)**	**colour arrangement**	**colour pattern**	**coloration pattern**	**combination of colours**
advancing coloration		+				+
alluring coloration		+		+		+
anticryptic coloration						+
apetetic coloration				+		+
aposematic coloration				+		+
combination coloration		+				
cryptic coloration					+	
directive coloration		+				
disruptive coloration				+		
epigamic coloration						+
episematic coloration		+				
procryptic coloration				+		+
protective coloration			+	+		
pseudepisematic coloration		+				
pseudoaposematic coloration				+		+
seasonal coloration	+					
sematic coloration		+				
warning coloration		+		+		

References

Chalker, Sylvia & Weiner, E. (1994). *The Oxford Dictionary of English Grammar*. London - New York - Sydney - Toronto: BCA.

Gordth, G. & Hendrick, D. H. (2001). *A Dictionary of Entomology.* Oxon: CABI Publishing.

Raţă, Georgeta. (2004). Agricultural Entomology: 'Colour' and/or 'Coloration'?. *4th International Scientific Days of Land Management in the Great Hungarian Plain, October 21-22, 2004, Mezötúr, Hungary*: 1-5.

Webster Encyclopedic Unabridged Dictionary of the English Language. (1996). New York: Gramercy Books.

NAMES OF FODDER PLANTS IN ENGLISH AND ROMANIAN

OANA BOLDEA

Introduction

This paper presents an analysis of the means of vocabulary enrichment. The idea it starts from and it attempts to prove is that English and Romanian have more in common than one might guess, if we think that one is a Germanic language and the other a Romance one. We try to see whether or not Latin plays a great part in English as well, where the popular names of plants are concerned. And the hypothesis we attempt to prove in our study is that most words can be traced back to Latin, either as the origin of the **borrowing** or at least as the source of the loan.

Material and Method

The approach used is the comparative etymological one. For this, the etymologies of the words were analyzed, in Romanian and in English as well, with greater emphasis on the English words. The reason is, on the one hand, that, for the Romanian ones, the etymology is easier to infer for a native speaker of this language and, on the other hand, because it is only natural that Romanian should have most words of Latin origin since it is a Romance language. The study focuses on borrowings as a means of enriching the vocabulary. The process by which one word starts being used in one language, after having previously been used in another, is called a **borrowing**. Such words are commonly called **loanwords**, although the term is by no means exact: the words are not given back to the language they were "borrowed" from. **Loan blend** denotes a word the meaning of which is borrowed but only part of the form is kept from the original language, while a **loan shift** refers to a word the meaning of which is borrowed but the form is nativized. Another concept used in the paper is that of **calque** or **loan translation**, a **borrowing** where the parts of the borrowed word are translated item by item into equivalent parts in the new language. The conclusions show that **calque** lies at the basis of most words where **borrowings** are concerned. When discussing **word borrowing**, distinction should be made between the terms **source of borrowing** (the language from which the word was taken into the target language) and **origin of borrowing** (the language to which the word can be traced). Thus, for many of the words analyzed here, the source is Latin, but it is not the origin, which in some cases may be Greek or some other language, as shown the conclusions. The material used in the study was a collection of names of fodder plants found in *Fodder Plants and Agriculture* (Rață *et al.* 2002). As the interest was in the names of the two plants in English and Romanian, the terminology in French in the abovementioned reference book was left aside. Other books studied provided the theoretical ground for the paper (**etymology** and **borrowings**).

Findings, Discussion and Conclusions

The findings of the study are extremely inciting. Looking at the etymology of the words, one can notice that our initial idea on the subject was correct: most of the names of fodder plants come from Latin in both languages. This is not unexpected, as all scientific names are in this language. One such example is *izma piperată*, the English equivalent being ***peppermint***, while the Latin is *mentha piperita*. Another such word is the Romanian *mei*, the English equivalent being ***millet***, while the Latin name is *Panicum milliaceum*. The different phonetic systems in the two languages under discussion gave two variants of the same word: *millium*. However, there is an even more interesting part, namely there are some English words clearly coming from Latin but whose Romanian equivalents have no such roots, while the expectation would be the opposite, since Romanian is a Romance language and English is a Germanic one. One such example is, for instance, *iarba grasă*, whose English equivalent is ***purslane***, while the Latin name is *Portulaca oleracea*. The English term has been used since Middle English, coming from Latin *via* French. Actually, the Romanian *grasă* also comes from Latin, but from another word (*grassus*), not the one denominating the plant in question. Naturally, this is not the only such example. The same happens with the English ***basil***, which clearly comes from the Latin *basilicum*, while in Romanian the term for the same plant is *busuioc* – not only it does not come from the same name of plant in Latin, but it does not come from Latin at all: it is traceable to a Bulgarian word. In the same way, the etymon of *păpădie* is Bulgarian, while the name of the same plant in English is ***dandelion***, coming *via* French from the Latin *dens leonis*. Another example is the

Latin *Berberis vulgaris*, which gave ***barberry*** in English, while the Romanian term is *dracilă*, having nothing to do with the aforementioned etymon; it comes from a Slavic word meaning 'bramble'.

Another aspect found through this study is that there are different means by which the words are derived. One class of words includes those which come directly from the Latin term, in which case there are some changes due to the phonetics of the language and to the influences from other languages. When this is the case, Latin is the origin of the **borrowing**. Such a term is *Cichorium*, which gave *cicoare* in Romanian and ***chicory*** in English.

Another English word having Latin as the origin language is ***rosemary***. Actually, with this term, we come across another means of enriching the vocabulary, namely **folk etymology**. The word appeared in Middle English, influenced by ***rose*** and the name ***Mary***, coming from the Latin *ros* ('dew') and *maris* ('sea'). The English felt the final *s* as the mark of the plural and eliminated it. This is how, what initially meant 'dew of the sea' turned into what it is nowadays. **Folk etymology** is also responsible for the English ***mandrake***: in Middle English, it appeared as a variant of *mandrage*, which was the shortened form of *mandragora*. Little by little, it changed into ***mandrake***, people taking it as coming from *man* and *drake* (probably due to the fact that it names a plant the root of which is thought to resemble people or monsters). For the abovementioned *mandragora* Latin plays a part, although the word does not come from it. In this case, Latin is the source of the **borrowing**, not its origin. The roots of this word must be searched for in the Greek *mandragoras*. This is only an example from the second class of names, those for which Latin is only the source. Another item belonging to this class is the English ***thyme***, coming from the Greek *thymon via* the Latin *thymus*. The same holds good for the word *carrot*, coming *via* French from Latin, where it was borrowed from Greek.

A third class of terms is represented by those which come from the Latin term indirectly, by means of translating the meaning of it, i.e. **calque**. Such are, for instance, the Romanian *floarea-soarelui* and the English ***sunflower***, both rendering the meaning of the Latin *Helianthus annus*. Or the Romanian *lupin alb* and the English ***white lupine***, both clearly translations of the Latin *Lupinus albus*. Another such term is the Latin *Vicia villosa*, which has been translated *măzăriche păroasă* into Romanian and ***hairy vetch*** into English. Other plants have got their name in different ways than those presented above. There are terms coming from other languages than Latin. One such an example is the English ***burdoch*** (the equivalent of the Romanian *brusture*, with unknown etymology), which comes from the Scandinavian term *burre* added to the Irish word *deoch*. ***Strawberry*** is another such term, which comes from the German words *stroh* and *beere*. Still other plants got their names due to the history of the region in which they are used. Such is, for instance, the Romanian *firuţă*, which, in English, is ***Kentucky bluegrass***, since, in the United States, Kentucky is the main region where it can be found in large amounts. Kentucky's popular name is "Bluegrass State", clearly showing the great abundance this type of fodder. ***Timothy*** is another term pertaining to this class: it was named after Timothy Handson, an American farmer who grew this grass and spread its cultivation in early 18th century.

References

Bădescu, Alice. (1984). *Gramatica limbii engleze*. [English Grammar]. Bucureşti: Editura ştiintifică şi enciclopedică.

Boldea, Oana. (2004). The Influence of Latin on the Names of Some Fodder Plants. *Proceedings of the 4th International Scientific Days of Land Management in the Great Hungarian Plain, Mezötúr, Hungary.*

Crystal, D. (1999). *The Penguin Dictionary of Language*. Harmondsworth: Penguin Books.

Dicţionarul explicativ al limbii române. [An Explanatory Dictionary of the Romanian Language]. (1998). Bucureşti: Editura Univers Enciclopedic.

Dictionary of American English. (1997). London: Longman.

Flavell, Linda & Flavell, R. (1995). *Dictionary of Word Origins*. London: Kyle Cathie Limited.

Leech, G. (1989). *An A-Z of English Usage*. London: Longman.

Leviţchi, L. (1975). *Limba engleză contemporană - Lexicologia*. [Contemporary English: Lexicology]. Bucureşti: Editura Academiei R.S.R..

Raţă, Georgeta, Moisuc, Al. & Samfira, I. (2002). *Fodder Crops and Agriculture*. Timişoara: Editura Orizonturi Universitare.

Raţă, Georgeta, Samfira, I., Diana-Andreea Boc-Sânmărghiţan & Butnariu, Monica. (2012). *Compendium of Forage Technical Terms in English, French and Romanian*. Newcastle upon Tyne: Cambridge Scholars Publishing.

Webster Encyclopedic Unabridged Dictionary of the English Language. (1996). New York: Gramercy Books.

ENGLISH AND ROMANIAN NAMES OF PLANTS IN FOOD ADDITIVES GUIDES

OANA BOLDEA

Introduction

The multitude of changes that have taken place in Romania in the past years have had a great impact upon the Romanian language itself. After 1989, our country has opened up towards Western countries, in an attempt to catch up with the new ideas and technologies the Romanians did not have access to before. This was the starting point of a great influx of objects ("object" here having the large meaning used in linguistics) into Romania. Together with the new products and technology imported into our country, we have had to solve the problem of the specific terminology. In the source language they had their name, but in Romanian they did not.

This is the problem people have had to find solutions for when the following questions arose:

What happens with those names?

How do we render them into Romanian?

There are two main ways to deal with this:

- solve it within the lexical resources the Romanian language offers;
- borrow new words.

The matter is an extremely complex one, encompassing a wide range of terms coming from various languages and belonging to all fields of activity. Still, English seems to be prevalent as language providing terms Romanian has to deal with. This is not surprising, on the one hand, because technology has been developing rapidly in the English-speaking countries and, on the other hand, because English has become an international language, spoken all over the world. The paper deals only with a small part of this general matter, trying to find ways of accomplishing the task of translating the English terminology appearing in the *Food Additives Guide*. We are trying to see how the terms can be adopted into Romanian without sounding awkward.

Results and Discussion

For the practical side of the study, we have used the *Food Additives Guide* (2002). This is a compendium of additives used in the Food Industry. The *Guide* is made up of several chapters, each presenting a different class of additives:

- unnumbered;
- colours (E100-E181);
- preservatives (E200-E290);
- antioxidants and acidity regulators (E296-E385);
- thickeners; emulsifiers and stabilizers (E400-E495);
- flavour enhancers (E620- E640);
- anticaking agents (E500-E585);
- glazing agents and sweeteners (E900-E1520).

It presents the names of the additives, their description and also the negative effects they can have on the health of the consumer. Their description includes the physical appearance (colour and aspect) as well as the material they are made of. The present study is restricted to those additives derived from plants. In the source mentioned previously, the additives come with their denomination, which can be a number, a "proper" name or both. Naturally, we have taken into consideration those which appear with names, and not represented by figures, as they are the ones to challenge a translator's inventiveness and other creative resources. Translating "E250" should be no problem.

For the theoretical background of this paper, books dealing with linguistics and specifically with lexicology proved extremely useful in understanding the general phenomenon of enriching the vocabulary, since word building is considered to be a boundary area between lexicology and grammar. Thus, we have read works written by outstanding Romanian researchers, like Leviţchi (1975) or Bădescu (1984). Of course, foreign books on the matter have also been consulted, in an attempt to have as complete an understanding of the phenomenon as possible. Such works are Quirk's *A University Grammar of English* (1973) and Swan's *Practical English Usage* (1995).

Another type of reference books used is that of the dictionaries, providing information either about the etymology of some names of plants, or about the plants themselves. *Webster's Encyclopedic Unabridged Dictionary* is one of the dictionaries displaying general information, while *Dicţionar botanic poliglot* (Vaczy 1980) and *Dicţionar etnobotanic* (Borza 1984) contain specific information about plants.

The method used is that of the analysis of the means of enriching the vocabulary. This analysis helps understanding how new words appear into the language. We have been interested especially in **derivational analysis**. This refers to the relation of morphemes within the word and interrelations (from the structural point of view) between different types of words, trying to establish the structural patterns words are built on. The two main means of word building are **derivation** and **composition**. **Derivation** is the means analysed here. Also called **affixation**, it consists in adding affixes to the stems of different words in order to form new words. Affixes may be classified from various points of view: according to their origin, to the parts of speech they serve to form, to the meaning they render, to the parts of speech they form, to their productivity, etc. We have been especially interested in the classification according to the part of the word they are added to (the beginning or the end, hence prefixes and suffixes), the meaning (if any) they bring to the word, and their origin. We have also dealt, to a certain extent, with **borrowings**, which refer exactly to the introduction of something from one language into another. In the case of our paper it is **loanwords** referring particularly to special terminology – names of additives in the food industry.

Conclusions

After analysing the names of the additives obtained from plants, we have seen that all those formed by affixation are actually made up with the help of suffixes. So, this is the first thing to note: no prefixation in the formation of English names of additives. The suffixes used are *-in*, *-ine*, -*ic* and, not so productive, *-an*. The problem appearing for the translator is what to do when having to render into Romanian the words ending in *-in* and *-ine*, respectively, since both of them are noun suffixes, so they are both added to nouns to form nouns. Moreover, they seem to be equally productive. We will analyse them one by one.

Thaumatin is the name of a sweetener derived from the tropical plant *Thaumococcus danielli*. It is an example of how the first mentioned suffix works. Formed in the same way, *curcumin* is an orange-yellow colouring agent derived from the root of the *curcuma* plant. Another example is *tannin*, derived from the nutgalls and twigs of oak trees, and occurring naturally in tea. Interestingly enough, its name comes from ME *tan* 'oak bark', coming from the Latin *tannum*.

There are quite a number of examples for the formation of words employing the second suffix, of which we mention here *caffeine* (coming from tea, kola nut or coffee and, etymologically, from *coffee*), *asparagine* (obtained from legumes, its etymon being *asparagus*) and *quinine* (coming from the cinchona bark, and etymologically from the word Quechua *kina* 'bark').

Both suffixes are of Greek origin and have approximately the same meaning: 'pertaining to', 'of the nature of'. It seems that a certain distinction of use is attempted, namely basic substances have the ending -*ine* rather than -*in* (*caffeine*, *quinine*), while -*in* is used for certain neutral compounds, glycerides, glucosides, and proteids (*palmitin*, for instance). In what the translation into Romanian is concerned, we are interested if we are to render the words by means of the masculine for -*in* and the feminine for -*ine*, since this seems to be the case with the English words. The translation into Romanian is usually done by -*ină*, as *Dicționarul explicativ al limbii române* (1998) does not mention a suffix -*in*. Thus, traditionally, -*ină* has been used for preserving, in Romanian, the derivational process used for the formation of the word in English. This is the case of most names of additives used in the *Guide*, especially those the Romanian translations of which have been used for a while, such as *caffeine* or *lecithin*. In the same way, we tend to supply words like *curcumin*, or *palmitin*. *Quinine* is a special case: since it is also called *quinina* in English, we could claim that, actually, we are rendering this term by the Romanian *chinină*, and not the other one, and in this case it is no longer felt as a **derivative word**. We could, then, say that the translator is merely observing an unwritten rule which states that the translator should tend to change as little of the word as possible (as allowed by the phonetic system of the target language) when introducing it as a novelty into the language. But this way of translating by -*ină* is by no means a set rule. The translation of *tannin* seems more appropriately done by the Romanian *tannin* (by the masculine, not the feminine). We believe this is possible because the word is no longer felt as a **derivative**.

Another suffix used is -*ic*. This is a suffix forming adjectives. It appears in such names as *citric acid* (coming from *citrus*), *fumaric acid* (coming from the name of the plant it is derived from, namely plants of the genus *Fumaria*) or *benzoic acid* (coming from *benzoin*, a resin exuded by trees native to Asia). It is also of Greek origin, and in Romanian it is transposed as it is, since the Romanian language allows it. Thus, the translations of the abovementioned words will be *acid citric*, *acid fumaric*, and *acid benzoic*. The case of ***aspartic acid*** is somewhat special, in that it has three other names: *aminosuccinic acid*, *asparaginic acid*, and *asparagic acid*. Although all names can be transposed into Romanian as they are, it would be best for one of them to become a commonly used term in order to avoid any possible misunderstandings.

The less productive suffix is *-an*. It appears in the name *carrageenan* coming from a sort of purplish-brown seaweed called *Irish moss* or *carrageen* (*Chondrus crispus*), or in *sorbitan*. Just like *-ic*, this is also an adjectival suffix. We do have this suffix in Romanian, but there is a problem while translating it, because in Romanian the suffix *-an* is an augmentative as in *bogătan* ('very rich') or in *băietan* ('lad'). That is why we believe the best way to render the word *carragenan* into Romanian is *caragină*, since *caragenan* would seem awkward, although *caragină* is also the Romanian for the plant the substance comes from, *carrageen*. Still, in what *sorbitan* is concerned, the most probable rendering in Romanian is by the same form as in English, as the *-an* ending can also be found in other chemistry-related Romanian such as *butan*, *etan*, *propan*.

There are also other types of names in the *Food Additive Guide* from the point of view of their formation and of their translation. The English language generally displays a certain tendency for **compound words**, and it is only natural for it to be easily observed here as well. The **compounds** are usually translated into Romanian by means of paraphrase. Thus, *sodium fumarate*, *guar gum*, and *quillaia extract* become *fumarat de sodiu*, *guma de guar*, *extract de quillaia* in Romanian.

The general tendency is for a translator to keep the form of the word, as much as the phonetical rules of the language allow it, when it denominates a new concept or object. No change happens with the following words denominating additives like *aronia* (derived from *Aronia arbutifolia*), *acacia* (derived from the sap of *Acacia Senegal*), or plants such as *curcuma*, *tara*, *guar*. And little change is made in others, where it is not possible to keep the form as it is, as in the case of the words ending in *-th* in English. Because Romanian does not have such a possibility, the *h* will drop in *amaranth* and *tragacanth*. But still the larger part of the word is maintained.

References

Bădescu, Alice. (1984). *Gramatica limbii engleze*. [English Grammar]. Bucureşti: Editura ştiintifică şi enciclopedică.

Boldea, Oana. (2004). Names of Plants in Food Additives Guide. *Lucrări ştiinţifice, Facultatea de Agricultură XXXVI*: 548-551.

Borza, Al. (1968). *Dicţionar etnobotanic*. [An Ethnobotanical Dictionary]. Bucureşti: Editura Academiei.
Crystal, D. (1999). *The Penguin Dictionary of Language*. Harlow: Longman.
Dicţionarul explicativ al limbii române. [An Explanatory Dictionary of the Romanian Language]. (1998). Bucureşti: Editura Univers Enciclopedic.
Flavell, Linda & Flavell, R. (1995). *Dictionary of Word Origins*. London: Kyle Cathie Limited.
Food Additives Guide. (2002). Online: FoodAG.com.
Leech, G. (1989). *An A-Z of English Usage*. London: Longman.
Leviţchi, L. (1975). *Limba engleză contemporană - Lexicologia*. [Contemporary English: Lexicology]. Bucureşti: Editura Academiei R.S.R..
Quirk, R. & Greenbaum, S. (1973). *A University Grammar of English*. Harlow: Longman.
Raţă, Georgeta. (1999). *English for Horticulturists*. Timişoara: Editura Mirton.
Salapina, Galina. (1974). *Limba engleză contemporană*. [Contemporary English]. Tipografia Universităţii din Timişoara.
Swan, M. (1995). *Practical English Usage*. Oxford: Oxford University Press.
Vaczy, C. (1980). *Dicţionar botanic poliglot*. [A Multilingual Botanical Dictionary]. Bucureşti: Editura Ştiinţifică şi Enciclopedică.
Webster Encyclopedic Unabridged Dictionary of the English Language. (1996). New York: Gramercy Books.

English Plant Names and Their German Counterparts

Astrid-Simone Groszler
and Biljana Ivanovska

Introduction

The purpose of this paper is to notice the way in which English plant names formed with animal names are rendered in the German language. By that we mean to notice not just whether or not these particular English plant names have German equivalents, but also how many of these equivalents observe the choice of the animal name instead of just giving a semantic equivalent.

As we have mentioned the term **equivalency**, we think it useful to extend on it. The concept of **textual equivalence** comes up, defined as the combination of similarities in source text and target text information flow and the cohesive roles of source text and target text devices in their respective texts. We may speak of functional equivalence when we deal with the translator's decision as to which consideration to be taken into account at any time.

Even if we cannot speak of actual text in the case of plant names, we do have to take into account the definitions presented above, since the same translation rules apply in the case of a text as in the case of mere plant or animal names. As we note later in the paper, the processes of both textual and functional equivalency are used in finding the German name for English plant names since, in several cases, there is no perfect equivalency between the English and German equivalents, even though we could establish some instances where perfect equivalency occurred.

We noticed that in some cases the animal names used in the two languages are the same, while in others they are not.

Material and Method

We have inventoried 78 English common plant names containing different animal names gathered from English language and specialised dictionaries and books. Then we have looked for the German counterparts of the English plant names analysing their equivalency and frequency. We have, thus, made a contrastive analysis of the concepts in English and German.

Results

The plant names we have analysed contain different animal names, ranging from insects (e.g. ***fly agaric***) to molluscs (e.g. ***oyster fungus***), birds (e.g. ***gooseberry***), and mammals (e.g. ***ox-eye daisy***). Below, we group the plant names according to the plant group they pertain to:

- ferns and mosses: ***adder's tongue (fern)***, ***hart's tongue (fern)***, ***stagshorn clubmoss*** (also *running ground pine* and *running clubmoss*);
- grasses, rushes and sedges: ***cocksfoot grass***, ***false fowl sedge***;
- mushroom names: ***fly agaric***, ***oyster fungus***, ***turkey tails***, ***wolf-fart puffball*** (also *pear-shaped puffball*), ***yellow stagshorn fungus***;
- seaweed and coastal plants: ***crab's eye lichen***;
- trees and shrubs: ***alder buckthorn***, ***bird cherry***, ***buckthorn***, ***cowberry*** (also *cranberry*, *red bilberry*, and *alpine cranberry*), ***crab apple***, ***dog rose***, ***dogwood***, ***goat willow*** (with the variants *great sallow* and *sallow*), ***gooseberry***, ***horse chestnut***, ***sea buckthorn***;
- waterplants: ***common duckweed***, ***common horsetail***, ***hare's tail cottongrass***, ***mare's tail***, ***narrow leaf cattail***, ***water horsetail***, ***wood horsetail***;
- wild flowers: ***alpine mouse-ear***, ***bear's foot***, ***bee orchid***, ***bird's-foot (trefoil)***, ***bird's-nest orchid***, ***black horehound***, ***bloody crane's-bill***, ***bristly ox-tongue***, ***buckshorn plantain***, ***catsear***, ***common fleabane***, ***common mouse-ear***, ***common storksbill***, ***common water-crowfoot***, ***cow parsley***, ***cowslip***, ***creeping crowfoot*** (or *creeping buttercup*), ***cuckoo flower***, ***cut-leaved crane's-bill***, ***dog's mercury***, ***fat hen***, ***field chickweed***, ***fly orchid***, ***foxglove*** (or *lady's glove*), ***frog orchid***, ***goatsbeard***, ***greater butterfly orchid***, ***harebell***, ***henbane***, ***hogweed***, ***horseshoe vetch***, ***houndstongue***, ***ivy-leaved toadflax***, ***lion's foot***, ***lizard orchid***, ***lousewort***, ***meadow crane's-bill***, ***monkey flower***, ***mouse-ear hawkweed***, ***ox-eye daisy***, ***pignut***, ***ragged robin***, ***red chickweed*** (also *scarlet pimpernel*), ***red goosefoot***, ***sea wormwood***,

sheep sorrel, ***smooth hawksbeard***, ***viper's bugloss*** (or *blueweed*), ***wood crane's-bill***, ***wormwood***, ***yellow horned-poppy***.

If we examine both English and German plant names, we can see there are several classes:

- perfect equivalency between English and German plant names: E ***adder's tongue*** – G *Natternzunge*, E ***bee orchid*** – G *Bienen-Ragwurz*, E ***bird cherry*** – G *Vogelkirsche*, E ***bird's-nest orchid*** – G *Vogel-Nestwurz*, E ***bloody crane's-bill*** – G *Blutstorchschnabel*, E ***common storksbill*** – G *Reiherschnabel*, E ***dog-rose*** – G *Hundsrose*, E ***fly agaric*** – G *Fliegenpilz*, E ***fly orchid*** – G *Fliegen-Ragwurz*, E ***houndstongue*** – G *Hundszunge*, E ***meadow crane's-bill*** – G *Wiesenstorchschnabel*, E ***monkey flower*** – G *Affenblume*, E ***oyster fungus*** – G *Austernpilz*, E ***red goosefoot*** – G *Roter Gänsefuß*, E ***wood crane's-bill*** – G *Waldstorchschnabel*, etc.;
- almost perfect equivalency between English and German plant names: E ***common fleabane*** – G *großes Flohkraut*, E ***goatsbeard*** – G *Bocksbart*, E ***hart's tongue*** – G *Hirschzungenfarn*, E ***horse chestnut*** – G *gewöhnliche Rosskastanie*, E ***lousewort*** – G *blattreiches Läusekraut*, E ***mouse-ear hawkweed*** – G *kleines Habichtskraut* (there is only the attribute of a mouse, G *klein* 'little'), E ***viper's bugloss*** – G *Natternkopf*, E ***yellow stagshorn fungus*** – G *Klebiger Hörnling*, etc.;
- no equivalency in German for English plant names: E ***bristly ox-tongue*** – G *Natternkopf-Bitterkraut* (E *ox* is replaced by G *Natter* 'adder', animals from different classes), E ***cat's ear*** – G *Ferkelkraut*, E ***fat hen*** – G *weißer Gänsefuß* (the animal name in German is 'goose'), E ***goat willow*** – G *Weidenkätzchen* (instead of *goat* there is *Kätzchen* 'kitty', still a mammal, but a different order), E ***lizard orchid*** – G *Bocksriemenzunge*), E ***mouse-ear*** – G *Hornkraut* (we are dealing with a different body part here, namely 'horn' instead of 'ear'), E ***stagshorn clubmoss*** – G *Keulen-Bärlap* (with the variant *Wolfsklaue*, still mammals, but *Bär* means 'bear' and *Wolf* 'wolf'), E ***turkey tails*** – G *Schmetterlingstramete* (the German name takes another class, namely insects, *Schmetterling* 'butterfly'), etc.;
- no animal name in German plant names(e.g. E ***alder buckthorn*** – G *Faulbaum*, E ***cow parsley*** – G *Wiesenkerbel*, E ***cowberry*** – G *Preiselbeere*, E ***crab apple*** – G *Zierapfel*, E ***dogwood*** – G *Hartriegel*, E ***fox glove*** – G *Fingerhut*, E ***gooseberry*** – G *Stachelbeere*, E ***ox-eye daisy*** – G *Wiesenmargerite*, E ***wolf-fart puffball*** – G *Birnenstäbling*, etc.;

- no German counterpart for the English plant name: E *crab's eye-lichen*.

Discussion

In the previous section, we brought forth an account of English common plant names containing animal names and their German counterparts. We have seen that all but one English plant name (i.e. 1%) entail an equivalent. Our present findings differ in this respect from our previous research of the Romanian equivalents of the English plant names. In that particular case, we found a larger percent of English common plant names containing an animal name which had no Romanian equivalent, namely 14%. This difference in percentage, 14% as opposed to 1%, can be explained by the fact that some plants common in the UK are not found in the native Romanian flora, whereas the German climate facilitates their presence in Germany.

If we look at all the numbers, we can see that from the 78 English names analysed:

- 34 (i.e. 44%) do not entail animal names;
- 26 (i.e. 33%) observe the English choice in animal names;
- 3 (4%) observe the same animal body part name;
- 2 (almost 3%) take a different animal name;
- only 1 (i.e. 1%) does not account for a German equivalent.

Thus, though almost all English common plant names containing an animal name entail a German equivalent, these counterparts do not always take an animal name.

Our research has clearly shown that though almost all English common plant names containing an animal name entail a German equivalent, these counterparts do not always take a German animal name, a fact which may be determined by the part of the plant we look at when choosing the respective animal name.

Also, when taking a look at our findings regarding their Romanian counterparts, we find out that the percentage of German plant names containing the same animal name is quite higher than that of Romanian plant names (i.e. 33% as opposed to 29%).

Conclusions

After having analysed the 78 common English plant names containing an animal name and their German counterparts, we have found that:

- 1% do not exhibit a German common plant name, as opposed to 14% of the Romanian counterparts;
- 3% take a different animal name;
- 4% observe the same animal body part name;
- 33% observe the English choice in animal names;
- 44% do not entail animal names.

Comparing the percentages, we notice that even though the percentage of plant names containing the same animal or animal body part name in German is larger than that of plant names containing a different animal or animal body part name (33%, respectively 4%, vs. 15%, respectively 3%), the greater percentage (i.e. 44%) of plant names do not contain any animal name.

References

Baker, M. (Ed.) (2001). *Routledge Encyclopaedia of Translation Studies*. New York: Routledge.

Bejan, D. (1991). *Nume româneşti de plante*. [Romanian Plant Names]. Cluj-Napoca: Editura Dacia.

Borza, Al. (1968). *Dicţionar etnobotanic*. [An Ethnobotanical Dictionary]. Bucureşti: Editura Academiei.

British Wild Flowers. Online: http://www.british-wild-flowers.co.uk/.

Englisch-Deutsch Wörterbuch. [English-German Dictionary]. Online: http://www.dict.cc/.

Gartendatenbank. [Garden Data Bank]. Online: http://www.gartendatenbank.de/.

Groszler, Astrid-Simone & Ivanovska, Biljana. (2010). English Common Names of Plants and Their Romanian Counterparts. In Georgeta Raţă (Ed.), *Teaching Foreign Languages: Languages for Special Purposes*. Newcastle upon Tyne: Cambridge Scholars Publishing. 36-41.

Groszler, Astrid-Simone & Ivanovska, Biljana. (2011). English Plant Names Containing an Animal Name and Their German Counterparts. In Georgeta Raţă (Ed.), *Academic Days of Timişoara: Language Education Today*. 36-41.

Groszler, Astrid-Simone. (2009). English Plant Names Containing an Animal Name and Their Romanian Counterparts. *Journal of Linguistic Studies 3 (2)*: 17-20.

Index der medizinischen Pflanzen. [Index of Medicinal Plants]. Online: http://www.samen-online.com/heilpflanzeninfos/pflanzenindex/index.html.

Lungu Badea, G. (2003). *Dictionary of Terms Employed in the Theory, Practice and Didactics of Translation.* Timişoara: Ed. Orizonturi.

Raţă, Georgeta, Grigoriu, Alma-Lioara & Imbrea, Ilinca-Merima. (2005). *Flora Marii Britanii. Nume de plante în limbile latină, engleză şi română* [British Flora. Names of Plants in Latin, English and Romanian]. Timişoara: Mirton.

Wiesenblumen / Waldblumen. (Namen und Angaben der Blumen gemäss FLORA HELVETICA). [Wood Flowers. Names and Uses of Flowers from *Flora Helvetica*]. Online: http://www.waldwiesenblumen.gabathuler.org.

ENGLISH PLANT NAMES AND THEIR ROMANIAN COUNTERPARTS

ASTRID-SIMONE GROSZLER AND BILJANA IVANOVSKA

Introduction

The purpose of this paper is to see how English plant names formed with the help of animal names are represented in the Romanian language. By that we mean to see not just whether or not these particular English plant names have a Romanian equivalent, but also how many of these equivalents observe the choice of the animal name instead of just giving a semantic equivalent.

In terms of **equivalency**, we think it useful to extend on the concept. In Baker (2001), there is a definition of equivalency as

> 'the relationship of a source text (ST) and a target text (TT) that allows the TT to be a translation of the ST in the first place. Equivalence relationships are also said to hold between parts of STs and parts of TTs. [...] equivalence is commonly established on the basis of: the source language (SL) and target language (TL) words supposedly referring to the same thing in the real world, i.e. on the basis of their referential or denotative equivalence; the SL and TL words triggering the same or similar associations in the minds of native speakers of the two languages, i.e. their connotative equivalence; the SL and TL words being used in the same or similar contexts in their respective languages, i.e. what Koller (1989) calls text-normative equivalence; the SL and TL words having the same effect on their respective readers, i.e. pragmatic (Koller 1989) or dynamic equivalence (Nida 1964); the SL and TL words having similar orthographic or phonological features.'

Another concept that comes up is **textual equivalence**, defined as the combination of similarities in ST and TT information flow and the cohesive roles of ST and TT devices in their respective texts. We may speak of **functional equivalence** when we deal with the translator's decision as to which consideration to be taken into account at any time.

Even though we cannot speak of actual text in the case of plant names, we still have to take into account the definitions presented above, because the same translation rules apply in the case of texts as in the case of mere plant or animal names. Later in the paper, we point out that the processes of both textual and functional equivalency are used in finding Romanian names for English plant names since there is no perfect equivalency between the English and Romanian plant names. We noticed that, in some cases, animal names used in the two languages are the same, whereas in others they are not.

Material and Method

We have inventoried 80 English common plant names containing different animal names, picked up from English language and English specialised dictionaries and books. Then we have looked for the Romanian counterparts of the English plant names analysing their equivalency and frequency. We have also studied 59 Romanian common plant names containing animal names and analysed their English counterpart. Thus, we have undertaken a contrastive analysis of the concepts in the two languages.

Results

The plant names we have analysed contain different animal names, ranging from insects (***fly agaric***) to molluscs (***oyster fungus***), birds (***gooseberry***), and mammals (***ox-eye daisy***). We have grouped these plant names according to the plant group they pertain to.

First, we studied mushroom names: ***fly agaric***, ***oyster fungus***, ***turkey tails***, ***wolf-fart puffball***, ***yellow stagshorn fungus***.

In plant names denominating ferns and mosses, we found: ***adder's tongue (fern)***, ***hart's tongue (fern)***, ***stagshorn clubmoss***.

In trees and shrubs, the research revealed: ***alder buckthorn***, ***bird cherry***, ***buckthorn***, ***cowberry***, ***crab apple***, ***dog rose***, ***dogwood***, ***goat willow***, ***gooseberry***, ***horse chestnut***, ***sea buckthorn***.

Wild flowers brought forth a large number of names: ***alpine mouse-ear***, ***bear's foot***, ***bee orchid***, ***beebread***, ***bird's-foot trefoil***, ***bird's-foot***, ***bird's-nest orchid***, ***black horehound***, ***bloody crane's-bill***, ***bristly ox-tongue***, ***buckshorn plantain***, ***cat's-ear***, ***common fleabane***, ***common storksbill***, ***common water-crowfoot***, ***cow clover***, ***cow parsley***, ***cowslip***, ***creeping crowfoot***, ***cuckoo flower***, ***cut-leaved crane's-bill***, ***dog's mercury***, ***fat hen***, ***field chickweed***, ***fly orchid***, ***foxglove***, ***frog orchid***,

goatsbeard*, *greater butterfly orchid*, *harebell*, *henbane*, *hogweed*, *horseshoe vetch*, *houndstongue*, *ivy-leaved toadflax*, *lion's foot*, *lizard orchid*, *lousewort*, *meadow crane's-bill*, *monkey flower*, *mouse-ear hawkweed*, *ox-eye daisy*, *pignut*, *ragged robin*, *red chickweed*, *red goosefoot*, *sea wormwood*, *sheep sorrel*, *smooth hawksbeard*, *viper's bugloss*, *wood crane's-bill*, *wormwood*, *yellow horned-poppy. Names denominating water plants are: ***common duckweed*, *common horsetail*, *duck's meat*, *hare's-tail cottongrass*, *horse pipes*, *mare's-tail / marestail*, *narrow leaf cattail*, *water horsetail*, *wood horsetail***.

Sea and costal plants revealed ***crab's-eye lichen***, while grasses, rushes and sedges revealed ***cocksfoot grass*, *false fox sedge***.

If we examine the Romanian plant names, we can see there are names in which occur the same animal as in the English name and plant names in which there are names of other animals.

The following pertain to common plant names exhibiting the same animal: E ***adder's tongue*** (R *limba şarpelui*), E ***bee orchid*** (R *albină*), E ***bird's-foot trefoil*** (R *unghia-găii* – *gaie* is *chough*, a bird related to the crow), E ***bird's-nest orchid*** (R *cuibul rândunelii*), E ***buckthorn*** (R *spinul cerbului*), E ***common storksbill*** (R *ciocul berzei* with the variant *ciocul-cucoarei*) – there is the same Romanian translation for ***cut-leaved / bloody crane's-bill***, *crane* meaning *cocor / cucoară* in Romanian, E ***cowslip*** (R *ţâţa vacii* with the variant *ciuboţica cucului*, *cuc* meaning 'cuckoo'), E ***cuckoo flower*** (R *scuipatul cucului*), E ***dog's mercury*** (R *buruiană câinească*), E ***fly agaric*** (R *muscariţă,* with the variant *pălăria-şarpelui* – 'snake hat'), E ***goat willow*** (R *salcie căprească*), E ***goatsbeard*** (R *barba caprei, ţâţa caprei*), E ***hart's tongue*** (R *limba cerbului*), E ***houndstongue*** (R *limba câinelui*), E ***lousewort*** (R *păducherniţă*), E ***mare's tail*** (R *coada calului*, with the variants *barba ursului*, *urs* meaning 'bear', and *coada iepei*, *iapă* meaning 'mare'), E ***ox-eye daisy*** (R *ochiul boului*), E ***red goosefoot*** (R *talpa gâştei*), E ***viper's bugloss*** (R *viperină*), E ***yellow stagshorn fungus*** (R *barba caprei*).

As for the plant names containing other animal names, they are as follows: E ***alder buckthorn*** (R *lemn câinesc* – instead of *buck* we have *câine* 'dog'), E ***black horehound*** (R *urechea porcului* – again the animal name in Romanian is 'pig', a mammal but from another order), E ***common duckweed*** (R *linte broştească* where 'duck' is replaced by *broască* 'frog', animals from different phyla), E ***horse chestnut*** (R *castan porcesc* – instead of *horse* there is *porc* 'pig', still a mammal, but again a different order), E ***oyster fungus*** (R *păstrăv de fag* – *păstrăv* meaning 'trout', still an aquatic animal but from another phylum), E ***stagshorn clubmoss*** (R *brânca ursului / barba ursului* – *urs* meaning 'bear').

There are also instances where Romanian common plant name does not include an animal name: E ***bird cherry*** = R *mălin negru*, E ***cowberry*** = R *merişor, smirdar, afin roşu, afin, coacăz de munte*, E ***crab apple*** = R *măr pădureţ*, E ***dog rose*** = R *cacadâr, măceş, răsură, rujă, rug sălbatic, trandafir sălbatic*, E ***dogwood*** = R *sânger,* E ***gooseberry*** = R *agriş*, E ***sea buckthorn*** = R *cătină albă*, E ***stagshorn clubmoss*** = R *pedicuţă*, E ***turkey tails*** = R *iască*, E ***wolf-fart puffball*** = R *pufai*, even though there are several Romanian names, none of them contains a colour name, etc.).

A last category of English common plant names containing animal names is the one with no Romanian counterparts whatsoever: ***common water-crowfoot***, ***fly orchid***, ***frog orchid***, ***greater butterfly orchid***, ***horseshoe vetch***, ***monkey flower***, ***pignut***, etc.

We also found 56 Romanian common plant names containing an animal name. Although all had an English counterpart, none of the English equivalents contained an animal name: R *băşica porcului* = E ***puffball***, R *băşina calului* = E ***giant puffball***, R *burete păstrăv* = E ***dryad's saddle***, R *buretele viperei* = E ***death cap***, R *cerga ursului* = E ***bracken***, R *cireaşa lupului* = E ***deadly nightshade***, R *coada rândunicii* = E ***liverwort***, R *coada şoricelului* = E ***yarrow***, R *coada-mâţei-de-baltă* = E ***sphagnum moss***, R *guşa porumbelului* = E ***bladder campion***, R *iarba limbricilor* = E ***milk vetch***, R *lemnul câinelui* = E ***spindle tree***, R *ochiul lupului, limba-boului* = E ***bugloss***, R *oiţe* = E ***wood anemone***, R *pălăria-şarpelui, burete şerpesc* = E ***parasol mushroom***, R *păstrăv de pădure* = E ***peppery milk cap***, R *pita vacii* = E ***cep***, R *porumbel* = E ***blackthorn***, R *porumbul şarpelui* = E ***lords and ladies***, R *spinarea lupului* = E ***lady fern***, R *struţişori* = E ***lesser clubmoss***. However two of them contained animal body parts (R *copita calului* = E ***hoof fungus***, R *untul vacii* = E ***green winged orchid***, ***lady orchid***). As for the animal names in the Romanian common plant names we analysed, they range from annelids (R *iarba limbricilor* – the animal name used here is 'round worm') to fish (R *limba peştelui* the animal name used here is 'fish') amphibians (R *izma-broaştei* – the animal name used here is 'frog'), reptiles (R *buretele viperei* – the animal name used here is 'viper'), to birds (R *coada rândunicii* – the animal name used here is 'swallow'), and mammals (R *ţolul lupului* – the animal name used here is 'wolf').

Discussion

In the previous section, we brought forth an account of English common plant containing an animal name and their Romanian counterparts. We have seen that most English plant names entail an equivalent, but a very small number of plant names still remains which bear no equivalent in the Romanian language. This can be explained by the fact that some of the plants common in the UK are not found in the native Romanian flora. The choice of the animal name coming up in the English and Romanian equivalents may be determined by the shape of the plant or by some kind of resemblance with the animal referred to.

If we look at numbers, we can see that of the 80 English plant names studied, only 7 do not account for a Romanian equivalent, which represents 14%. Deepening our investigation, we could find out that, from the remaining 73 Romanian counterparts, 32 (i.e. 40%) do not entail animal names, that 22 of the Romanian common plant names (i.e. 29%) observe the English choice in animal names, while the other 14 (17%) take different animal names. Thus, though most English common plant names containing an animal name entail a Romanian equivalent, these counterparts do not always take a Romanian animal name.

As for the 56 additional Romanian common plant names containing an animal name, 2 of them (3.5 %) entailed animal body parts.

Conclusions

After having analysed the 80 common English plant names containing an animal name and their Romanian counterparts, we have found that:

- 7 English plant names do not have a Romanian counterpart, which may be determined by the fact that the respective plants are not native to Romania;
- 14 English plant names contain a different animal name in Romanian;
- 22 English plant names have the same animal names as in Romanian;
- 32 English plant names do not have animal names in the Romanian counterpart.

Comparing the percentages, we can see that even though the percentage of plant names containing the same animal name in Romanian is larger than that of plant names containing a different animal name (29% vs. 17%), the greater percentage (i.e. 40%) of plant names do not contain any animal name at all.

After extending our research on Romanian common plant names containing animal names, we found not only English common plant names whose Romanian counterparts do not contain animal names, but also Romanian common plant names containing an animal name, whose English counterparts do not entail animal names.

References

Baker, M. (Ed.) (2001). *Routledge Encyclopaedia of Translation Studies*. New York: Routledge.

Bejan, D. (1991). *Nume româneşti de plante*. [Romanian Plant Names]. Cluj-Napoca: Editura Dacia.

Borza, Al. (1968). *Dicţionar etnobotanic*. [An Ethnobotanical Dictionary]. Bucureşti: Editura Academiei.

British Wild Flowers. Online: http://www.british-wild-flowers.co.uk/.

Groszler, Astrid-Simone & Ivanovska, Biljana. (2010). English Common Names of Plants and Their Romanian Counterparts. In Georgeta Raţă (Ed.), *Teaching Foreign Languages: Languages for Special Purposes*. Newcastle upon Tyne: Cambridge Scholars Publishing. 36-41.

Groszler, Astrid-Simone. (2009). English Plant Names Containing an Animal Name and Their Romanian Counterparts. *Journal of Linguistic Studies 3 (2)*: 17-20.

Lungu Badea, G. (2003). *Dictionary of Terms Employed in the Theory, Practice and Didactics of Translation*. Timişoara: Ed. Orizonturi.

Raţă, Georgeta, Grigoriu, Alma-Lioara & Imbrea, Ilinca-Merima. (2005). *British Flora. Names of Plants in Latin, English and Romanian*. Timişoara: Ed. Mirton.

COMMON NAMES OF YELLOW WILD FLOWERS IN ENGLISH AND ROMANIAN

OANA BOLDEA

Introduction

Plants have played an important role in people's lives, whether it was for their nutritional values, their medicinal uses, or, simply, for their beauty. Even from times immemorial, people started to recognize them and to keep them in mind for one or more of their characteristics. Many of the names given to the plants in the old times, the so-called "common names" or "vernacular names", can tell us many things about the plants in question.

It is extremely interesting to study wild plants rather than garden plants from the point of view of their vernacular names. Since garden plants have very well-known names, it seemed more intriguing to focus our attention on wild flowers, which are commonly met in Romania and Great Britain. People gave the flowers names they considered suitable, and we were extremely incited by the different approaches people used in the process of naming. We wanted to see whether these names tell us different things about the flowers and about the people who named them. Another aim was to identify possible similitudes between the processes of naming.

The idea of the study seemed of interest to us as a teacher of English teaching students in Horticulture. When teaching a language, one surely teaches also about the culture of that country. And we found that this process of naming wild plants, process which started a very long time ago, tells a lot about the way people think. Sometimes, the name provides information as to what practices were used involving that plant (that is the function of the plant, which can be medicinal, religious, practical or food-related). In other cases, the name gives a description of the flower in question, or tells about its whereabouts.

Material and Method

The material used in conducting this study comprised both Romanian and English books on the subject. The English book we focused on was *The Oxford Book of Wild Flowers* (Nicholson *et al.* 1982), which provided the English common names of the flowers. As it also contains pictures of the flowers themselves, it was a lot easier to identify them and then look for their Romanian names, as well.

The Romanian book which proved to be of utmost importance in our research was *Dicţionar etnobotanic* by Borza (1968). Fortunately, this dictionary includes both Romanian common names and Latin names of the plants, and also gives the common names in various languages. Usually, the languages given are languages spoken in Romania by minorities (German, Hungarian, and Russian). Still, as sometimes English names are also given, it was of real help in the study.

The etymology of the English words was found also in the *Webster Encyclopaedic Unabridged Dictionary of the English Language* (1996).

The material analyzed comprised 150 flower names. As we wanted to limit, in a way, the number of names under analysis, we chose to deal only with yellow flowers. But although the number of the plants is the one mentioned above, the number of names is far larger: for each plant, we found at least 3 names in each language. Sometimes we found even 10 names for the same plant. The explanation can be found in the fact that the names were all used in different parts of the country.

Romanian and English names were the interest of this paper, but sometimes Latin names proved important too, either because it was similar to the English or to the Romanian name, or because it brought some new information about the plant. That is why the Latin name is also sometimes mentioned and analyzed.

Results and Discussion

The results of the study are very inciting. Just as we supposed at the beginning, the names can tell many things about the plant itself. We can differentiate several classes of names of plants based on the process of naming.

First of all, there are the names which describe the **appearance of the flower**. Thus, there are names referring to the colour of the plant (either of the leaves or of its inflorescence). Such an example is the English ***silverweed***, named this way because of the colour of its leaves. The Romanian equivalent is *argentină*, with its variants *arginţică* and *arginţel*, clearly referring to the same thing as its English counterpart. We only found one English name of this plant, but eight Romanian names, not all referring to the colour of the plant. Some of them refer to its use: *buruiana scrântiturii*, *scrântitoare*, *scrintea*, *scrinteală*. All these names refer to the fact that the plant was often used in folk medicine to treat hurt ankles, or simply placed in shoes to keep the feet comfortable when walking long distances. Interestingly enough, although we found the same use in English, the name does not reflect it. This is not all: the Romanians also gave another name to this plant, *iarba gâştii*, which refers to the place where the plant can usually be found: on closely-grazed land, such as goose greens, hence the name. Actually, this name resembles very much the Latin name, which is *Potentilla anserina* L. Besides the names referring to the colour of the plant, there are those referring to the colour of its inflorescence. Thus, there is ***goldilocks***, which in Latin refers to the same characteristic: *Ranunculus auricomus* L., but which in Romanian is totally different: *buruiana de nouă daturi* refers to its use rather than to its appearance, while *talpa cocoşului* refers to the shape of its leaves. Another class is that of names giving information about the shape of the plant. We are talking about the shape of the leaves, as in *herb tuppence*, also known as ***pennywort***, which owes its name to the fact that the leaves are almost round, as a coin. The Latin name also has the same meaning: *Lysimachia numularia* L. Also, the shape of the inflorescence has been taken into consideration when naming other plants, such as *globe flower*, with the perfect Romanian equivalent *bulbuci*. It is clear that both refer to the globe-like inflorescence of the plant. Other names refer to the whole plant, as in ***creeping Jenny***, which is a plant that "creeps" on the ground. Although the name usually refers to the leaves, to the flower or to the entire plant, sometimes other parts of the plant were taken into consideration. Such an example is a plant whose all three names (English, Latin and Romanian) refer to the supposed resemblance of its fruiting receptacle to a mouse's tail: E ***mouse tail***, *Myosurus minimus* L., R *codiţucă*. Though less common than the ones already mentioned, there are instances when other characteristics are emphasized by the name of the plant. Thus, there is a plant named ***stinkweed*** obviously because of its strong smell. This characteristic seemed to strike the Romanians too, since the Romanian name is *puturoasă* 'stinky one'. The taste is also one of the features taken into account when naming plants. Such a plant is *wall pepper*, with the Latin name *Sedum acre* L., both referring to the bitter, peppery taste of its tiny swollen leaves. However, the Romanian name has no bearing on this characteristic: it refers to the fact that, in the past, it was used in folk medicine in treating haemorrhoids: *buruiana de trânji*, *trânjoaică* (as *trânji* is the vernacular name used in some parts of Romania for the abovementioned medical condition). There are also other physical characteristics underlined by the names. One such example is ***touch-me-not***, which owes its name to the fact that the ripe seeds explode when touched, shooting out their seeds. The Latin name refers to the same thing: *Impatiens carpensis noli-tangere* L. Obviously, the Romanians were not struck by this characteristic, since the Romanian name gives information about the use of this plant, rather than about anything else: *slăbănog, buruiana celor slabi*. The plant was used in folk medicine. ***Bladderwort*** is a good example for the class including other characteristics of the plants emphasized by their names. Its name comes from the fact that it has small bladders on its leaves. The Romanian name refers to something else, namely to the place where it grows, i.e. on water: *otrăţel de apă*. Although the English name of ***wayside cudweed***, also known as ***marsh cudweed***, refers to the fact that it grows in damp places and on waysides, the Romanian name of the same plant underlines one of its physical aspects: the fact that the leaves are woolly on both sides. Thus, the Romanian names are *iarbă flocoasă, iarbă lânoasă, lânărică*. ***Honeysuckle***, a well-known plant both in Romania and Great Britain, has been given this interesting name because its light colour and strong scent attract moths and bees that pollinate it while gathering nectar. We have presented so far names of plants referring to one of their physical characteristics. But more often than not, the name refers to more than one such characteristic. For instance, the name ***buttercup*** refers to both the colour and the shape of the flower. The Romanian name does not refer to the same thing: it is called *piciorul cocoşului* 'cock's foot', due to the shape of its leaves.

Besides the special characteristics of one plant or another, the names also refer to the **places where they grow**. These names make up the second large class of names classified according the naming process. Thus, there are names such as ***rock rose***, due to the fact that it grows on chalk and limestone. As a matter of fact, it is not a rose at all, but it resembles one, and it was given its Latin name (*Helianthemum chamaecistus* L.), which means 'sunflower', probably because the flowers open fully only in bright sunshine. The Romanian name keeps the meaning of the Latin one: *ruja soarelui*. ***Woodbine*** (better known as ***honeysuckle***) is yet another example for this class. The name refers to its habit of twining lightly round the trees and shrubs on which it climbs. The already-mentioned ***wayside cudweed*** or ***marsh cudweed*** falls into the same category.

As stated above, many plant names refer to more than one characteristic. Here are some that combine **one physical characteristic with the place where the plant grows**. The ***yellow stonecrop*** refers both to the colour of its flowers and to the fact that it forms mats of short stems in dry places such as walls, roofs, sand dunes, and even grassland. It is also known as ***wall pepper*** and ***biting stonecrop***, both names clearly referring to the same habit of growing on walls; the second characteristic emphasized, however, is not the same as before, so we are not talking about the colour, but about the taste. *Pepper* and *biting* both indicate the peppery taste of the leaves. The same holds true for ***yellow waterlily***, which gives us information about the place where the plant is to be found (on water) and also about the colour of the flower (yellow).

A third large class of names is composed of the names referring to the **use people gave these plants**. Several under classifications are possible. First of all, there are the plants used in folk medicine and the names of which are given according to the disease they treat. In some cases, the plant is no longer used for the purpose stated by its name. This is what makes the study the more interesting, since by looking at a plant's name one can learn a lot about the past of the people that named it. Such a plant is the already-mentioned ***silverweed***, whose English name says nothing about the use, but whose Romanian name *buruiana scrântiturii* tells us that it used to be important in treating sore ankles. In the same way, the plant which in English bears the name ***creeping cinquefoil*** (referring to the fact that the flowers are made up of five petals each and also to the way it grows) has a Romanian name which indicates that it was used in folk medicine for the illness called *apucătură* (an illness characterised by sleepiness, spasms and convulsions). The Romanian name is *buruiana apucatului*: it indicates both what it is used for and also the fact that it grows wild, that it is not a garden plant. There is another plant the names of which refer to its use in traditional medicine: the plant called ***tormentil*** (word coming from *torment* 'pain, anguish, annoyance'), which clearly reflects that it was used against something that was troubling sick people. Actually, in Great Britain its roots used to be boiled in milk to treat diarrhoea in children and calves, thus putting a stop to an annoying situation, as the name suggests. Interestingly enough, although the plant was recognized as a medicinal plant in Romania, too, its use is totally different, as the name lets us know. It is called *strâmsurea*, and it is still used in folk medicine for the disease called *strânsura la coş*, a disease of the lungs. The ***yellow loosestrife***, the English name of which is a mistranslation of the Latin *Lysimachia* (the plant is said to have been named after a certain *Lysimachos*), has four distinct Romanian names, all providing different information about the plant, depending on the part of the country they come from. Thus, the generic name *gălbenele* refers merely to its colour. Another name, *iarba sângelui*, is a proof that in traditional medicine the plant was used for diseases of the blood (both in humans and in animals). The third name found for this plant (*Lysimachia vulgaris* in Latin) is *floare de lungoare*. As *lungoare* is the common Romanian name for *typhoid fever*, we easily understand another use this plant had in folk medicine. And it would be misleading to believe that the fourth name found for the same plant, *buruiana de argint*, refers to the colour of the plant, especially since the plant does not resemble silver at all. In order to really understand its meaning, we have to remember that *argint* has the meaning of 'syphilis' in the Romanian countryside. Then the mystery is solved: the plant was used in treating syphilis. This interpretation is supported by another name of the flower: *ruşinici* '(*Approximately*) shame on you'. If we think of the fact that this disease was considered shameful, everything makes sense.

A series of wild plants were also used for **religious or even superstitious reasons**. We placed the religious reasons in the same class with superstitious since, more often than not, these two are intertwined, as we shall see below. One example is ***St. John's wort***, which was associated (in many countries, not only in Britain) with Saint John the Baptist. It used to be hung in doors and windows to prevent the access of evil spirits. It has the same name in Romanian too: *floarea lui Ioan* ('John's flower'), *iarba lui Sfântu Ion* ('St. John's grass'). Somewhat related to these names are other two names: *iarba crucii* and *iarba spaimei* – which again tell us about the mixture between popular beliefs and religion. In Romania, it has the same use already stated for Britain: it is believed to protect against evil spirits. Still, in Romania we found other various names of the same plant which account for its being used in circumstances related to medicine. Thus, it is also known as *iarba sângelui*, because it is a traditional remedy for the cattle which eliminate blood in their urine. It is also called *pojarniţă* (< *pojar* 'measles') due to its use in treating measles. Another name for ***cowslip*** is ***paigle*** because the heads of nodding flowers are thought to symbolize St. Peter's bunch of keys. ***Bedstraw*** is known in Romanian as *floarea lui Sântion*, again related to Saint John. It is the best example of how religion and superstition intertwine (another name for it is *sânziene*, or *drăgaică*), as the same flower has names which refer to religion (St. John the Baptist's birth) but also names indicating feasts having nothing to do with religion, but with superstition. It is true that the two celebrations follow shortly one after the other. The plant is thought to bring luck, keep the cattle in good health, wealth, and a long and happy life. That is why people used to hang it at their gates, bring it in their homes and make wreaths of it which they threw on the roofs of their houses. Yet another Romanian name of this plant has nothing to do with beliefs, whether religious or not. It refers to a much more practical use – curdling milk to make cheese: *închegătoare* or *smântânică*.

The latter name is to be included in the last class of names we will deal with here: the class of names indicating what we called **practical uses**. The English name of one and the same plant, ***yellow bedstraw*** or ***lady's bedstraw***, refers exactly to such a use, since in old days people used to sleep on a mattress of dried bedstraw and other plants covered with a sheet. Such a mattress smelt pleasant and could be easily burnt and renewed when it became soiled. ***Wormwood*** owes its name to the fact that its aromatic scent made people use it to keep away moths and other insects. There is another plant the names of which are the perfect translation from one language into the other: ***broom***, with the Romanian equivalent *mături*. It is clear that this plant has been used for the same thing in both countries, i.e. in broom-making. Even the Latin name reflects the same thing. It is called *Sarothamus scoparius* L., the first word meaning 'a shrub growing like a broom', and the second word having the meaning of 'a broom made of twigs'. Another use of this plant was that of keeping witches away (probably because the link existing in superstition between brooms and witches). The name ***dyer's greenweed*** tells a lot about the use of the flower. The plant was once used because a greenish-yellow dye was obtained from the leaves and stems – as both English and Latin names suggest (Latin *Genista tinctoria* L., where *tinctoria* means 'used in dyeing'). Although it is no longer used for this purpose, the name perpetuates its history. In Romania, it was used for the same reason: yet, the name tells us nothing about that. Instead, one of its names, *grozamă*, indicates that it was used in folk medicine against what the Romanians called *poceală* 'paralysis'.

Conclusions

As predicted at the beginning, the study of the naming processes in English and Romanian presented inciting results concerning not only the mechanisms of thought when naming the plants, but also the beliefs they had and the traditional ways of using the plants.

According to the process of naming, we could classify the names in three main classes (all having their own subdivisions):

- one including names which take into account the characteristics of the plant;
- another comprising names which indicate the places where the plants grow;

- names suggesting the uses that the plants have, or had at some time in the past.

This classification is by no means a clear-cut one, since many plant names can be found which indicate two or even all three of the abovementioned features of plants (e.g., ***dyer's greenweed***).

References

Bejan, D. (1991). *Nume româneşti de plante.* [Romanian Plant Names]. Cluj-Napoca: Editura Dacia.

Boldea, Oana. (2004). On the Common Names of Yellow Wild Flowers in Romanian and English – a Contrastive Approach. *Lucrări ştiinţifice. Seria agronomie 48.*

Borza, Al. (1968). *Dicţionar etnobotanic.* [An Ethnobotanical Dictionary]. Bucureşti: Editura Academiei.

Dicţionarul explicativ al limbii române. [An Explanatory Dictionary of the Romanian Language]. (1998). Bucureşti: Editura Univers Enciclopedic.

Flavell, Linda & Flavell, R. (1995). *Dictionary of Word Origins.* London: Kyle Cathie Limited.

Nicholson, B. E., Ary, S. & Gregory, M. (1982). *The Oxford Book of Wild Flowers.* Oxford: Oxford University Press.

Raţă, Georgeta. (1999). *English for Horticulturists.* Timişoara: Editura Mirton.

Vaczy, C. (1980). *Dicţionar botanic poliglot.* [A Multilingual Botanical Dictionary]. Bucureşti: Editura Ştiinţifică şi Enciclopedică.

Webster Encyclopedic Unabridged Dictionary of the English Language. (1996). New York: Gramercy Books.

TRUE AND FALSE BERRIES IN ENGLISH AND ROMANIAN

GEORGETA RAȚĂ

Introduction

Many English words have multiple meanings, which are all uses of the same word that have grown apart over time. These meanings are ultimately related. Thus, the word ***berry*** designates both 'any small, usually stoneless, juicy fruit, irrespective of botanical structure, as the *huckleberry, strawberry,* or *hackberry* ' and, in botany, 'a simple fruit having a pulpy pericarp in which the seeds are embedded, as the *grape, gooseberry, currant,* or *tomato'.* Or, this makes it a very misleading term from a botanical point of view: there are a lot of names of fruits that end in *-berry* owing their names to the fact that the fruits are small.

A ***berry*** develops from an ovary containing one or more carpels containing one or more ovules, so ***berries*** typically contain more than one seed (e.g. *grapes* and *goose**berries**).* A *tomato* also is classified as a ***berry****:* cut in half, it displays distinct sections, each representing a separate carpel with many seeds.

There are several different types of ***berries****:* a 'true' ***berry*** has a relatively soft pericarp with a thin exocarp or skin (e.g. *tomato, pepper, eggplant, grape,* and *persimmon);* a 'pepo' is a ***berry*** with a comparatively thick exocarp, or rind (e.g. *cantaloupe, watermelon, pumpkin, cucumber,* and *squash);* a 'hesperidium' is a ***berry*** with a leathery skin containing oils (e.g. citrus fruits, including *oranges, lemons,* and *grapefruits).*

Some fruits with the word ***berry*** in their names, such as *rasp**berry*** and *straw**berry**,* develop differently and are not really ***berries*** at all.

How to translate such English common names into another language (Romanian, for example), is a good question to ask, if we take into account that specialists such as agronomists or horticulturists need to do so mainly when dealing with specialised texts (literature). Can present dictionaries be of real help or not? And if they are, in what measure?

Material and Methods

We have inventoried 68 names of plant and/or fruit ending in *-berry* using some of the best encyclopaedias and dictionaries in the world. We then have tried to see if these names have Romanian equivalents or not, or if there is any other way lexicographers managed to help a non-native English-speaking specialist.

Results and Discussion

We have inventoried a number of 53 *'-berry'* plant names designing ***true berries***. These 53 common names have only 24 Romanian equivalents. They belong to the following 11 families of plants:

- ERICACEAE – 30 common names in English, with only 16 corresponding Romanian equivalents (the symbol Ø stands for 'no equivalent'): E ***American cranberry*** (*Vaccinium macrocarpon,* syn. *Oxycoccus macrocarpon*) large, tart, red fruit = R Ø, E ***bear huckleberry*** (*Gaylussacia ursina*) unpalatable reddish-black fruits = R Ø, E ***bearberry*** (*Arctostaphyllos uva-ursi, Rhamnus purshiana*) = R ***cruşân american, lemn-câinesc,*** *E* ***bilberry*** (*Vaccinium myrtillus*) edible sweet and black berries = R ***afin, E black huckleberry*** (*Gaylussacia baccata*) edible black fruits = R Ø, E ***blue berry*** (*Vaccinium myrtillus*) = *R* ***afină, coacăză neagră***, E ***bog whortleberry*** (*Vaccinium oxycoccos*) = *R* ***răchiţele***, E ***buckberry*** (*Gaylussacia ursina*) unpalatable reddish-black fruits = R Ø, E ***cowberry*** (*Vaccinium vitis-idaea*) = *R* ***merişor****, E* ***cranberry*** (*Vaccinium macrocarpum*) = *R* ***merişor American****, E* ***cranberry*** (*Vaccinium oxycoccos*) = *R* ***răchiţele****, E* ***cranberry*** (*Vaccinium vitis idaea*) = *R* ***merişor, merişoară****, E* ***dangleberry*** (*Gaylussacia frondosa*) dark-blue sweet fruits = R Ø, E ***European cranberry*** (*Vaccinium oxycoccos*) = *R Ø*, E ***foxberry*** (*Vaccinium vitis-idaea*) = *R* ***cruşân american, lemn-câinesc****, E* ***highbush blueberry*** (*Vaccinium corymbosum*) sweet blue or black berries = R Ø, E ***huckleberry*** (*Vaccinium myrtillus*) (*Gaylussacia*) = *R* ***afin****, E* ***June berry*** (*Sorbus aucuparia*) = *R* ***scoruş-de-munte****, E* ***large cranberry*** (*Vaccinium macrocarpon,* syn. *Oxycoccus macrocarpon*) large, tart, red fruit = R Ø, E ***lowbush blueberry*** (*Vaccinium angustifolium*) sweet blue or black berries = R Ø, E ***moss berry*** (*Vaccinium oxycoccos*) = *R* ***răchiţele****, E* ***mountain cranberry*** (*Vaccinium vitis-idaea*) = R Ø, E ***red bilberry*** (*Vaccinium vitis idaea*) = *R* ***merişor****, E* ***red huckleberry*** (*Vaccinium vitis-idaea*) = *R* ***merişor****,*

E ***red whortleberry*** (*Vaccinium vitis-idaea*) = *R* ***merişor-de-munte***, *E* ***small cranberry*** (*Vaccinium oxycoccos*) = R Ø, E ***southern gooseberry*** (*Vaccinium macrocarpon,* syn. *Oxycoccus macrocarpon*) large, tart, red fruit = R Ø, E ***swamp blueberry*** (*Vaccinium angustifolium*) sweet blue or black berries = R Ø, E ***whinberry*** (*Vaccinium myrtillus*) edible sweet and black berries = R Ø, E ***whortleberry*** (*Vaccinium myrtillus*) edible sweet and black berries = R ***afină***;

- ROSACEAE – 6 common names in English, with only 2 Romanian equivalents: E ***black chokeberry*** (*Aronia melanocarpa*) brilliantly, glossy black small fruits = R Ø, E ***Juneberry*** (*Amelanchier* sp.) = R Ø, E ***purple chokeberry*** (*Aronia prunifolia*) small fruits = R Ø, E ***red chokeberry*** (*Aronia arbutifolia*) brilliantly, glossy red small fruits = R Ø, E ***service-berry*** (*Amelanchier canadensis*) = *R* ***(specie de arbust)***, *E* ***small cranberry*** (*Vaccinium oxycoccos*) = *R* ***răchiţele****;*
- BERBERIDACEAE – 3 common names in English, with no Romanian equivalent: E ***(common) barberry*** (*Berberis vulgaris*) edible, strongly acidic bright coral red, cylindrical fruit = R Ø, E ***Japanese barberry*** (*Berberis thunbergii*) red fruit = R Ø, E ***Thunberg barberry*** (*Berberis thunbergii*) red fruit = R Ø;
- CAPRIFOLIACEAE – 4 common names with 1 Romanian equivalent: E ***American cranberry bush*** (*Viburnum americanum, Viburnum opulus* var. *americanum, Viburnum trilobum*) red berries = R Ø, E ***highbush cranberry*** (*Viburnum americanum, Viburnum opulus* var. *americanum, Viburnum trilobum*) red berries = R Ø, E ***sheepberry*** (*Viburnum lentago*) = R Ø, E ***sheepberry*** (*Viburnum opulus*) = R ***călin****;*
- GROSSULARIACEAE – 4 common names in English, with 2 Romanian equivalents: E ***American gooseberry*** (*Ribes hirtellum*) acidic berries = R Ø, E ***blackberry*** (*Ribes nigrum*) = *R* ***coacăz negru***, *E* ***European gooseberry*** (*Ribes uva-crispa,* syn. *Ribes grossularia*) acidic berries = R Ø, E ***gooseberry*** (*Ribes grossularia*) = *R* ***agrişă***;
- ACTINIDIACEAE – 1 common name in English with no Romanian equivalent: E ***Chinese gooseberry*** (*Actinidia chinensis, Actinidia deliciosa*) edible small, oval brownish-green fruit with a distinctive green flesh, with tiny purplish seeds surrounding a white core, better known under the name *kiwi* = R Ø;
- MELIACEAE – 1 common name in English with no Romanian equivalents: E ***Chinaberry*** (*Melia* sp.) yellow fruit = R Ø;
- PHYTOLACCACEAE – 1 common name in English with no Romanian equivalents: E ***pokeberry*** (*Phytolacca americana*) attractive, though often poisonous, rounded berries = R Ø;

- RANUNCULACEAE – 1 common name in English with 1 Romanian equivalent: E ***baneberry*** (*Actaea spicata*) fruits are white, red, or black berries = R ***arbalț***;
- ULMACEAE – 1 common name in English with 1 Romanian equivalent: E ***hackberry*** (*Celtis occidentalis*) = *R* ***celtis occidental, sâmbovină-americana***;
- VERBENACEAE – 1 common name in English with 1 Romanian equivalent: E ***pigeonberry*** (*Duranta erecta, Duranta plumieri, Duranta repens*) small yellow fruit = R Ø. Other 32 *'-berry'*-named plants design ***false berries***.

They have only 8 Romanian equivalents and they belong to the following 2 families:

- ROSACEAE – 25 common names in English, with only 5 Romanian equivalents: E ***American red raspberry*** (*Rubus idaeus* subsp. *strigosus*) black through purple and red to yellow drupes = R Ø, E ***beach strawberry*** (*Fragaria chiloensis*) the fruit is an aggregate of numerous nutlets = R Ø, E ***black raspberry*** (*Rubus occidentalis*) black through purple and red to yellow drupes = R Ø, E ***blackberry*** (*Rubus fruticosus*) a fruit which is an aggregate of small, purplish-black drupes = R ***mur***, *E* ***boysenberry*** (*Rubus ursinus, Rubus macropetalus*) a fruit which is an aggregate of small, purplish-black drupes = R Ø, E ***cloudberry*** (*Rubus chamaemorus*) edible orange-yellow berries = R ***(un fel de mur)***, E ***dewberry*** (*Rubus caesius, Rubus flagellaris*) a fruit which is an aggregate of small, purplish-black drupes = R ***mur***, ***rug***, *E* ***European red raspberry*** (*Rubus idaeus*) black through purple and red to yellow drupes = R Ø, E ***flowering raspberry*** (*Rubus odoratus*) = R Ø, E ***garden strawberry*** (*Fragaria x ananassa*) = R Ø, E ***loganberry*** (*Rubus ursinus, Rubus macropetalus*) a fruit which is an aggregate of small, purplish-black drupes = R ***(hibrid de zmeură şi mure)***, *E* ***meadow strawberry*** (*Fragaria virginiana*) the fruit is an aggregate of numerous nutlets = R Ø, E ***ornamental raspberry*** (*Rubus odoratus*) = R Ø, E ***raspberry*** (*Rubus idaeus*) black through purple and red to yellow drupes = R ***zmeur***, ***zmeură***, E ***red raspberry*** (*Rubus idaeus*) black through purple and red to yellow drupes = R Ø, E ***sand strawberry*** (*Fragaria chiloensis*) the fruit is an aggregate of numerous nutlets = R Ø, E ***sow-teat blackberry*** (*Rubus allegheniensis*) a fruit which is an aggregate of small, purplish-black drupes = R Ø, E ***strawberry*** (*Fragaria elatior*) = R ***căpşun***, E ***strawberry*** (*Fragaria vesca*) = R ***frag***, ***fragă***, ***fragi-de-pădure***, ***fragi-iepureşti***, E

thimbleberry (*Rubus odoratus*) = R Ø, E ***western dewberry*** (*Rubus ursinus*, *Rubus macropetalus*) a fruit which is an aggregate of small, purplish-black drupes = R Ø, E ***white blackberry*** (*Rubus* sp.) a fruit which is an aggregate of small, white drupes = R Ø, E ***wood strawberry*** (*Fragaria vesca*) the fruit is an aggregate of numerous nutlets = R Ø, E ***woodland strawberry*** (*Fragaria vesca*) the fruit is an aggregate of numerous nutlets = R Ø, E ***youngberry*** (*Rubus ursinus*, *Rubus macropetalus*) a fruit which is an aggregate of small, purplish-black drupes = R ***(specie de mur american)***;

- MORACEAE – 7 common names in English with 3 Romanian equivalents: E ***black mulberry*** (*Morus nigra*) compound fruits = R ***agud negru***, ***dud negru***, E ***fruitless mulberry*** (*Morus alba*, *Morus bombycis*) compound fruits = R Ø, E ***mulberry*** (*Morus*) = R ***dud***, E ***red mulberry*** (*Morus rubra*) compound fruits = R ***dud roşu american***, E ***silkworm mulberry*** (*Morus alba*, *Morus bombycis*) compound fruits = R Ø, E ***Texas mulberry*** (*Morus microphylla*) compound fruits = R Ø, E ***white mulberry*** (*Morus alba*, *Morus bombycis*) compound fruits = R Ø.

Conclusions

The 53 common names of *true berries* have 24 Romanian equivalents (45.28%), which is better than in the case of the 32 common names of *false berries* for which there are only 8 Romanian equivalents (25%). The total percentage reaches 37.64% (32 Romanian equivalents for 85 English common names ending in *-berry*).

This means that the specialist has either to render the English common name by a scientific (Latin) name (e.g. E ***American cranberry*** = R *Vaccinium macrocarpon*, syn. *Oxycoccus macrocarpon*, with additional description of the plant or fruit: large, tart, red fruit), or by an explanatory phrase: E ***service-berry*** (*Amelanchier canadensis*) = R ***(specie de arbust 'a bush species')***.

References

Bejan, D. (1991). *Nume româneşti de plante*. [Romanian Plant Names]. Cluj-Napoca: Editura Dacia.

Benson, M., Benson, Evelyn & Ilson, R. (1990). *The Combinatory Dictionary of English*. Amsterdam – Philadelphia: John Benjamins Publishing Company.

Borza, Al. (1968). *Dicţionar etnobotanic*. [An Ethnobotanical Dictionary]. Bucureşti: Editura Academiei.

Chalker, Sylvia & Weiner, E. (1994). *The Oxford Dictionary of English Grammar*. London - New York - Sydney - Toronto: BCA.

Leviţchi, L. & A. Bantaş. (1995). *Dicţionar englez-român*. [English-Romanian Dictionary]. Bucureşti: Ed. Teora.

Leviţchi, L. (Ed.). (2004). *Dicţionar englez-român*. [English-Romanian Dictionary]. Bucureşti: Ed. Univers Enciclopedic.

Longman Dictionary of English Language and Culture (2003). London: Longman.

Raţă, Georgeta. (2006). A Case of Polysemy in the English of Horticulture: True and False *Berries*. *Language and the media – One language: many worlds*. Zagreb – Split: HDPL. 603-612.

Raţă, G. (2005). True and False *Berries* in English and Romanian: A Contrastive Approach. *Scientific Works L (6), Plovdiv, Bulgaria*: 411-414.

The Random House Dictionary of the English Language. (1968). New York: Random House.

Webster Comprehensive Dictionary. (1995). Chicago: J. G. Ferguson Publishing Company.

Webster Encyclopedic Unabridged Dictionary of the English Language. (1996). New York: Gramercy Books.

Webster's Universal Dictionary & Thesaurus (2003). New Lanark: Geddes & Grosset.

METAPHORS IN COMPOUND ENGLISH AND ROMANIAN PLANT NAMES

ALINA-ANDREEA DRAGOESCU

Introduction

Metaphor is highly used in common speech and plant name coinage is especially abundant in metaphorical processes. This figure of speech associates nature to human beings by means of characteristic features or salient properties. Hence, it may be argued that plant names originating in anthropomorphic notions build upon the complementarities between nature and human nature in a harmonized view of the living world. It is our aim to analyse the role that this specific type of metaphor plays in scientific, as well as vernacular denominational practices. Thus, anthropomorphized plant names illustrate the major theme of anthropomorphism, which is undoubtedly a breeding ground for metaphors. Being closely connected with myth and fable, personification also steps into play, as a trope by which non-human beings are ascribed human characteristics, endowed with human qualities, represented as possessing human form, or spoken of in anthropomorphic language. Besides explaining such analogies between nature and human nature, another aim of this study is to find explanations for the choice of some plant names which are often related or analogous in English and Romanian, being calqued after the Latin denomination. However, names are sometimes surprisingly diverse and distinctive from a semantic point of view, revealing justifications for their chosen names at the same time. The interpretation of metaphorical names often results in the attribution of surfacing properties, which are not normally associated with that which is signified, according to the rules of **semantic composition** *a posterior* devised by linguists. Thus, Bejan (1991) explains that metaphor – the basic process of denominational practice in popular botanical vocabulary – is based on two elements: an identification morpheme of the plant class and a specific variable which specifies different plant properties or marks an opposition within the same class. Nonetheless, the present interpretation of metaphor primarily aims at explaining how these properties are derived by their imaginative inventors, but also how metaphor is created and understood. We argue that this resourceful process of imagination involves no special interpretive mechanism that pragmatics may provide, but stands to reason in natural ways. Our main concern is with the 'emergent property' issue, typical of pragmatic approaches to metaphor, as well as the truth-conditional pragmatic approach.

Material and Method

We have drawn an inventory of metaphorical constructions, advancing explanations for various associations between plants and human beings, with the help of etymology and pragmatic inference. English botanical and vernacular plant names, as well as their Romanian counterparts are analysed in order to reveal symbolic connotations of names scientists or simple people have attributed to plants. The contrastive analysis aims at a comparative study of equivalencies between Romanian and English, as well as of their etymological source and of the connection between their origin and meaning. Latin scientific plant names are dissected where scientific etymologies are significant and the metaphorical function of the popular plant name concurs with the Latin scientific name. Therefore, the applied linguistics approach of the analysis makes use of etymological and contrastive analysis tools.

Results

We have analysed English common plant names and their Romanian and scientific correspondents, grouped them into three categories – parts of the human body, beings and proper names, human moral traits and other abstractions – based on the same anthropomorphic metaphor which has been the working hypothesis of the paper.

Discussion

1. Parts of the Human Body

By analysing a large variety of plant names, it may easily be noticed that a particular feature is often associated with the encoded concept metaphors activate. For example, the feature 'bearded' appears in numerous plant names, whose Latin epithet is *barbata* or *barbatus*, meaning 'bearded', on account of the plant's hairy aspect. The English vernacular name ***beard-tongue*** and the Romanian *barba limbii* 'the tongue's beard' are almost perfect equivalents, both possibly calqued after the botanical name *Penstemon barbatus*. The Latin genus name comes from the Greek words *penta* 'five' and *stemon* 'stamen', while the species name alludes to the flower, which resembles a bearded mouth or a hairy tongue pulled out. Likewise, the Romanian name of *Sinningia barbata* is *Sinningia cu barbă* 'bearded Sinnigia', due to its covering of hairs, which is why it is closely associated with, hence activated by, the encoded concept *beard*. By a similar mechanism, two species of willow (*Salix babylonica*) are anthropomorphized in Romanian: *salcia plângătoare* (E ***weeping willow***), which has typical drooping branches and *salcia pletoasă* 'hairy willow', whose branches are strongly pendulous, such as hair tresses. The scientific name *babylonica* derives from Linnaeus' misunderstanding that it was the tree described in Psalm 137, whereas the popular name is originally and directly activated by the weeping-like aspect of its leaves and branches. Several plant names contain the epithet *capitata* (< L *capitatus* 'having a head'), given the shape correspondence: for instance, *Primula capitata*, whose common name *capitata primrose* is a **linguistic calque** of the scientific name. The scientific genus name is also maintained in Romanian and it is shortened from L *Primula veris* 'firstling of spring', feminine of L *primulus*, diminutive of *primus*, because it is one of the first flowers to bloom (OED). *Butia capitata* is a palm whose vernacular names in English (***palm***) and Romanian (*palmier*) have switched to the large group of hand-related etymologies. Interestingly, both these names derive from the Latin *palma* 'palm tree', originally meaning 'palm of the hand'. The tree was so called due to the shape of its leaves, like the fingers of a hand. Despite building upon a queer zoomorphic metaphor, ***foxgloves*** have equivalent anthropomorphized names in both Romanian (*degețelul roșu* 'little red finger') and Latin (*Digitalis purpurea*), the former being the exact translation of the latter. However, the English zoomorphic name might also be a corruption of an anthropomorphic metaphor, originally ***folksgloves***. This possibility is also confirmed by its other vernacular names – ***fairy-folks-fingers*** in Wales and ***bloody fingers*** in Scotland – alluding to the same metaphor as the scientific name. The generic part of the scientific name means 'finger-like', in reference to the tubular-shaped bell flowers of *Digitalis purpurea*, which could easily fit a human fingertip (WE). There is also an English variant of the popular plant name – ***lady's glove*** – which is closely related semantically to the Romanian and Latin names, as 'glove' and 'finger' both activate the common referent 'hand', which sets them in close propinquity. Closely connected from a semantic standpoint, *Boesenbergia pandurata* has been rendered analogously in both English (***fingerroot***) and

Romanian (*degetarița* – a feminized form derived from *degetar* 'thimble', from *deget* 'thumb'). This name was inspired by the rhizome of the plant which resembles a handful of fingers, as it belongs to the ginger family. The genus-name *panduratus* derives from Gk *pandoura* 'fiddle', after the god Pan, who used to play that instrument. The comparison obviously hints at the fiddle-shaped stamina. Moreover, ***five-finger fern***, also called ***maidenhair fem***, and its Romanian counterpart *ferigă* possibly refer to 'threadlike or hair-like fern', from L *fili-* 'thread', which also gave R *fir* 'thread', according to the etymologist's speculation. A more bizarre Romanian fern name for *Adianthum capillus-Veneris* is *buricul-Vinerei* 'Friday's bellybutton', which has deformed *capillus* 'hair' to *buric* 'bellybutton' and the name of goddess Venera into Vineri 'Friday', a mythological feminine being in Romanian fairytales. Other synonymic names for the same fern are the more common *părul-fetei* 'girl's hair', but also *părul Maicii-Domnului*, which corresponds to the English ***black maidenhair fern*** in reference to Virgin Mary. All this collection of names related to hair has so been called by reference to the gracefully arching leaves like maidens' hair tresses. Several other plants have been named in analogy with human hair (e.g. ***hairy bitter cress***, ***hairy stereum***, ***tufted hair grass***, ***wavy hair grass***) by translating the Latin epithets *hirsute*, *hirsutum* 'hairy', also due to the aspect of the plant. Numerous other plants are named after organs belonging to the human body, the majority of which can be explained by analogy to either their shape or, more rarely, their properties. For example, ***liverwort*** (*Marchantia polymorpha*) is similarly rendered in Romanian as *fierea pământului* 'gall of the earth', suggesting bitterness, gall or liver conditions. However, this erroneous association originated in the 9th century, when the pre-scientific medieval concept of the Doctrine of Signatures held that the outward appearance of plants and their similarity to other bodies indicated its uses. The concept was maintained by the herbalists of the Renaissance who concluded that the red sap of the ***bloodroot*** (*Sanguinaria canadensis*) would cure blood diseases or the fused leaves of ***boneset*** (*Eupatorium perfoliatum*) could heal broken bones. By the same token, the lobed liver-like appearance of liverworts led to the belief that it might function as a remedy for liver diseases. However, they do not appear to be of any use in the treatment of liver problems, as morphological plant signatures have not actually led to the discovery of any therapeutic value. The etymology of ***spleenwort*** (L *Asplenium adiantum* < Gk *a-* 'not' + *splen* 'spleen') presents us with an akin mystification. The confusing reference to ***spleenwort*** as a cure for spleen and liver problems originates in antique medicine. The ending -*wort* comes from OE *wyrt* meaning 'herb'. Likewise, *Hepatica acutiloba* is a type of liverwort whose leaves have acute tips. The genus name is derived from the Greek *hepar* 'liver' and the epithet means 'sharply pointed lobes'. This plant was believed, according to the Doctrine of Signatures, to treat liver diseases because the colour and shape of leaves resembles that of the organ (Armitage 2006). Another wort species, ***toothwort*** (*Lathraea squamaria*), has a very similar Romanian counterpart, *colţişor* 'little fang' (*Dentaria bulbifera*). Both the English and Romanian names relate to the foliage of the plant whose deeply cut lobes resemble human teeth. Other vernacular names include *muma pădurii* or *mama pădurilor* 'the mother of forests', which designate a hideous mythological being, and also *buricul pământului* 'bellybutton of the Earth'. In addition, the ***toothed wrack***, also called ***serrated wrack***, are adaptations of the Latin name *Fucus serratus*, the species name meaning 'notched like a saw' (< L *serra* 'saw'), displaying the resemblance with either teeth or saw teeth. The word *wrack* carries the sense of 'seaweed or cast up on shore', as in shipwreck: calqued after the Latin *focus*, it means 'seaweed, sea wrack, tangle'. Another anthropomorphised fern name, ***hart's tongue (fern)*** (*Phyllitis scolopendrium*), is adapted to Romanian as *limba vecinei* 'neighbour's tongue', perhaps alluding to the gossipy 'tongues' inspired by the long leaves. It is the narrow leaves that also suggested the English vernacular name *tongue*, as if the earth were sticking its tongue out. The word *hart* should not be mistaken for *heart*, as the OED cites its origin in the OE *heorot* 'male deer'. The zoomorphic image of a deer's rounded horns is much more suitable, despite the abandonment of one more anthropomorphic example from our study. Other similar names are often self-explanatory: ***dead (man's) tongue*** has a paralysing effect on the organs of speech (Grieve 1995), ***devil's tongue*** (L *amorphophallus* < Gk *amorphos* 'without form' and *phallus* 'phallus'), ***mother-in-law's tongue*** has very long leaves, ***painted tongue*** (< L *salpiglossis*) has very colourful, tongue-like petals – they allude to the shape of the prominent spadix which unwraps from the stem like a pulled out tongue or phallus. Romanian counterparts of these names are sometimes perfect translations: *limba soacrei* for ***mother-in-law's tongue***. Some of them combine parts of the body with names of human beings.

2. Beings and Proper Names

A large number of flowers are associated with Christian figures in English, Romanian, French, as well as in other European languages spoken in Christian countries. Botanists, folklorists and lexicographers have documented numerous plants named in symbolic association with

Virgin Mary, Jesus Christ, and the saints. Before the age of Christianity, many flowers had been associated with pagan deities, nymphs and other fabulous creatures. However, with the advent of Christendom, religious names amply replaced older names. This profuse flower symbolism illustrates popular faith, while drawing on the metaphor of garden as paradise, which was applied to actual plants in European rural traditions. Due to the process of Christianization, a number of feminine names changed from scientific names alluding to ancient or medieval mythology into names with religious connotation. For instance, *Daphne mezereum* becomes ***lady laurel*** and *Andromeda polifolia*, from the Latinized version of the Greek mythological character Andromede, becomes ***bog rosemary*** in folk imagination. Such is also the case of ***(Our) Lady's-mantle*** or ***Mary's mantle***, whose scientific name *Alchemilla acutiloba* builds on a medieval alchemists' metaphor. Such vernacular names incorporate an additional connotation of Mother Mary's ceaseless protection of the faithful from evil, just as any mother protects her offspring. The botanical genus name of *Thelypteris quelpaertensis* comes from Gk *thelys* 'female' and *pteris* 'fern', also maintaining a feminine metaphor in the English names ***maiden fern*** or ***queen's-veil***, with an added pious undertone. Moreover, the symbolical notion of Mary's loving motherhood was generically rendered by means of flower names such as ***sweet Mary***, ***Mary-love***, ***Mary's hand***, suggesting her mercy and protection for devout sons and daughters. The names ***Mary's milkdrops*** and ***Our Lady's milk herb*** are the religious variants for the more familiar ***lungwort*** (*Pulmonaria officinalis*), activating images of the Nursing Madonna. Some scientific names already contained religious metaphors: ***Virgin's bower*** (*Clematis virginiana*). An orchid from the genera *Cypripedium* is called ***Our Lady's shoes / slipper*** and ***Lady-slipper*** based on the fable that it sprung up under Mary's feet as she visited Elizabeth. Another related species, ***showy / Queen Lady's-slipper*** (*Cypripedium reginae*) derives its scientific name from Kypris, an early Greek name of Aphrodite, the goddesses of love, born on the island of Cypress: *pedilon* 'slipper' and *reginae* 'of the queen' (Armitage 2006). In addition, the shape of its inflorescence is very illustrative of an elegant lady's shoe. The botanical name of ***Lady's veil*** or ***baby's breath*** (*Gypsophila paniculata*) means 'lover of chalk', in reference to the type of soil in which this plant thrives. Romanians have associated the delicate white inflorescence with bridal purity, hence the name *floarea miresei*, which may also be considered a religious metaphor alluding to the Virgin bride. By the same marital metaphor, ***bachelor's buttons*** translates perfectly into Romanian as *butonii burlacului*, but it can also be encountered as ***Mary's crown*** in English. Its botanical name

Centaurea cyannis derives from L *centaurus* < Gk *kentauros* 'centaur', a half man-half horse creature from Greek mythology, and probably the centaur named Chiron, famous for his knowledge of the medicinal uses of plants. Furthermore, Mary's queenly attributes are activated by the plant name ***Mary's crown***, just as the ***Virgin lily*** or ***Madonna lily*** (*Lilium candidum*) call to mind the same imperial metaphor. Another common medieval name for the *Amaryllis belladonna* lily was ***beautiful lady***, the literal translation of the Latin epithet, in reference to the Virgin's features. Mary's mediation on behalf of humans is symbolized by flowers such as ***Our Lady's keys***, ***ladder to heaven***, or ***Mary's candle***. Her sorrow for the human race is evoked by other names like ***Mary's tears***, ***Queen's tears***, or ***Our Lady's tears*** for the ***lily-of-the-valley***. Their Romanian counterpart, the neutral diminutive *lăcrimioare* 'small tears', also reflects the Christian legend that the flowers have blossomed from Mary's tears beside the cross. On the other hand, the Romanian name *lacrima Maicii Domnului* ('Our Lady's tear') is a totally different plant, the ***waxplant*** or ***waxflower***, given this metaphoric name because the abundant nectar they produce suggests tear shedding. Thus, some flowers preserve a religious connotation only in Romanian. Another example would be the name *mâna Maicii-Domnului* 'Mother Mary's hand', a heath spotted orchid, partly sharing the hand metaphor inspired by the L *Dactylorhiza* (< Gk *daktylos* 'finger' and *rhiza* 'root'), so named due to its finger-like root nodules. Moreover, *palma-Maicii-Domnului* 'Mother Mary's palm' is another type of orchid which only has an original religious connotation in Romanian. The Latin name *Orchis maculata*, like all orchids, derives the genus name from Gk *orkhis* (< *orkheos*) meaning 'orchid', literally 'testicle', so called because of the shape of its root. Another explanation would be the fact that the most beautiful part of the flower that we smell and admire represents, in fact, the plant's genitalia. The Romanian names for *Asplenium adianthum*, *părul-Maicii-Domnului* 'Virgin Mary's hair', *părul-fetei* 'girl's hair', or *părul-orfanei* 'orphan girl's hair' belong to a set of semantic equivalents for the English ***maidenhair fem***, discussed above. The scientific name comes from the Gk *adiantos* 'unwettable', as its fronds are particularly water-repellent, and *nigrum* 'black' describing the scales colour. Even though ***field bindweed*** (*Convolvulus arvensis*) displays an anthropomorphic metaphor only in the Romanian name *poala Maicii-Domnului* 'Virgin Mary's Lap', its medieval name was nonetheless ***Our Lady's little glass***. ***Costmary*** (*Tanacetum balsamita*) has two comparable Romanian forms – *izma Maicii-Preciste* 'Virgin Mary's mint' or *busuiocul-sfintelor* 'Saints' basil'.

The cases where Romanian plant names differ from both the English and the Latin names display the originality of Romanian creators of metaphors, which have not always been translated or calqued after foreign or scientific names. In addition, the proliferation of miraculous legends of the Madonna often involved plants. Thus, flowers such as ***Our Lady's bedstraw*** were said to have bloomed when Jesus was laid on them, testifying to his divinity. The Romanian names of the plant are *sânziene* and *floarea lui Sfântul Ion* 'Saint John's flower', in association to Christian fables. In Romanian folklore, *noaptea de sânziene* 'lady's bedstraw night' (June 23rd) is considered magical because the gates of heaven open up and fairies called 'sânziene', such as the flowers, descend upon earth. The scientific name of the genus, *Gallium* (< Gk *gala* 'milk'), was given to the property of the plant, which was used to curdle milk. Several 'milk' plants were said to have received white spots from the drops of the nursing Madonna's milk. In a reverse direction, botanists gave the ***milk thistle*** its scientific name, *Silybum Marianum*, based on the religious legend. While the common name merely refers to the white veins on the leaves resembling spilled milk, the species name *Marianum* clearly honours Virgin Mary. Other metaphors pointing to the glorification of Mary are related to an array of ornaments. For example, ***jewelweeds*** (*Impatiens wallerana*) is also known by the name ***Our Lady's earrings***, metaphoric jewellery fancied by the faithful for the adornment of Mary. The common name is determined by the shape of flowers, which hang like precious earrings, whereas the scientific genus name is probably a semantic metaphor of the violent seed dispersal (Armitage 2006). Thus, 'Mary'-flowers came to metaphorically connote a diversity of abstract symbols of Mary's virtues, evoking her immaculate maidenly spirituality. They have often received their names by association with features like purity, whiteness, delicacy, and their general characteristics as flowers. Additionally, healing plants especially esteemed were honoured with the name of Mary, whose role was to heal the wounds of the world. For example, the common name of *Calendula officinalis*, ***marigold***, is derived from 'Mary's gold' connoting the shining glory of Virgin Mary in Christian tales. Although legendary stories often provide the explanation for these and other names, the key is sometimes missing (Stokes 1984). Despite the vast set of religious names, many plants have preserved names referring to beings from ancient Greek and Roman mythology. For instance, *Atropa belladonna*, commonly known as ***belladonna*** or ***deadly nightshade*** in English, is similarly calqued in Romanian as *beladonă*, while also occurring profusely in anthropomorphic imagery, such as *doamna-codrului* 'forest lady' and *doamnă-mare* 'big lady'. The scientific name of this poisonous plant is a **blend** based on the name of Atropos, one of the three Fates in Greek mythology – significantly the one who cuts the thread of life, as the plant is highly poisonous – and the Italian phrase meaning 'beautiful lady'. It acquired this name due to the purpose it was used for, as its juice dilates pupils, therefore being employed as a cosmetic enhancer in the past. A final example, the English (*Venus flytrap*) and the Romanian (*Capcana lui Venus*) names of the plant *Dionaea muscipula* are metaphors connoting more than the carnivorous plant's capturing capabilities, as it entraps insects. In turn, Venus is part of the name not because of its beauty, but because she is the Goddess of love and beauty, but because the two red lobes surrounded by hairs and sensitive to touch reminded those who coined the name of female genitalia. While the genus name originate in the Gk *Dione*, a metronymic name for Aphrodite, the specific name *muscipula* does not mean 'flytrap' in Latin, but 'mousetrap' (< L *mus* 'mouse' + L *capere* 'to seize, grasp'). Thus, it seems that the venereal metaphor of the Latin name was intended as a metaphor for the love-goddess' gripping device that captures incautious mammals. Finally, a wide range of plants have been given nasty names, by association with human genitalia. For want of space and discomfiture, the plentiful inventory could not be listed here, as this area is infinitely resourceful.

3. Human Moral Traits or Other Anthropomorphic Abstractions

Plant names are sometimes metaphorically christened as if they were endowed with human feelings or even abilities. For instance, ***walking ferns*** (*Camptosorus rhizophyllus*) have pointed fronds that root at the tip, which suggest the fact that they are in motion. The Latin etymology is derived from the Greek *kamptos* 'bent', given the curving appearance and the flexion of the fern. The epithet *rhizophyllus* 'root-leaved' also alludes to the fact that the leaves soon become roots in their turn. ***Sneezewort*** (*Achillea ptarmica*), also called ***bastard pellitory***, ***fair-maid-of-France***, or ***sneezeweed***, is closely related to the Romanian *strănutătoare* 'sneezing plant', which implies the ability to sneeze, as if possessed by the plants themselves. Nonetheless, the plant does provoke sneezing, as verified by the species name *ptarmica* (< Gk *ptairo* 'sneeze'), meaning 'causing sneezing'. The genus name *Achillea* originates from Linnaeus (< L *achillea* < Gk *Akhilleus* 'Achilles', because, as a curative plant, it was associated with his healing of Telephus in the legends of Olympus. Flowers seemed to also have feelings, as people have often connected to nature at times of grief or distress. The ***melancholy thistle*** (*Cirsium* *helenioides*) probably got its name from the way the solitary flower-head positioned at the top of a branchless stem hangs downwards. *Cirsium* comes from both the Greek words *kirsion* 'thistle' and *cirsos* 'swollen vein, varices', which thistles were faultily thought a remedy for. *Floarea-jelei* 'the flower of sorrow' and *iarba urâciunii* 'the grass of ugliness' (*Linaria acutiloba*) have a zoomorphic English correspondent – ***toadflax***. For odd reasons, it has acquired some more vernacular names containing anthropomorphic metaphors among which some are at odds: ***brideweed***, ***bridewort***, and ***dead men's bones***. The first two connote the start of a new life, while the latter implies the opposite. The closely connected *floarea-suferinței* 'flower of sufferance' or *floarea pasiunii* (*Passiflora incarnata*) has a perfect English equivalent, ***passion flower***, and they are both related semantically to human sufferance rather than carnal passion. However, ***passion flowers*** are often given a mystifying sense of romantic or even sexual innuendo. As early as the 15th century, Christian missionaries noticed the distinctive structure of this plant, which they associated symbolically to the sufferings of Christ on the Cross. For instance, the radial filaments of the flower mimic the round crown of thorns, the ten petals and sepals represent the apostles (St. Peter the denier and Judas the betrayer being excluded for lack of more petals), the stigmata represent the nails, and the anthers stand for Jesus' wounds, etc. This symbolism is also authenticated by the etymology of the word *passion* (< L *passionem* 'suffering, enduring' < *pati* 'to suffer, endure'. By analysing English and Romanian counterparts of scientific plant names, we have found that an impressive number play upon anthropomorphic metaphors. Furthermore, the samples discussed above are semantically connected more often than not. This can be explained by the fact that their shape or properties inspired people to give them similar names. In some cases, as illustrated above, the vernacular choice name was also informed by the awareness of the scientific Latin name of the plant. This reveals the fact that we deal with learned and knowledgeable people who possessed accurate botanical information rather than making simple analogies, as in the case of most vernacular metaphors. Thus, numerous Romanian names have been found to be adaptations of the Latin name and (partial) correspondents of the English name. In extremely rare cases, the English plant names have no connection whatsoever with the Romanian ones, being original innovations in both languages. While some names are anthropomorphic in one language, their equivalents are zoomorphic in the other. To illustrate both cases, we may note ***hart's tongue*** (< OE *heorot* 'deer') which is, in Romanian, *limba vecinei* 'neighbour's tongue'. Few names are downright contradictory (e.g. *floarea-jelei* 'flower of sorrow' or *iarba urâciunii* 'the grass of ugliness' are in English ***brideweed***, ***bridewort***, but strangely also ***dead men's bones*** at the same time). Some plants have undergone alteration, losing their anthropomorphic reference in either English or Romanian, though they had originally been akin. For instance, the ***field bindweed*** (*Convolvulus arvensis*) has preserved an anthropomorphic metaphor only in Romanian (*poala Maicii-Domnului*), while its medieval English name was initially ***Our Lady's little glass***. However, few plant names differ significantly, while most resemble and call to mind analogous metaphors, as we have seen above.

Conclusions

Although too few plant names have been analysed for want of space, it may still be noticed that most of them display the pervasiveness of the mimesis metaphor. In addition, they often take their English and Romanian names after the Latin (scientific) name, upholding Latin as the language of universal understanding, which opens up doors for taxonomic information. The analysis of plant name metaphors has revealed that most plant names are the ingenious result of well-informed naturalists or, in some cases, the work of most inspired simple people, displaying an infinite grasp of the natural world. Plant names disclose essential information about plants, especially if they are curative or toxic and they may enlighten us on a series of properties which could be of use. For that reason, understanding denominational practices and their etymologies is essential, as certain metaphors activate specific aspects or features of the plant. Most significantly, few words fail to live up to their names, whereas most names have at least one convincing reason to bear certain words. We have yet to discover the genuine importance of names and their full connotations, for truth and destiny lie within them. Vernacular metaphors have undeniably enriched botanical nomenclature, thus being collected by specialists and maintained alongside scientific names. Moreover, local names play a very important role in the ethno-botanical study of other aspects of life, pointing to the vast knowledge people had about their surrounding universe. Ultimately, anthropomorphised metaphors indicate the connection between nature and human nature, which points to a harmonized view of life as interconnectedness.

References

Armitage, M. A. (2006). *Armitage's Native Plants for North American Gardens*. Portland, Oregon: Timber Press.

Bejan, D. (1991). *Nume româneşti de plante*. [Romanian Plant Names]. Cluj-Napoca: Editura Dacia.

Borza, Al. (1968). *Dicţionar etnobotanic*. [An Ethnobotanical Dictionary]. Bucureşti: Editura Academiei.

Dragoescu, Alina-Andreea. (2010). A Study of Metaphors in Compound Plant Names. In Georgeta Raţă (Ed.), *Teaching Foreign Languages: Languages for Special Purposes*, Newcastle upon Tyne: Cambridge Scholars Publishing. 42-54.

Grieve, M. (1995). *A Modern Herbal*. Online: http://botanical.com/.

Online Etymology Dictionary. Online: http://www.etymonline.com/.

Raţă, Georgeta, Grigoriu, Alma-Lioara & Imbrea, Ilinca-Merima. (2005). *British Flora. Names of Plants in Latin, English, and Romanian*. Timişoara: Mirton.

Stokes, John. S. Jr. (1984). *Flowers of the Virgin Mary*. The Marian Library/International Marian Research Institute. Dayton, Ohio. Online: http://campus.udayton.edu/mary/resources/m_garden/JS-FOT VM-AVE.html.

The American Heritage Dictionary of the English Language. (2008). Boston: Houghton Mifflin.

The Columbia Electronic Encyclopaedia. (2004). Online: http://www.reference.com/browse/Doctrine+of+Signatures.

NAMES OF PESTS IN ENGLISH AND ROMANIAN

OANA BOLDEA

Introduction

The present paper deals with the problem of crop pests in a different manner than it has been discussed before. This time it is not a specialist in plant protection that approaches the matter, but a linguist. Teaching English to future agriculturists, we could not help but notice that English and Romanian terms referring to the same pest are not always perfect equivalents. This made us want to search more into the matter in an attempt to identify the differences among the terms in the two languages and, perhaps, even point out to the reason why they occur.

In this way, we hoped to clarify the subject for our students, willing to help them learn **agricultural English** more successfully. Also, the subject is of interest for the people that find themselves in the situation of having to translate from one language to another, whether they are professional translators or not.

Methods

The approach used in the paper is that of the comparative analysis. What we did was to compare English and Romanian terms and underline the differences in translation. Moreover, where it was possible to get at the root of the differences, we did, and we tried to explain why they occur. The material we have used consists of a number of books from various fields. Naturally, books on entomology proved to be extremely useful, such as Pălăgeşiu's *Curs de entomologie agricolă*, or *Quarantine Pests for Europe*, edited by Smith *et al.* (1995). Still, dictionaries were of utmost importance for the paper, whether they were "general" ones, such as *Dicţionarul explicativ al limbii române* (1998) and *Webster's Encyclopedic Unabridged Dictionary of the English Language* (1996), or specialized ones, like *The Wordsworth Dictionary of Botany* (1996) and *Dicţionar botanic polyglot* (1980). The corpus we worked on is based on the terms that appear in *Plant Protection* written by our colleagues (2002).

Results

What we found was that most terms are perfect equivalents. This means that they can be translated from one language into the other without any problem. Such terms are, for instance, *Ditylenchus dipsaci* Kuhn. whose equivalent is ***bulb and stem nematode*** in English and ***nematodul tulpinilor şi bulbilor***, in Romanian, or *Taeniothrips inconsequens* Uzel. with the English ***pear thrips*** and the Romanian ***tripsul părului***. Much in the same way, *Taeniothrips gladioli* Hall. is translated into English as ***gladiolus thrips*** and into Romanian as ***tripsul gladiolelor***, while *Acyrtosiphon onobrychidis* B.F. becomes ***green pea louse*** in English and ***păduchele verde al măzării*** in Romanian.

Another aspect we came across was that some of the Latin terms do not have equivalents in Romanian, as they have in English. This is what happens with *Aulacaspis pentagona*, for instance, for which the English term is ***papaya scale***, but which lacks the Romanian translation. The same holds true for *Parlatoria blanchardi*, with the English ***date palm scale***. The lack of Romanian equivalents in these cases is easy to account for: the reason is that the plants these pests attack are not grown in Romania. Since we do not have the plants, we do not have the pests in question. What the translator must do in such a circumstance is simply use the Latin name of the pest.

Discussion and Conclusions

What seemed very inciting when researching the names of pests was that when differences do occur, Romanian terms seem to be more concrete than English ones: they seem to give more information about the organism they name. For instance, *Anguina tritici* Steinb. has the English equivalent ***wheat nematode***, which indicates that it is a pest attacking wheat. The Romanian equivalent comes with new information. Thus, ***nematodul boabelor de grâu***, in addition to the plant the pest attacks, also indicates the part of the plant they feed on, namely the grains. The same happens with *Pratylenchus pratensis* De Man. While the English term ***meadow nematode*** lets us know that the pest appears on plants growing on meadows, the Romanian term, ***nematodul rădăcinilor de graminee***, clarifies the part of the plant attacked (i.e., the roots). For *Phylloxera vastatrix* Planch., the English ***grape phylloxera*** establishes that it is a pest attacking grapes, while the Romanian ***filoxera rădăcinilor de vie*** is much more exact, informing that actually it is not the grapes but the vine which is attacked, and not just any part of the plant, but the roots.

But it would be wrong to conclude that the new information brought about by the Romanian term has always to do with the part of the plant attacked, as discussed above. A good example is *Cryptomyzus ribis* L., for which the English term is ***red currant aphid***, establishing which plant is attacked. The Romanian ***păduchele galben al coacăzului roşu*** also gives the colour of the pest, thus differentiating it from any other aphid that might attack the same plant.

If the Romanian terms seem to be more exact, in English the same pest has often two or more names, which does not happen in Romanian. Synonymy, in this respect, seems to be a characteristic of the English language. Thus, *Heterodera rostochiensis* Woll. is ***nematodul auriu al cartofului*** in Romanian, and both ***golden nematode of potato*** and ***potato root eelworm***, in English. This is possible also because in English there are two terms for ***nematode***, while in Romanian there is but one. Another such pest is *Heterodera schachtii* Schmidt., which has one name in Romanian, ***nematodul sfeclei***, but no less than three in English: ***sugar beet nematode***, ***beet eelworm***, and ***beet cyst nematode***. Also, for *Abraxas grossulariata* L., there are three English terms (***currant moth***, ***currant worm***, ***magpie moth***), but only one Romanian term (***cotarul agrişului***).

Another interesting finding is that the same pests have often different names in English and Romanian because of the different aspects taken into consideration when naming them. Thus, for *Helix pomatia* L., the English term refers to the place where it can be found, ***grapevine snail***, while the Romanian term points to physical characteristics of the pest, ***melcul cu cochilie***. Much in the same way, the English term for *Piesma quadrata* Fieb. points to the part of the plant attacked, ***beet leaf bug***, while the Romanian one refers to the physical appearance of the bug, namely to the fact that it is small: ***ploşniţa mică a sfeclei***. This is exactly what happens with *Brachychaudus helichrisi* Kalt., for which the English term describes the effect the bug has on the plant, ***leaf-curling plum aphid***, while the Romanian name describes the physical appearance of the bug: ***păduchele mic al prunului***.

What seemed striking to us was that sometimes the names of pests differ so much from one language to the other that they indicate different plants as being attacked by the same pest. For instance, *Doralis fabae* Scop. has the English ***black bean aphid***, while the Romanian term is ***păduchele negru al sfeclei***. While the same physical characteristic is kept in both names (the black colour), the English term points out that it attacks bean plants, while the Romanian name points to the beet as the damaged plant. The same holds true for *Eulecanium corni* Bche., for which the English term is ***European peach scale*** and the Romanian ***păduchele ţestos***

al prunului points to a different plant attacked, namely the plum tree. A third example is *Eriosoma lanuginosum* Hartig: while the English term, ***woolly elm aphid***, establishes that it attacks elms, the Romanian one tells us that it is currants it damages: ***păduchele lânos al agrişului şi coacăzului***. A reason for this might be that the name was chosen in each language depending on the plant most often attacked by the pest.

References

Boldea, Oana. (2008). Plant Names in Romanian and English: A Contrastive Approach. *Scientific Papers. Faculty of Agriculture XL*: 157-162.

—. (2005). Names of Pests: A Contrastive Approach. *The 2nd International Scientific Conference "Rural Development". Proceedings II, 17-19 November, 2005, Lithuanian University of Agriculture, Kaunas, Lithuania*: 54-56.

Dicţionarul explicativ al limbii române. [An Explanatory Dictionary of the Romanian Language]. (1998). Bucureşti: Editura Univers Enciclopedic.

Pălăgeşiu I. (2002). *Curs de entomologie agricolă. Partea generală*. [A Course in Agricultural Entomology. General Part]. Timişoara: Ed. Eurobit.

—. (2002). Curs de entomologie agricolă. Partea specială. [A Course in Agricultural Entomology. Special Part]. Timişoara: Ed. Eurobit.

Raţă Georgeta, Manea, D., Giuchici, Camelia & Grozea, Ioana. (2002). *Plant Protection. A Compendium of Technical Terms in Latin, English, French and Romanian*. Timişoara: Editura Mirton.

Raţă, Georgeta. (1999). *English for Horticulturists*. Timişoara: Editura Mirton.

Smith, I. M., McNamara, D. G., Scott, P. R. & Holderness, M. (Eds.). (1995). *Quarantine Pests for Europe*. Wallingford: CAB International.

Usher, G. (1996). *The Wordsworth Dictionary of Botany*. Reston, VA: Wordsworth Editions Ltd.

Vaczy, C. (1980). *Dicţionar botanic poliglot*. [A Multilingual Botanical Dictionary]. Bucureşti: Editura Ştiinţifică şi Enciclopedică.

Webster Encyclopaedic Unabridged Dictionary of the English Language. (1996). New York: Gramercy Books.

COMMON NAMES OF PLANT DISEASES IN ENGLISH AND FRENCH

GEORGETA RAȚĂ, IONEL SAMFIRA
ND CAMELIA GIUCHICI

Introduction

A good knowledge of the plant disease denomination system can help both students and academics acquire languages for special purposes more efficiently. This paper attempts to provide both categories with a practical method of learning English for Special Purposes or French for Special Purposes based on the knowledge of the mechanisms that rule this type of denomination.

Material and Method

The comparative study of common names for plant diseases has been carried out on a corpus of 513 common names of plant diseases (of which 273 English and 240 French) selected among the data supplied by the HYPP Project, an encyclopaedic database on plant protection stored on CD-ROM whose main subject areas are dealt: Study of weeds, Plant pathology, and Agricultural zoology. HYPP exists in six languages: German, English, Spanish, French, Italian and Portuguese. The data are not simply translated but, in general, adapted to the biological situation of each country. HYPP has two aims: to give more information to a user who already has one or several hypotheses on the problems of a crop, using the index and the database; and to help in identifying a crop problem. HYPP is a tool designed for teaching, training and also for farmers and their development advisers. Teachers and their pupils find a source of important information for courses of phyto-technology or plant biology. HYPP is also of interest for farmers, development advisers, technicians of cooperatives and phyto-sanitary industries because it forms a unique work summarising plant protection in the whole of western and Mediterranean Europe. As it is multilingual, HYPP can also be used as a dictionary for anyone having translation difficulties. We understand by **diseased** 'the state of a plant which has been infected by a pathogen'.

Results and Discussion

The comparative analysis shows that there are four ways of naming plant diseases in both English and French: *using plant disease names indicated as such by the dictionaries*; *using plant disease names explicitly indicated as belonging to the field of pathology in general or implicitly indicated as such by their endings*; *using generic names of diseases*; and *using descriptive and/or explanatory names emphasising the aspect of the plant and/or effect of the disease on the plant*. The figures between brackets below show the number of occurrences.

1. **Names of plant diseases** indicated as *Plant Pathol.*:

- E (170): ***anthracnose*** (< F) (11): ***anthracnose***, ***anthracnose*** *of almond / soybean, lucerne* ***anthracnose***; ***blackleg*** (3); ***black rot*** (2): ***black rot***, *false* ***black rot*** *of grapevine*; ***blight*** (27): ***blight***, *Ascochyta / Coryneum / Mycospharella / Pinodella* ***blight***, *chestnut / peach / pepper / pisi / spinach* ***blight***, *leaf / pear blossom / pod /spur / stem* ***blight*** *(of maize)*, *bacterial / early / fire / late* ***blight***; ***blotch*** (4): ***blotch***, *Septoria nodorum* ***blotch***, *net* ***blotch*** *of barley*; ***canker*** (6): ***canker***, *Eutypa / Nectria* ***canker***; ***curl*** (1): *peach leaf* ***curl***; ***die(-back)*** (6): ***die-back***, ***die-back*** *of asparagus / hazel / hazelnut, shoot* ***die-back***; ***leaf rust*** (1); ***leaf spot*** (4); ***mildew*** (23): ***mildew***, *American gooseberry / gooseberry / cucurbit* ***mildew***, *downy / powdery* ***mildew***; ***mosaic*** (5): *cauliflower / cucumber / potato / tobacco* ***mildew***, *rugose* ***mildew*** ; ***pox*** (2): *plum* ***pox***; ***rot*** (36): ***rot***, *Ascochyta / Botrytis / Didymella* ***rot***, *lettuce* ***rot***, ***rot*** *on fruits, bunch / collar / crown / foot / root / stem / tuber* ***rot***, *bacterial / bird's eye / brown / charcoal / firm / grey / soft / violet / white* ***rot***; ***rust*** (13): ***rust***, ***rust*** *of asparagus / wheat, beet / raspberry* ***rust***, *white pine* ***rust***, *crown / stem* ***rust***, *black / blister / brown / stripe / yellow* ***rust***; ***scab*** (4): ***scab***, *apple / pear* ***scab***, ***scab*** *of cucurbits*; ***scald*** (1): ***scald*** *of barley*; ***scorch*** (1): *leaf-* ***scorch*** *of strawberry*; ***smut*** (4): ***smut***, ***smut*** *of leek / onion*, ***head*** *smut*; ***wilt*** (16): ***wilt***, *Fusarium / Verticillium* ***wilt***, ***wilt*** *of cucumber / melon / watermelon, melon* ***wilt***, *bacterial* ***wilt***;
- *F* (102): ***anthracnose*** (15): ***anthracnose***, ***anthracnose*** *de l'amandier / du groseiller / de la luzerne / du noyer / de l'olivier*, ***anthracnose*** *maculée;* ***black rot*** (< E) (1); ***blanc*** (4): ***blanc***, ***blanc*** *des Curcubitacées / du lin / du pommier;* ***charbon*** (3): ***charbon*** *à Sphacelotheca / Ustilago*, ***charbon*** *de l'oignon;* ***cloque*** (1): ***cloque*** *du*

pêcher; ***échaudage*** (1); ***enroulement*** (1): ***enroulement*** *de la pomme de terre;* ***flétrissement*** (3): ***flétrissement*** *du lin,* ***flétrissement*** *des bouquets floraux,* ***flétrissement*** *bactérien;* ***fonte*** (2): ***fonte*** *du collet,* ***fonte*** *blanche;* ***frisolée*** (1): ***frisolée*** *de la pomme de terre;* ***gale*** (1): ***gale*** *rugueuse;* ***mildiou*** (< E) (14): ***mildiou, mildiou*** *de la laitue / de l'oignon / de la pomme de terre / de la tomate / de la vigne;* ***mosaïque*** (9): ***mosaïque, mosaïque*** *du choufleur / du concombre / de la laitue / de la pomme de terre,* ***mosaïque*** *du virus du tabac,* ***mosaïque*** *commune;* ***piétin*** (4): ***piétin, piétin****-verse;* ***pourridié*** (3): ***pourridié, pourridié*** *laineux de la vigne,* ***pourridié*** *des racines;* ***pourriture*** (20): ***pourriture*** *des capitules / du collet / des racines (de l'artichaut),* ***pourriture*** *bactérienne / blanche / charbonneuse / grise / noble / phoméenne;* ***rot*** (< E) (2): ***rot*** *blanc / livide;* ***rouille*** (12): ***rouille*** *de l'ail / de l'asperge / du blé / de la betterave / du framboisier / des graminées / de l'orge / des Ribes,* ***rouille*** *des glumes,* ***rouille*** *brune / couronnée / jaune / naine / noire,* ***rouille*** *à colonnettes;* ***tavelure*** (2): ***tavelure*** *du poirier / du pommier;* ***verse*** (3): ***verse*** *des tiges, piétin-****verse****.*

2. **Names of diseases** (indicated as *Pathol.* and/or by their specific ending: *-mania*, *-ose*, *-osis*, in English, and *-ine* and *-ose*, in French):

- E (12): ***apoplexy*** (1); ***black measles*** (1); ***Botrytis*** (1): ***Botrytis***; ***cancer*** (1): *bark* ***cancer*** *of fig*; ***cercosporiose*** (1): ***cercosporiose*** *of olive*; ***ergot*** (< F) (1); ***excoriosis*** (2); ***gangren(e)*** (1); ***Pestalotiopsis*** (1); ***rhizomania*** (1); ***tuberculosis*** (1): ***tuberculosis*** *of olive-tree*;
- F (86): ***alternariose*** (4); ***ascochytose*** (1); ***bactériose*** (6): ***bactériose, bactériose*** *du fraisier / du noisetier / du noyer*; ***cercospor(i)ose*** (2); ***chancre*** (10): ***chancre*** *à Cytospora / Didymella / Fusicoccum / Nectria,* ***chancre*** *du figuier / du pêcher / du poirier / du pommier,* ***chancre*** *de l'écorce / du tronc,* ***chancre*** *bactérien / commun / européen*; ***cladosporine*** (1); ***cladosporiose*** (1); ***cylindrosporiose*** (1); ***dégénérescence*** (1): ***dégénérescence*** *infectieuse*; ***dépérissement*** (1): ***dépérissement*** *bactérien à Pseudomonas*; ***eutypose*** (1); ***excoriose*** (1); ***fusariose*** (6); ***Fusarium*** (1); ***gangrène*** (1); ***graisse*** (2): ***graisse*** *de l'artichaut / du pois*; ***helminthosporiose*** (4): ***helminthosporiose, helminthosporiose*** *de l'orge*; ***kabatiellose*** (2); ***mastigosporiose*** (1); ***meunier*** (1): ***meunier*** *de la laitue*; ***moniliose*** (2); ***nécrose*** (1): ***nécrose*** *du collet*; ***oïdium*** (9): ***oïdium, oïdium*** *du groseiller / du lin,* ***oïdium*** *brun*; ***Pestalotiopsis*** (1); ***phomopsis*** (3): ***phomopsis, phomopsis*** *du tournesol*; ***plomb*** (1): ***plomb*** *parasitaire*; ***Pseudopeziza*** (1);

ramulariose (1); ***rhizocton(i)e*** (3); ***rhizomanie*** (1): ***rhizomanie*** *de la betterave*; ***sclérotiniose*** (4); ***septoriose*** (1); ***suie*** (1): ***suie*** *du maïs*; ***variole*** (2); ***verticilliose*** (6); ***virose*** (1): ***virose*** *à énations de la luzerne.*

3. **Generic names of diseases**:

- E (39): ***disease*** (21): *Cercospora / Cylindrosporium / Fusarium / phomopsis / Ramularia / Verticillium / Xanthomonas* ***disease***, *anthracnose / smut* ***disease***, *chocolate spot / dead-arm / ink / mild / peacock / shot-hole / silver-leaf* ***disease***, ***disease*** *of onion / potato, St. John's* ***disease***; ***virus*** (18): *artichoke / barley / beet / cucumber / lactuca / lettuce / lucerne / pea / potato / Prunus* ***virus***, *enation / fizzletop / latent / leafroll / leaf-rolling / mild yellowing / mosaic / mottled crinkle / necrotic ringspot / severe yellowing / yellow / yellow dwarf* ***virus***, ***virus*** *yellows*, ***virus*** *A*;
- F (21): ***maladie*** (17): ***maladie*** *de l'encre / de la pourriture des racines / du plomb parasitaire*, ***maladie*** *de la tache annulaire nécrotique du pommier / des taches brunes / chocolat / noires*, ***maladie*** *des racines ligneuses / des stries foliaires de l'orge / des taches brunes de l'orge / des taches communes de la luzerne / des taches pourpres du fraisier / des tiges noires de la luzerne / des tumeurs de l'olivier*, ***maladie*** *en œil de paon des feuilles d'olivier*, ***maladie*** *criblée*; ***virus*** (4): *artichoke mottled crinkle* ***virus***, *mosaïque du* ***virus*** *du tabac, pea seed-borne mosaic* ***virus***, ***virus*** *latent de l'artichaut.*

4. **Descriptive and/or explanatory names of diseases**:

- E (46): ***banding*** (1): *strawberry vein* ***banding***; ***black stem*** (1): ***black stem*** *of lucerne*; ***corky root*** (1); ***drop*** (1): *lettuce* ***drop***; ***flavescence dorée*** (1); ***fleck*** (2): *leaf* ***fleck*** *(of pasture grasses)*; ***leaf roll*** (1): *potato* ***leaf roll***; ***marmor lactucae*** (1); ***mould*** (14): ***mould*** *of hazelnut / soybean, fruit / leaf / silique* ***mould***, *grey / snow / white* ***mould***; ***red core*** (1): ***red core*** *of strawberry root*; ***shot-hole*** (1); ***spot*** (18): ***spot*** *of cereals / lucerne / maize, leaf / pod / ring* ***spot***, *angular / bird's eye / black / brown / chocolate / common / eye* ***spot***; ***streak*** (2): *potato leaf-drop / stipple* ***streak***; ***stripe*** (1): *barley leaf* ***stripe***;
- F (20): ***banding*** (< E) (1): *strawberry vein* ***banding***; ***bigarrure*** (1): ***bigarrure*** *de la pomme de terre*; ***brûlure*** (2): ***brûlure***, ***brûlure*** *des dards*; ***corky root*** (1); ***court-noué*** (1); ***criblure*** (1): ***criblure*** *à Coryneum*; ***feu*** (1): ***feu*** *bactérien*; ***flavescence dorée*** (1); ***jaunisse*** (3): ***jaunisse*** *de la betterave / de l'orge*, ***jaunisse*** *grave / modérée /*

nanisante; ***moisissure*** (1): ***moisissure*** *grise*; ***moucheture*** (1): ***moucheture*** *brune du Dactyle*; ***nervation*** (1): ***nervation*** *noire des Crucifères*; ***panachure*** (1); ***pied noir*** (1); ***rayure*** (1): ***rayure*** *réticulée de l'orge*; ***rougeot*** (1): ***rougeot*** *parasitaire*; ***tacheture*** (1): ***tacheture*** *foliaire*. (Figure 7-2)

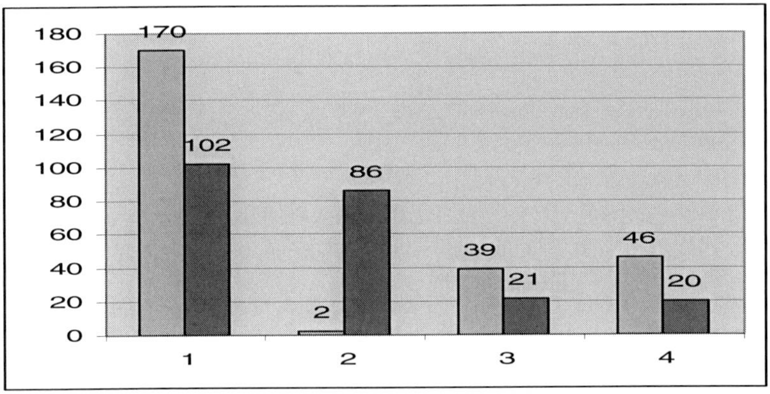

Figure 7-2. Common names of plant diseases in English and French: 1 - plant pathology; 2 - pathology; 3 - generic; 4 - descriptive/explanatory

Conclusions

The following conclusions can be drawn:

- the names of plant diseases of the *Plant Pathology-type* are much numerous in English than in French (170 *vs.* 102);
- the names of plant diseases of the *Pathology-type* (including names derived with specific suffixes) are seven times less numerous in English than in French (12 *vs.* 86);
- the names of plant diseases of the *Generic Name-type* is almost two times bigger in English than in French (39 *vs.* 21);
- the names of plant diseases of the *Descriptive Name-type* are more than two times bigger in English than in French (46 *vs.* 20).

This seems to indicate:

- a great concern for precision on behalf of both English (170) and French (102) languages (with the only difference that English is far richer than French in synonyms of this type);

- a more productive **derivative system** in French (86) than in English (12);
- a greater tendency towards descriptive (46) and explanatory (39) names of plant diseases in English in comparison to French ones (20 and 21, respectively).

Borrowings from one language to the other are exceptional, French translating English names for plant diseases in most cases.

References

HYPP. Version 1.0 copyright © 1994

Robert, P. (1996). *Dictionnaire alphabétique & analogique de la langue française*. [An Alphabetical and Analogical Dictionary of the French Language]. Paris: S.N.L.

Rață, Georgeta & Giuchici, Camelia. (2002). Common Names of Plant Diseases: A Comparative Approach (English and French). *Proceedings of the 6th International Symposium "Interdisciplinary Regional Research Hungary – Romania – Yugoslavia", University of Novi Sad (Yugoslavia), 3-4 October, 2002.*

The Random House Dictionary of the English Language. (1968). New York: Random House.

COMMON NAMES OF PLANT DISEASES IN ROMANIAN AND ENGLISH

GEORGETA RAŢĂ, MARIA-ADRIANA PROCA AND CAMELIA GIUCHICI

Introduction

Approaching a specialized text in any field of knowledge is a true challenge for anybody, be it a researcher, a professor, a translator, a student, or a teacher of foreign languages in a technical university, for instance. Despite its international character, using a specialised vocabulary can sometimes be troublesome. This is also the case of common names of plant diseases in languages such as English and French, on the one hand, and Romanian, on the other hand.

Material and Method

We inventoried a number of 82 Romanian common names of plant diseases in our *Plant Protection. A Compendium of Technical Terms in Latin, English, French and Romanian* (2002). We then picked up their English equivalents and counted their occurrences in the same source. The method we used in doing so is the comparative one.

Results and Discussion

The 82 Romanian common names of plant diseases are as follows: ***alternarioză***, with 6 English equivalents (*black rot*, *black spot*, *blight*, *early blight*, *mould*, and *spot*) and 12 occurrences; ***antracnoză***, with 7 equivalents (*anthracnose*, *anthracnose disease*, *bird's eye rot*, *blight*, *blotch*, *ring spot*, and *spot*) and 18 occurrences; ***arsură***, with 3 equivalents (*blight*, *blotch*, and *scald*) and 3 occurrences; ***arsura bacteriană***, with 6 equivalents (*bacterial blight*, *blight*, *die-back*, *spot*, *tuberculosis*, and *Xanthomonas disease*) and 9 occurrences; ***ascochitoză***, with 4 equivalents (*anthracnose*, *Ascochyta rot*, *black stem*, and *blight*) and 4 occurrences; ***băşicare***, with 1 equivalent (*curl*) and 1 occurrence; ***boala petelor roşii***, with 1 equivalent (*rot brenner*) and 1 occurrence; ***boala plumbului***, with 4 equivalents (*apoplexy*, *black measles*, *esca*, and *silver-leaf disease*) and 5 occurrences; ***brunificarea şi frângerea***, with 1 equivalent (*Phomopsis disease*) and 1 occurrence; ***cancerul***, cu 2 equivalents (*blight* and *cancer*) and 2 occurrences; ***cancerul bacterian***, with 1 equivalent (*bacterial canker*) and 1 occurrence; ***cancerul deschis***, with 1 equivalent (*canker*) and 1 occurrence; ***cărbunele***, with 2 equivalents (*anthracnose* and *bird's eye rot*) and 2 occurrences; ***cercosporioza***, with 2 equivalents (*Cercospora disease* and *cercosporiose*) and 2 occurrences; ***ciuruirea***, with 5 equivalents (*blight*, *Coryneum blight*, *shot-hole*, *shot-hole disease*, and *spot*) and 5 occurrences; ***cornul***, with 1 equivalent (*ergot*) and 1 occurrence; ***eutipoza***, with 2 equivalents (*die-back* and *Eutypa canker*) and 2 occurrences; ***excorioza***, with 3 equivalents (*disease*, *excoriosis*, and *false black rot*) and 4 occurrences; ***făinarea***, with 2 equivalents (*Oidium* and *powdery mildew*) and 9 occurrences; ***făinarea americană***, with 2 equivalents (*American mildew* and *mildew*) and 2 occurrences; ***focul bacterian***, with 1 equivalent (*fire blight*) and 1 occurrence; ***fuzarioza***, with 7 equivalents (*die-back*, *disease*, *Fusarium*, Fusarium *disease*, *Fusarium wilt*, *rot*, and *wilt*) and 14 occurrences; ***îngălbenirea aurie***, with 1 equivalent (*Flavescence dorée*) and 1 occurrence; ***îngălbenirea moderată***, with 1 equivalent (*mild yellowing virus*) and 1 occurrence; ***înnegrirea bacteriană***, with 1 equivalent (*bacterial wilt*) and 1 occurrence; ***înnegrirea şi putregaiul umed***, with 1 equivalent (*blackleg*) and 1 occurrence; ***mana***, with 5 equivalents (*blight*, *downy mildew*, *mildew*, *red core*, and *rot*) and 18 occurrences; ***monilioza***, with 1 equivalent (*brown rot*) and 2 occurrences; ***mozaicul***, with 6 equivalents (*blight*, *enation virus*, *marmor*, *mosaic*, *mosaic virus*, and *virus*) and 9 occurrences; ***mozaicul comun***, with 1 equivalent (*mild disease*) and 1 occurrence; ***mozaicul „Y"***, with 1 equivalent (*rugose mosaic*) and 1 occurrence; ***mucegaiul de zăpadă***, with 2 equivalents (*rot* and *snow mould*) and 2 occurrences; ***mucegaiul violet***, with 1 equivalent (*violet rot*) and 2 occurrences; ***nervaţiunea neagră***, with 3 equivalents (*black rot*, *die*, and *die-back*) and 3 occurrences; ***pătarea***, with 3 equivalents (*brown rot*, *fleck*, and *spot disease*) and 3 occurrences; ***pătarea brună***, with 7 equivalents (*black spot*, *blight*, *brown spot*, *common spot*, *mould*, *scab*, and *spot*) and 8 occurrences; ***pătarea brună-violacee***, with 1 equivalent (*blight*) and 1 occurrence; ***pătarea cafenie***, with 1 equivalent (*mould*) and 1 occurrence; ***pătarea inelară***, with 1 equivalent (*ringspot virus*) and 1 occurrence; ***pătarea în ochi***, with 1 equivalent (*eyespot*) and 2 occurrences; ***pătarea neagră***, with 2 equivalents (*black leg* and *rot*) and 3 occurrences; ***pătarea reticulată***, with 1 equivalent (*net blotch*) and 1 occurrence; ***pătarea roşie***, with 1 equivalent (*leaf-scorch*) and 1 occurrence; ***piticirea și îngălbenirea***, with 1 equivalent (*yellow dwarf virus*) and 1 occurrence; ***plum-pox***, with 1 equivalent (*plum-pox*) and 2 occurrences; ***putregaiul***, with 6 equivalents (*brown rot*, *crown rot*, *die-back*, *ink disease*, *rot* and *white rot*) and 9 occurrences; ***putregaiul alb***, with 2 equivalents (*rot* and *white rot*) and 5 occurrences; ***putregaiul alb cu scleroți***, with 6 equivalents (*drop*, *soft rot*, *white mould*, and *white rot*) and 9 occurrences; ***putregaiul amar***, with 2 equivalents (*anthracnose* and *anthracnsoe disease*) and 3 occurrences; ***putregaiul bacterian***, with 1 equivalent (*bacterial rot*) and 1 occurrence; ***putregaiul brun***, with 1 equivalent (*brown rot*) and 1 occurrence; ***putregaiul cenușiu***, with 3 equivalents (*grey mould*, *grey rot*, and *rot*) and 8 occurrences; ***putregaiul uscat***, with 2 equivalents (*gangren* and *rot*) and 2 occurrences; ***putrezirea***, with 1 equivalent (*charcoal rot*) and 1 occurrence; ***rapănul***, with 1 equivalent (*scab*) and 2 occurrences; ***rincosporioza***, with 2 equivalents (*blotch* and *scald*) and 2 occurrences; ***rizoctonioza***, with 1 equivalent (*eyespot*) and 2 occurrences; ***rizomania***, with 1 equivalent (*rhizomania*) and 1 occurrence; ***rugina***, with 1 equivalent (*rust*) and 4 occurrences; ***rugina brună***, with 2 equivalents (*brown rust* and *rust*) and 2 occurrences; ***rugina coronată***, with 1 equivalent (*crown-rust*) and 1 occurrence; ***rugina cu coloane***, with 1 equivalent (*blister rust*) and 1 occurrence; ***rugina galbenă***, with 2 equivalents (stripe rust and yellow rust) and 2 occurrences; ***rugina naegră***, with 1 equivalent (black rust) and 1 occurrence; ***scurt-nodarea***, with 1 equivalent (fanleaf) and 1 occurrence; ***septorioza***, with 2 equivalents (late blight and Septoria nodorum blotch) and 2 occurrences; ***sfâșierea***, with 1 equivalent (stripe) and 1 occurrence; ***suberificarea***, with 1 equivalent (corky-root) and1 occurrence; ***ștricul***, with 1 equivalent (streak) and 1 occurrence; ***tăciunele***, with 1 equivalent (smut disease) and 1 occurrence; ***tăciunele comun***, with 1 equivalent (smut) and 1 occurrence; ***tăciunele prăfos***, with 1 equivalent (head smut) and 1 occurrence; ***uscarea cenușie***, with 1 equivalent (canker) and 1 occurrence; ***verticilioza***, with 1 equivalent (wilt) and 1 occurrence; ***veștejirea***, with 1 equivalent (wilt) and 4 occurrences; ***virescența***, with 1 equivalent (vein banding) and 1 occurrence; ***virusul „A"***, with 1 equivalent (virus A) and 1 occurrence; ***virusul încrețirii***, with 1 equivalent (crikle virus) and 1 occurrence; ***virusul îngălbenirii***, with 2 equivalents (yellow virus and yellowing virus) and 2 occurrences; ***virusul latent***, with 1 equivalent (latent virus) and 1 occurrence; ***virusul mozaicului***, with 1 equivalent (mosaic) and 1 occurrence; ***virusul răsucirii***, with 4 equivalents (fizzletop virus, mosaic virus, roll, and virus) and 4 occurrences.

To these 82 Romanian common names of plant diseases correspond

166 English common names of plant diseases, which indicates a solid settlement of such names in English due to the long-lasting tradition in growing plants in English-speaking countries. The 166 English names have an impressive number of occurrences (645) compared to the 185 Romanian ones, i.e. 3.52 times more names in English than in Romanian.

Conclusions

Taking all this into account, we are inclined to believe that both English and Romanian common names of plant diseases are polysemantic as the multiple meanings of a word are all uses of the same word that have grown apart over time. The difference is that English has developed a greater number of nouns to denote plant diseases as it settled long before Romanian even became a language. The fact that a lot of Romanian names of plant diseases have English equivalents that do not always have the same Romanian equivalents is, however, an issue.

References

Chalker, Sylvia & Weiner, E. (1994). *The Oxford Dictionary of English Grammar*. London - New York - Sydney - Toronto: BCA.

Raţă, Georgeta, Proca, Maria-Adriana & Giuchici, Camelia. (2004). On Common Names of Plant Diseases in English and Romanian. *Lucrări ştiinţifice. Facultatea de Agricultură XXXVI*: 614-617.

Raţă Georgeta, Manea, D., Giuchici, Camelia & Grozea, Ioana. (2002). *Plant Protection. A Compendium of Technical Terms in Latin, English, French and Romanian*. Timişoara: Editura Mirton.

DOG BREEDS IN CROATIAN AND ROMANIAN

ANICA PERKOVIĆ, GEORGETA RAȚĂ AND MARTINA PERKOVIĆ

Introduction

The purpose of this research is to see if there are great similarities or differences between the lexicographical fields of two different languages such as Croatian (a Slavic language) and Romanian (a Romance language).

The hypothesis of the research was that in certain lexicographic fields (e.g. dog breeds, horse breeds, swine breeds, etc.) dealing with animal breeds developed within the Anglo-Saxon area, names are in most cases either borrowed as such or adapted at different levels (phonological, morphological, semantically, etc.).

Our background information consisted in systematic information from different similar lexicographical fields, which we corroborated with the information supplied by a recent *Dictionary of European Anglicisms* (Gorlach 2005), on the one hand, and with other lexicographical works, on the other hand.

Material and Methods

In our comparative approach of dog breeds in Croatian and Romanian, we have analysed dog breed names that have become Anglicisms in a number of European languages including Croatian and Romanian, such as supplied by Manfred Gorlach's *Dictionary of European Anglicisms. A Usage Dictionary of Anglicisms in Sixteen European Languages* (2005). The names of dog breeds inventoried by Gorlach (2005) were compared with dog breed names of English origin in Croatian and Romanian comprehensive and language dictionaries.

The method we have used in our analysis is a comparative one: it consists in studying and comparing two modern languages - in our case, Croatian and Romanian - from a specific point of view - in our case, names of dog breeds - in search of similarities and differences between the two languages.

Results and Discussion

We have inventoried in Gorlach's *Dictionary of European Anglicisms* (2005) a number of 31 dog-related Anglicisms (29 nouns and 1 verb) (words designating dog-related fields such as: dog breeds, dog types, dog shows, etc.): ***Airedale (terrier)***, ***basset***, ***beagle***, ***best in show***, ***bobtail***, ***boxer***, ***bull terrier***, ***bulldog***, ***chow-chow***, ***cocker (spaniel)***, ***dingo***, ***dog***, ***dog cart***, ***fox terrier***, ***golden retriever***, ***greyhound***, ***harrier***, ***husky***, ***mastiff***, ***Newfoundland***, ***pincher***, ***pit bull terrier***, ***pointer***, ***pup***, ***puppy***, ***setter***, ***sled dog***, ***spaniel***, ***terrier***, ***trim***, and ***trimming***.

Of these 31 dog-related Anglicisms, only 17 (55%) designate dog breeds: ***Airedale (terrier)*** 'a breed of dog', ***basset*** 'a breed of short-legged dog', ***beagle*** 'a breed of hound', ***bobtail*** 'a breed of dog', ***boxer*** 'a dog with a smooth brown coat and pug like face', ***bull terrier*** 'a breed of short-haired dog', ***bulldog*** 'a dog of a sturdy powerful breed with large head and smooth hair', ***chow-chow*** 'a dog of a Chinese breed with long hair and bluish-black tongue', ***cocker (spaniel)*** 'a small breed of dog with a silky coat', ***dog*** 'a special breed of dog', ***fox terrier*** 'a short-haired terrier', ***golden retriever*** 'a breed of dog with a thick golden-coloured coat', ***greyhound*** 'a breed of dog often used for racing', ***mastiff*** 'a dog of a large strong breed', ***Newfoundland*** 'a dog of a very large breed with a thick coarse coat', ***pincher*** 'a dog with a cropped tail', and ***pit bull terrier*** 'a dog of an American variety of bull terrier, noted or its ferocity'.

All these dog-breed related names can be grouped into 3 categories: *names borrowed by both languages (Croatian and Romanian)*, *names borrowed only by Romanian*, and *names not borrowed by either language*.

1. Of the 17 Anglicisms designating dog breeds, only 13 (76%) have been borrowed by both Croatian and Romanian, as shown in Table 7-4 (dog breed names in both italic and bold fonts are names that have preserved the English spelling intact in both Croatian and Romanian).

None of the Romanian language dictionaries we have consulted mentions ***Airedale (terrier)*** or ***Newfoundland***, which leads to the conclusion that the author was either misinformed or there has been some misinterpretation in the process.

As for the rest of dog breed names: *baset* (pl. *baseţi*) is mentioned by DOOM 2 for the spelling and by DEX '98, NODEX, DN, and MDN as an indirect borrowing through French; *beagle* is mentioned by MDN as a direct borrowing from English; *boxer* is mentioned by DN as an indirect borrowing through French and by MDN as an indirect borrowing through

French (Fr) or German (G); *bull terrier* (spelled *bull-terrier*) is mentioned by DN as an indirect borrowing through French; *buldog* (pl. *buldogi*) is mentioned by DOOM 2 for the spelling, by DN and MDN as a direct borrowing from English, and by DEX '98, Sinonime, DN, and NODEX as an indirect borrowing through French (< *bouledogue*); *cocker (spaniel)* and *cócher* (pl. *cocheri*) are mentioned by DOOM 2 for the spelling, by DN, DEX '98, and NODEX as direct borrowings from English, and by DEX '98 and NODEX as an indirect borrowing through French; *dog* (pl. *dogi*) is mentioned by DOOM 2 for the spelling, and by DEX '98, NODEX, DN, and MDN as a both direct borrowing from English and an indirect borrowing through French (< *dogue*); *foxterier* (pl. *foxterieri*) is mentioned by Sinonime and DOOM 2 for the spelling, and by DEX '98, DN, MDN, and NODEX as indirect borrowing through French; *greyhound* is mentioned by MDN as a direct borrowing from English; *pincher*, spelled as *pinscher*, is mentioned by MDN as a direct borrowing from German (< *Pinscher*).

Table 7-4. Dog breed names borrowed by both Croatian and Romanian

Dog breed name		
in English	**in Croatian**	**in Romanian**
airedale (terrier)	*erdel terijer*	*airedale (terrier)*
basset	*basset*	*baset* (pl. *baseți*)
beagle	*beagle*	*beagle*
boxer	*bokser*	*boxer*
bull terrier	*bull terrier*	*bull terrier*, *bull-terrier*
bulldog	*bulldog*	*buldog* (pl. *buldogi*)
cocker (spaniel)	*koker spanijel*	*cocker (spaniel)*, *cócher* (pl. *cocheri*)
dog	*doga*	*dog* (pl. *dogi*)
fox terrier	*foksterijer*	*foxterier* (pl. *foxterieri*)
greyhound	*greyhound*	*greyhound*
newfoundland	*njufaundlend*	*(câine) terra-nova*
pincher	*pinć*	*pincher*, *pinscher*
pitbull terrier	*pitbul*	*pitbul terier*

Are considered direct borrowings from English: *beagle*, *buldog*, *cocker / cocher*, *dog*, and *greyhound* (38% of the total dog breed names); are considered indirect borrowings: *baset*, *boxer*, *bull-terrier*, *buldog*, *cocker / cocher*, *dog*, and *foxterier* (< Fr) (54% of the total dog breed names), *boxer* (< Fr or G) (8% of the total dog breed names), and *pinscher* (< G) (8% of the total dog breed names).

Together with forms identical to the English ones - *beagle*, *boxer*, *bull terrier*, *cocker (spaniel)*, and *dog greyhound* (representing 46% of the total dog breed names) - there are also forms adapted to the Romanian spelling - *baset*, *bull-terrier*, *bulldog*, *cócher*, *foxterier*, and *pitbul terrier* (representing 46% of the total dog breed names).

The plural forms of 5 of these dog breed names - *baset* (pl. *baseţi*), *buldog* (pl. *buldogi*), *cócher* (pl. *cocheri*), *dog* (pl. *dogi*), and *foxterier* (pl. *foxterieri*) - shows that these names have been integrated to the phonetic and orthographic system of the Romanian language and that they are no longer felt as foreignisms.

As for *Newfoundland*, it is rendered in Romanian not by the same noun, but by some kind of paraphrase - *(câine) terra-nova* (*terra-nova* is an approximate translation of *Newfoundland*) - while *pincher* proves to be a misspelled form of *pinscher*.

2. A single dog breed name was borrowed by Romanian alone, as shown in Table 7-5 (names in both italic and bold have preserved the English spelling intact).

Table 7-5. Dog breed names borrowed only by Romanian

Dog breed name		
in English	**in Croatian**	**in Romanian**
chow-chow	-	*chow-chow*

Chow-chow is mentioned by DN and MDN as an indirect borrowing through French.

3. According to Gorlach (2005), other 3 dog breed names have been borrowed by other European languages, but not by Croatian or Romanian: *bobtail* (by German, Dutch, French, Spanish, Italian, Russian, Bulgarian, Finnish, and Hungarian), *golden retriever* (by German, Dutch, Norwegian, Icelandic, French, Spanish, Russian, Bulgarian, and Finnish), and *mastiff* (by Dutch, Norwegian, French, Spanish, Italian, Russian, Polish, Bulgarian, Finnish, Hungarian, and Greek), as shown in Table 7-6.

Mastif(f) (pl. *mastifi*) is mentioned by DOOM 2 for its spelling and by MDN as a direct borrowing from English.

Table 7-6. Dog breed names borrowed by neither Croatian nor Romanian

Dog breed name in English	in Croatian	in Romanian
bobtail	-	-
golden retriever	-	-
mastiff	-	-

Conclusions

Dog breeds names borrowed by both Croatian and Romanian show different tendencies in the two studied languages. Thus, Croatian has borrowed only 4 dog breed names as such (*basset*, *beagle*, *bull terrier*, and *greyhound*) compared to the 8 dog breed names borrowed by Romanian (*Airedale (terrier)*, *beagle*, *boxer*, *bull terrier*, *cocker (spaniel)*, *dog*, *greyhound*, *pincher*), preferring to adapt the rest of the names to the linguistic norms of the language, mainly in spelling (*erdel terijer*, *bokser*, *buldog*, *koker spanijel*, *doga*, *foksterijer*, *njufaundlend*, *pinč*, and *pitbul*). On the other hand, Romanian has both intact English forms and adaptations for the same dog breed name (*bull terrier* / *bull-terrier*, *cocker (spaniel)* / *cócher* (pl. *cocheri*), and *pincher* / *pinscher*), with Romanian plurals for the forms it adapted (*baset* (pl. *baseţi*), *buldog* (pl. *buldogi*), *cócher* (pl. *cocheri*), *dog* (pl. *dogi*), *foxterier* (pl. *foxterieri*)), which shows a wider appetite for Anglicisms in Romanian. These plural forms show that these names have been integrated to the phonetic and orthographic system of the Romanian language and are no longer felt as foreignisms, be they direct or indirect borrowings. There is a single dog breed name borrowed only by Romanian - *chow-chow* - but this name alone is not enough to draw a conclusion about the borrowing trends in the two studied languages. As for the dog breeds names borrowed by neither Croatian nor Romanian according to Gorlach (2005) - *bobtail*, *golden retriever*, and *mastiff* - the German author seems misinformed, since the last name is mentioned by a Romanian language dictionary.

The hypothesis of the research that in certain lexicographic fields (e.g. dog breeds, horse breeds, swine breeds, etc.) dealing with animal breeds developed within the Anglo-Saxon sphere, names are in most cases either borrowed as such or adapted at different levels (phonological, morphological, semantically , etc.) proved to be correct, with this mention that Romanian is more inclined to do so than Croatian, maybe because in most cases dog breed names have come into Romanian through French, which facilitated their absorption under the cover of French borrowings.

As additional research, we have in mind the study of other similar lexicographic fields such as the field of horse breed names, poultry breed names, etc.

References

Aniš, V. (2006). *Veliki rječnik hrvatskog jezika*. [Great Dictionary of Croatian Language]. Zagreb: Novi Liber.

Dicţionar ortografic, ortoepic şi morfologic al limbii române. [An Orthographic, Orthoepic, and Morphological Dictionary of Romanian Language]. (2005). Bucureşti: Editura Univers Enciclopedic. (DOOM 2)

Dicţionarul explicativ al limbii române. [An Explanatory Dictionary of the Romanian Language]. (1998). Bucureşti: Editura Univers Enciclopedic. (DEX '98)

Gorlach, M. (2005). *A Dictionary of European Anglicisms. A Usage Dictionary of Anglicisms in Sixteen European Languages. Oxford*: Oxford University Press.

Marcu, F. & Constant, M. (1986). *Dicţionar de neologisme*. [A Dictionary of Neologisms]. Bucureşti: Editura Academiei. (DN)

Marcu, F. (2000). *Marele dicţionar de neologisme*. [The Great Dictionary of Neologisms]. Bucureşti: Editura Saeculum. (MDN)

Noul dicţionar explicativ al limbii române. [The New Explanatory Dictionary of the Romanian Language]. (2002). Bucureşti: Editura Litera Internaţional. (NODEX)

Raţă, Georgeta. (2010). Nume de rase canine în limba română: o abordare etimologică. [Names of Dog Breeds in Romanian: An Etymological Approach]. *Agrobuletin AGIR 4*: 129-133.

Perković, Anica, Rata, Georgeta & Perković, Martina. (2009). Dog Breeds in Croatian and Romanian: A Comparative Approach. *Proceedings of the 44th Croatian and 4th International Symposium on Agriculture*: 747-751.

Seche, M. & Seche, L. (2002). *Dicţionar de sinonime*. [A Dictionary of Synonyms]. Bucureşti: Editura Litera Internaţional. (Sinonime).

VERBS OF ANIMAL COMMUNICATION IN ENGLISH, FRENCH, AND ROMANIAN

GEORGETA RAŢĂ AND IASMINA IOSIM

Introduction

The **verbs of animal communication** are verbs denoting the way in which people perceive sound communication in animals. The analysis of their origin and language use levels in different languages shows a very dynamic evolution in this category of verbs.

Material and Method

We have used in the inventory of the verbs of animal communication some of the best English, French, and Romanian language dictionaries.

The method we have used in the study of these verbs is the comparative one – a method based on the resemblances and differences between root words, affixes, basic forms and variants, explanations, etc.

Results and Discussion

1. English verbs of animal communication. A number of 57 English verbs of animal communication have been inventoried: ***baa***, ***bark***, ***bawl***, ***bay***, ***bell***, ***bellow***, ***bleat***, ***boom***, ***bray***, ***burble***, ***cackle***, ***call***, ***caw***, ***chatter***, ***cheep***, ***chirp***, ***chuck***, ***chuckle***, ***cluck***, ***coo***, ***croak***, ***crow***, ***cry***, ***gabble***, ***gaggle***, ***growl***, ***grunt***, ***hiss***, ***honk***, ***hoot***, ***howl***, ***low***, ***meow / mew / miaou / miaow / miaul***, ***moo***, ***murmur***, ***neigh***, ***peep***, ***pur / purr***, ***quack***, ***rattle***, ***roar***, ***scream***, ***sing***, ***snort***, ***squeal***, ***thrill / trill***, ***tinkle***, ***twitter***, ***utter***, ***wail***, ***warble***, ***whinny***, ***whistle***, ***yap yelp***, and ***yodel / yodel***. From an etymological point of view, 37 of these verbs have an attested English origin (Old English and Middle English: ***bark***, ***bawl***, ***bell***, ***bellow***, ***bleat***, ***boom***, ***burble***, ***call***, ***chatter***, ***chirp***, ***chuck***, ***cluck***, ***croak***, ***crow***, ***cry***, ***growl***, ***grunt***, ***hiss***, ***hoot***, ***howl***, ***low***, ***murmur***, ***neigh***, ***peep***, ***quack***, ***rattle***, ***roar***, ***scream***, ***sing***, ***snort***, ***squeal***, ***tinkle***, ***twitter***, ***utter***, ***whistle***, and ***yelp***), 3 have an attested French origin (Old French and Old North French: ***bay***, ***bray***, and ***warble***), 2 have an attested German origin (Germanic and German: ***gaggle*** and ***yodel / yodel***), 1 is Italian (***thrill / trill***), and 1 is Scandinavian (***wail***).

As the verbs of animal communication show the way humans perceive sounds produced by animals, there are a considerable number of imitative verbs (13) resulted from a process in which a word belonging to one word class (in our case, mimetic words) gets to be used as part of another word class without the addition of an affix – a process called **conversion**, **reclassification**, or **functional shift**: ***baa***, ***cackle***, ***caw***, ***cheep***, ***chuckle***, ***coo***, ***gabble***, ***honk***, ***meow / mew / miaou / miaow / miaul***, ***moo***, ***pur / purr***, ***whinny***, and ***yap***.

2. French verbs of animal communication. A number of 81 French verbs of animal communication have been inventoried: ***aboyer*** 'bark, yowl, bay, cry, woof, howl, wail', ***babiller*** 'twitter, babble, prattle, chatter; murmur', ***barrir*** 'trumpet', ***bégueter*** 'to utter the characteristic cry of a goat', ***béler*** 'baa, beat out, bleat', ***beugler*** 'bawl, roar', ***blatérer*** 'to utter the characteristic cry of a sheep', ***bourdonner*** 'buzz, hum; ring; drone', ***braire*** 'bray, heehaw', ***bramer*** 'bell', ***cacaber*** 'to utter the characteristic cry of a partridge', ***cacarder*** 'gaggle', ***cajoler*** 'to utter the characteristic cry of a magpie', ***cancaner*** 'gosip, tattle', ***caqueter*** 'cackle', ***carcailler*** 'to utter the characteristic cry of a quail', ***chanter*** 'sing, chant; chirp', ***chuinter*** 'cry', ***clabauder*** 'yelp, bark, yap', ***clapir*** 'squeal, emit a sharp shrill cry, squawk', ***coasser*** 'croak', ***crailler*** 'scream', ***craqueter*** 'to utter the characteristic cry of a crane', ***criailler*** 'cry', ***crier*** 'cry, scream, shout, yell; call, clamour; crow, shriek, sing out', ***croasser*** 'caw, croak', ***gazouiller*** 'chirp, warble', ***gémir*** 'whine', ***glapir*** 'bark, squeak, yelp', ***glatir*** 'to utter the characteristic cry of an eagle', ***glouglouter*** 'bubble, gurgle', ***glousser*** 'chuckle, cluck, chortle, gobble, squawk', ***grailler*** 'croak', ***grisoller*** 'warble, sing with trilled and modulated voice', ***grogner*** 'grumble, grunt, rumble, murmur, plain, mump, snort', ***grommeler*** 'grumble, mutter; grunt, murmur', ***gronder*** 'rumble, growl, roar, grumble', ***hennir*** 'neigh, whinny', ***huer*** 'hoot', ***hululer / ululer*** 'hoot, screech, ululate', ***hurler*** 'shriek, screech, yell; howl, roar, bellow, squeal; wail, bawl; shrill, squall', ***jaboter*** 'chatter, prattle; babble, gossip; tattle, prate', ***jacasser*** 'chatter, prattle; jabber, twitter', ***japer*** 'yap, chatter, nag', ***jargonner*** 'jabber, chatter, prattle; mumble, babble', ***jaser*** 'chatter, prattle; babble, gossip; tattle, prate', ***lamenter*** 'lament, weep, wail; moan, mourn', ***margauder / margot(t)er*** 'to utter the characteristic cry of a quail', ***meugler*** 'moo, make a mooing sound (like a cow)', ***miauler*** 'mew, meow; caterwaul', ***mugir*** 'bellow, roar, boom, blow', ***nasiller*** 'snuffle, speak with a nasal, speak with a twang', ***pépier*** 'chirp, tweet, twitter, chirrup, peep', ***piailler*** 'squeak, squeal', ***piauler*** 'cheep, squeak', ***raire*** 'growl', ***râler / raller*** 'moan, growl, grumble', ***ramager*** 'to utter the characteristic cry of a dove', ***rauquer*** 'growl, make a low guttural sound (of an animal)', ***réer*** 'growl, make a low guttural sound (of an animal)', ***renâcler*** 'grumble; balk', ***ronronner*** 'purr, whir, drone', ***roucouler*** 'coo', ***rugir*** 'roar', ***siffler*** 'whistle, blow, wheeze, hiss', ***striduler*** 'stridulate', ***triller*** 'trill', ***trompeter*** 'trumpet', and ***vagir*** 'cry, wail, lament'.

From an etymological point of view, 25 of these verbs have an attested Latin origin (***barrir***, ***béler***, ***beugler***, ***blatérer***, ***cacaber***, ***chanter***, ***coasser***, ***gémir***, ***glapir***, ***glatir***, ***grogner***, ***gronder***, ***hennir***, ***hululer / ululer***, ***hurler***, ***lamenter***, ***meugler***, ***mugir***, ***pépier***, ***raire***, ***réer***, ***rugir***, ***siffler***, ***striduler***, and ***vagir***), 20 have an old and new attested French and Provençal etymon (***bégueter***, ***bourdonner***, ***bramer***, ***cajoler***, ***cancaner***, ***clabauder***, ***craqueter***, ***criailler***, ***glouglouter***, ***grailler***, ***jaboter***, ***jacasser***, ***margauder / margot(t)er***, ***râler***, ***ramager***, ***rauquer***, ***renâcler***, ***ronronner***, ***triller***, and ***trompeter***), and 1 has an attested Dutch etymon (***grommeler***). A considerable number of these verbs (19) have formed on French "soil", with the most productive affixes being *-(et)ter* (***bourdonner***, ***cancaner***, ***clabauder***, ***craqueter***, ***glouglouter***, ***grailler***, ***jaboter***, ***jargonner***, ***nasiller***, ***râler / raller***, ***ramager***, ***rauquer***, ***ronronner***, ***triller***, and ***trompeter***), *-ailler*, *-asser*, and *-iller* (***criailler***, ***jacasser***, and ***nasiller***) and a single prefix, *re-* (***renâcler***).

Though, as a rule, the verbs of animal communication show the way humans perceive sounds produced by animals, there are relatively few imitative verbs (16): ***babiller***, ***cacarder***, ***caqueter***, ***carcailler***, ***chuinter***, ***crailler***, ***croasser***, ***gazouiller***, ***grisoller***, ***huer***, ***japer***, ***jaser***, ***miauler***, ***piailler***, ***piauler***, and ***roucouler***.

3. Romanian verbs of animal communication. A number of 47 Romanian verbs of animal communication have been inventoried: ***a bate*** 'bark, yap', ***a behăi*** 'baa', ***a boncăi / boncălui*** 'low, bellow', ***a cârâi*** 'croak', ***a cârcâi*** 'cluck', ***a ciripi*** 'chirp', ***a chelălăi*** 'yap', ***a chiţăi / chiţcăi*** 'squeak', ***a cloncăi*** 'cluck, chuck', ***a cloncăni*** 'croak', ***a corcorozi*** 'cluck', ***a cotcodăci*** 'cluck', ***a croncăi / croncăni*** 'croak', ***a cucuriga / cucurigi*** 'to utter the characteristic cry of a cock or rooster', ***a fluiera*** 'whistle', ***a gâgâi*** 'gaggle', ***a gânguri*** 'coo', ***a gunguni / gunguri*** 'coo', ***a grohăi*** 'grunt', ***a hămăi / hâmâi*** 'yap', ***a lătra*** 'bark, yap', ***a măcăi*** 'to utter the characteristic cry of a duck', ***a măcăni*** 'to utter the characteristic cry of a duck', ***a mârâi*** 'snarl', ***a mehăi*** 'baa', ***a mierlăi / miorlăi / miorlâi*** 'meow', ***a mieuna / mioni / miuna*** 'meow', ***a mormăi*** 'growl, grumble', ***a mormorosi*** 'growl, grumble', ***a mugi*** 'moo', ***a necheza*** 'neigh, whinny', ***a***

orăcăi 'croak', ***a piscui*** 'chirp', ***a piui*** 'chirp', ***a rage / răgi*** 'bray, roar', ***a râncheza*** 'neigh, whinny', ***a sâsâi*** 'hiss', ***a schelălăi*** 'yap', ***a scheuna*** 'yelp, yap', ***a striga*** 'cry, call', ***a şuiera*** 'hiss', ***a toarce*** 'pur', ***a ţârâi*** 'chirp', ***a ţâţâi*** 'chirp', ***a ţipa*** 'cry', ***a ugui*** 'coo', and ***a zbiera*** 'cry, bellow'.

From an etymological point of view, 6 of these verbs have an attested Latin etymon (***a bate***, ***a lătra***, ***a mugi***, ***a rage / răgi***, ***a şuiera***, and ***a toarce***) and 1 verb has an attested Serbo-Croatian etymon (***a boncăi***).

Most of these verbs (27) have formed on Romanian "soil", with the most productive suffixes being *-ăi* (***a behăi***, ***a chiţăi / chiţcăi***, ***a cloncăi***, ***a grohăi***, ***a hămăi***, ***a măcăi***, ***a mehăi***, ***a mierlăi / miorlăi***, and ***a orăcăi***), *-âi* and *-i* (***a cârâi***, ***a hâmâi***, ***a mârâi***, ***a miorlâi***, ***a ţârâi***, and ***a ciripi***, ***a cotcodăci***, ***a cucurigi***, ***a mioni***, ***a piui***, respectively), *-a* and *-ăni* (***a cucuriga***, ***a fluiera***, ***a mieuna***, and ***a cloncăni***, ***a croncăni***, ***a măcăni***, respectively), and *-ui* (***a piscui***,), and a single prefix, *s-* (***a schelălăi***).

Though, as a rule, the verbs of animal communication show the way humans perceive sound produced by animals, there are relatively few imitative formations (7): ***a cârcâi***, ***a chelălăi***, ***a corcorozi***, ***a gâgâi***, ***a gânguri***, ***a gunguni***, and ***a sâsâi***.

Conclusions

As shown in Figure 7-3, the following conclusions can be drawn from what we have presented above:

- as far as the verbs of animal communication with attested etymon are concerned, the French language has 48 such verbs, English has 41, and Romanian has only 7;
- as far as the verbs of animal communication produced through **derivation** are concerned, the Romanian language has 27 such verbs, French has 18, while English has none;
- as far as the imitative verbs of animal communication are concerned, the French language ahs 16 such verbs, English has 13, and Romanian has only 7.

From the point of view of a Romanian student in animal husbandry, English verbs of animal communication seem to be easier to understand and to learn than French ones, as most of them are (al)most identical with the sound(s) uttered by the animals.

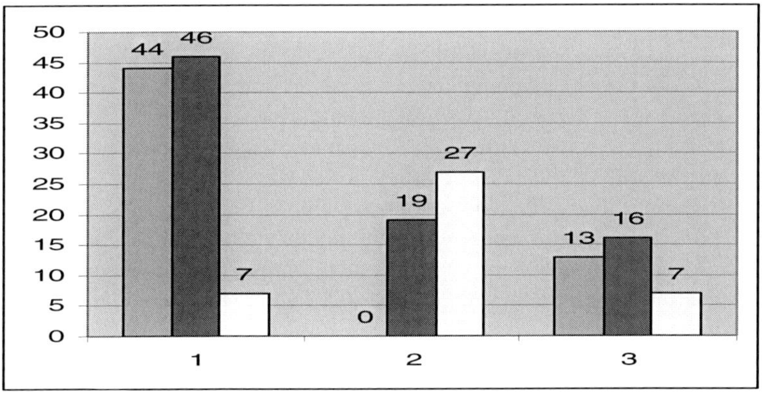

Figure 7-3. Comparison between English, French, and Romanian verbs of animal communication from the point of view of their attested etymon (1), derivational character (2), and imitative character (3)

References

Alcock, J. (1989). *Animal Behaviour: An Evolutionary Approach*. Sunderland, MA: Sinauer Associates, Inc.

Dicţionarul explicativ al limbii române. [The Explanatory Dictionary of the Romanian Language]. (1998). Bucureşti: Editura Univers Enciclopedic.

Raţă, Georgeta & Radanov, Iasmina. (2003). Verbs of Animal Communication in English, French and Romanian: A Comparative Approach. *Scientific Papers, Faculty of Agriculture XXXV*: 473-478.

Raţă, Georgeta. (1997). Le verbe et le groupe verbal en français contemporain. [The Verb and the Verbal Group in Contemporary French]. Bucureşti : E.D.P. R.A.

—. (1997). Les verbes de communication non-conventionnels. Essai de classification. [Unconventional Verbs of Communication. A Tentative Classification]. *Studia Universitatis Babeş-Bolyai. Philologia 4:* 67-71.

—. (1998). Comunicarea animală: o abordare lingvistică. [Animal Communication: A Linguistic Approach]. *Lucrări ştiinţifice. Medicină veterinară XXXI*: 263-268.

—. (2001). *Contribuţii la teoria comunicării.* [Contributions to the Theory of Communication]. Timişoara: Editura Mirton.

—. (2001). Verbele de comunicare animală în limba engleză. Încercare de clasificare. [Verbs of Animal Communication in English: A Tentative

Classification]. *Lucrări ştiinţifice. Zootehnie şi Biotehnologii XXXIV:* 297-303.

—. (2001). Verbele de comunicare animală în limba franceză. Încercare de clasificare. [Verbs of Animal Communication in French: A Tentative Classification]. *Economia de piaţă la etapa actuală: realizări şi perspective. Lucrări ştiinţifice II.* Chişinău: Universitatea Agrară de Stat din Moldova. 138-141.

—. (2001). Verbele de comunicare animală în limba română. Încercare de clasificare. [Verbs of Animal Communication in Romanian: A Tentative Classification]. *Lucrări ştiinţifice. Zootehnie şi Biotehnologii XXXIV:* 416-419.

Robert, P. (1996). *Dictionnaire alphabétique & analogique de la langue française.* [Alphabetical and Analogical Dictionary of the French Language]. Paris: S.N.L.

The Random House Dictionary of the English Language. (1968). New York: Random House.

Webster Comprehensive Dictionary. (1995). Chicago: J. G. Ferguson Publishing Company.

ENGLISH ANIMAL IDIOMS AND THEIR ROMANIAN EQUIVALENTS (I)

ANDREEA VARGA
AND ASTRID-SIMONE GROSZLER

The purpose of this research paper is to establish semantic and cultural similarities and differences when translating **animal idioms** from English into Romanian.

Idioms are terms or phrases whose meaning cannot be inferred simply from the meaning of each of its words, but they comprise a figurative significance which is known through common use. (*Idioms*. Online: http://idioms.thefreedictionary.com/) The definition of **idioms** can be narrowed down to words collocated together which become fossilized and which, eventually, alter the meaning of the words that exist.

Collocations "consist of two or three lexical (sometimes called full, descriptive, substantial) words, usually linked by grammatical (empty, functional, relational) words. The collocates within a collocation define and delimit each other by eliminating at least some of their other possible meaning; the defining may be mutual and equally balanced, but more often it is closer for one to collocate than for the other." (Newmark 1980) A **collocation** can be sundered into syntagmatic or horizontal, thus relaying on a common structure, or paradigmatic or vertical, residing in words pertaining to the same semantic field which may function as substitutes for each other or be semantic opposites. These fall within their category and become **collocations** only when arrayed syntagmatically.

In effect, **idioms** are colloquial metaphors requiring some foundational cognizance, information or experience, employed within a culture where the interlocutors must have a common reference point. Perforce, they constitute more than just a semantic part of language, but rather a totem of a particular culture. Nonetheless, **metaphorical idioms** can be construed as more universal since animals occur in all cultures as embodiments of particular features.

Another operating concept that bolsters up the thesis of this paper is **equivalence**: defined under a larger scope, it constitutes the result of a translation process revolving around the identity relation between two terms pertaining to two different languages and possessing almost the same connotation and denotation. Reduced to a more narrow meaning, **equivalence** is a translation technique which presupposes rendering a term or set phrase in the source language into a different term in the target language, but preserving the correspondence with the same reality. **Equivalences** are established between texts integrating in a certain culture, in a particular communication context, and which are perceived as consequences of the interaction between the translator and the text. Several types of **equivalence** can be distinguished: cognitive (engendered by the semantics of the text and the knowledge brought forth by the translator, based on real or unreal experiences), emotional (intuitive), denotative (conveying the information advanced by the source language text), connotative (preserving the style, the language register, the sociolect, the geographical outreaching of the expressions), standard or normative (according to the genre of the text for translation, for example, a textbook, an advertisement, etc.), pragmatic or dynamic (adapted to the knowledge of the reader in order to be comprehended) (Lungu-Badea 2003).

The plethora of animal **idioms** can be subdued to manicheistic segregation, embodying either the good or the bad. Herd animals, in general, horses, cattle, are associated with fertility as inheritance from the Celts and, thus, gain positive coordinates. However, they preserve specific features of characteristic behaviour which are emphasized and employed at a metaphorical level: ***a cash cow*** (a profitable business), ***a dark horse*** (a candidate who is little known to the general public), ***a horse of a different colour*** (something totally separate and different), ***a sacred cow*** (a belief or system that is treated with much respect and is not usually criticized), ***as gentle as a lamb*** (very gentle), ***as innocent as a lamb*** (having no guilt, naïve), ***as meek as a lamb*** (quiet, docile, meek), ***as strong as a horse / ox*** (very strong), ***by shank's mare*** (surprise, excitement),***holy cow*** (surprise, excitement), ***horse sense*** (common sense, practical thinking), ***horse trade*** (to bargain in a hard and skilful way), ***like lambs to the slaughter*** (quietly and without complaining about the dangers that may lie ahead), ***on horseback*** (on the back of a horse), ***straight from the horse's mouth*** (directly from the person who said something, directly from a dependable source), ***the black sheep of the family*** (the worst or the most unpopular member of a family), ***to back the wrong horse*** (bet on the wrong horse), ***to change horses in midstream*** (to make new plans or choose a new leader in an activity), ***to eat like a horse*** (to eat a lot), ***to flog / beat a dead horse*** (to continue fighting a battle that has been won), ***to get off one's high horse*** (to begin to be humble and agreeable), ***to get on one's high horse*** (to behave with arrogance), ***to hold one's horses*** (to wait, to be patient), ***to***

hoof it (to walk or run), ***to horse around*** (to play around in a rough way), ***to lock the barn door after the horse is gone*** (to try to deal with something after it is too late), ***to look a gift horse in the mouth*** (to complain if a gift is not perfect), ***to put (someone or something) out to pasture*** (to retire someone or something, just as you would put a horse that is too old to work out to pasture), ***to put the cart before the horse*** (to do things in the wrong order), ***to ride herd on (someone)*** (to watch closely and strictly supervise someone), ***to separate the sheep from the goats*** (to divide people into two group), ***to work like a horse*** (to work very hard), ***until the cows come home*** (for a very long time), ***wild horses could not drag (someone away)*** (there is nothing that will force someone to go somewhere or do something), ***you can lead a horse to water (but you can't make it drink)*** (you can give someone the opportunity to do something but you cannot force him or her to do it if they do not want to). On the other hand, bad animals are usually associated with the ***snake*** or the ***wolf***: ***a snake in the grass*** (a treacherous or underhand person), ***a snake-pit*** (any arena where people are viciously fighting each other for power). ***Serpents*** symbolize trouble corollary to strife and infertility, whereas ***wolfs***, though paragons of loyalty, success, perseverance, stability, thought, intuition, learning, occur in **idioms** with negative connotations: ***a lone wolf*** (someone who prefers to spend time alone and has few friends), ***a wolf in sheep's clothing*** (a person who pretends to be good), ***to cry wolf*** (to give a false alarm, to warn of a danger that is not there), ***to keep the wolf from the door*** (to maintain oneself at the most basic level), ***to keep the wolves at bay*** (to fight against some kind of trouble), ***to throw (someone) to the wolves*** (to send someone into danger without protection, to sacrifice someone), ***to wolf down (something)*** (to gulp down something, to eat something quickly).

Idioms and **collocations** can be posited as having a Romanian equivalent and sharing the same common feature of the animate subject. Such an eloquent example is ***to hold one's horses*** (R *a ţine caii în frâu*), which elicits the power, strength, freedom, wildness of the horse. However, if we are to consider the **idiom** ***to bet on the wrong horse***, it is most obvious why it has no Romanian equivalent. Horse racing is so deeply inveterate in English tradition that it has almost become an iconic sport. But not in Romanian culture. The **idiom** ***to look a gift horse in the mouth*** has a perfect equivalent in *calul de dar nu se caută la dinţi*, which underscores that horses were an expensive, wealthy gift in both traditions. The lamb is cogently perceived as meek and innocent in both of the aforementioned languages, the **idioms** having literal translations in the latter language: ***as gentle as a lamb*** preserves its structure based on a simile in Romanian and is translated by *blând ca un mieluşel*, ***like lambs to slaughter*** is rendered as *ca mielul la tăiere*, ***the black sheep of the family*** becomes *oaia neagră a familiei* (maybe also due to the biblical reference). ***Swine*** are universally tropes of filth, ordure, dung and inferiority: E ***to cast pearl before swine*** – R *a arunca perle la porci*, E ***as fat as a pig*** – R *gras ca un porc*. Nonetheless, they are associated in the Chinese tradition with being highly intelligent, scholarly, easily angered, easily swayed, and affected by emotions (which is fully true). To further continue our analysis on equivalents, the ***bull*** is another universal paragon of wealth, potency, beneficence, generative force, male procreative strength, kingship, taming of the masculine and animal nature. These traits are shed light upon in the following **idioms**: ***as strong as a bull*** (R *puternic ca un taur*) and ***to take the bull by the horns*** (R *a apuca taurul de coarne*). The ***mouse*** is subsumed under the same category of universally acknowledged animals for a characteristic feature illustrated in the **idiom *as poor as a mouse church***, its Romanian counterpart being *sărac ca un şoarece de biserică*, which explains the pervasiveness of the church in all cultures.

On balance, **animal idioms** may revolve around the same trait of characteristic behaviour which is inherent in the animal, they may be attached features eloquent for a particular geographic space, or they may inherit traits *a posteriori* as a consequence of their social, historical or religious involvement and significance in these events. **Idioms** enucleate a cornucopia of details in the history of a language explaining for alterations and providing with depictions of certain cultures.

References

Idioms. Online: http://idioms.thefreedictionary.com/

Lungu Badea, G. (2003). *Dictionary of Terms Employed in the Theory, Practice and Didactics of Translation*. Timişoara: Ed. Orizonturi.

Newmark, P. (1980). *Approaches to Translation*. Phoenix, AR: ELT.

Perković, Anica, Raţă, Georgeta, Perković, Marina & Turalija, Alka. (2009). Animal Idioms in Different Languages. *Proceedings of the Second International Conference "Research People and Actual Tasks on Multidisciplinary Sciences" Lozenec, Bulgaria 3*: 120-124.

Varga, Andreea & Groszler, Astrid-Simone. (2008). Animal Idioms and Their Romanian Equivalents. *Scientific Papers, Faculty of Agriculture 40 (3)*: 369-373.

English Animal Idioms and Their Romanian Equivalents (II)

Astrid-Simone Groszler

Introduction

The purpose of this paper is to reveal the way in which English animal **idioms** are put into Romanian. By that we mean to observe not just whether or not English animal **idioms** have a Romanian equivalent, but also how many of these equivalents stick to the choice of the animal name instead of just giving a semantic equivalent.

We know that **idioms** are terms or phrases whose meaning cannot be inferred simply from the meaning of each of its words, but they comprise a figurative significance which is known through common use. We can say that **idioms** are colloquial metaphors requiring some foundational cognizance, information or experience, employed within a culture where the interlocutors must have a common reference point. **Idioms** with a metaphorical meaning can be construed as more universal since animals occur in all cultures as embodiments of particular features.

We have shed some light upon the term **idiom**. Another term that must be subdued to our attention is **equivalency**. In Baker (2001), there is a definition of equivalency as 'the relationship of a source text (ST) and a target text (TT) that allows the TT to be a translation of the ST in the first place. Equivalence relationships are also said to hold between parts of STs and parts of TTs. […] equivalence is commonly established on the basis of: the source language (SL) and target language (TL) words supposedly referring to the same thing in the real world, i.e. on the basis of their referential or denotative equivalence; the SL and TL words triggering the same or similar associations in the minds of native speakers of the two languages, i.e. their connotative equivalence; the SL and TL words being used in the same or similar contexts in their respective languages, i.e. what Koller (1989) calls text-normative equivalence; the SL and TL words having the same effect on their respective readers, i.e. pragmatic (Koller

1989) or dynamic equivalence (Nida 1964); the SL and TL words having similar orthographic or phonological features.'

Also the concept of **textual equivalence** comes up, defined as the combination of similarities in ST and TT information flow and the cohesive roles of ST and TT devices in their respective texts. We may speak of functional equivalence when we deal with the translator's decision as to which consideration to be taken into account at any time.

Material and Method

We have inventoried 82 English **animal idioms** gathered from English language and specialised dictionaries. Then, we have looked for the Romanian counterparts of the English **animal idioms** analysing their equivalency and frequency.

Results

The **idioms** we have analysed include bird and insect **idioms**. The animals' characteristic physical or behavioural features are put to display in the following: *like a **bat** out of hell* ('moving very quickly'), *like a **bear** with a sore head* ('very irritable and bad-tempered'), *eager **beaver*** ('hardworking and enthusiastic, overzealous'), *have a **bee** in one's bonnet* ('have an obsessive idea'), *the **bee**'s knees* ('exceptionally good'), ***birds** of a feather* ('very similar in many ways'), *kill two **birds** with one stone* ('succeed in doing two things at the same time'), *like a red flag to a **bull*** ('sure to make someone very angry'), *take the **bull** by the horns* ('dealing with a difficult situation or problem'), *a social **butterfly*** ('a person who has a lot of friends and acquaintances and likes to flit from one social event to another'), *a **cat** in gloves catches no mice* ('being too careful and polite prevents you from obtaining what you want'), *a **cat** can look at a king* ('nobody is so important that an ordinary person cannot look at or be curious about them'), *a fat **cat*** ('rich and powerful person misusing his/her money and power'), *herding **cats*** ('difficulty of coordinating a situation which involves people who all want to act independently'), *let the **cat** out of the bag* ('reveal a secret, often not intentionally'), *like a **cat** on hot bricks* ('very nervous or restless'), *like something the **cat** dragged in* ('dirty, untidy or generally unappealing'), *play **cat and mouse*** ('treat someone alternately cruelly and kindly, so that they do not know what to expect'), *raining **cats and dogs*** ('raining very heavily'), *wait for the **cat** to jump* ('delay taking action until you see how events will turn out'), ***chicken** out of something* ('not doing something because of fear'), *like a* *headless* ***chicken*** ('without thinking or analyzing the situation carefully'), *till the* ***cows*** *come home* ('long time'), ***crocodile*** *tears* ('false tears showing insincere grief'), *eat* ***crow*** ('admit that you were wrong about something and apologize'), *as the* ***crow*** *flies* ('distance measured in a straight line'), ***dog's*** *life* ('a situation or job which you find unpleasant or unsatisfactory'), ***dog*** *eat* ***dog*** ('intense competition and rivalry in pursuit of one's own interests, with no concern for morality'), *a* ***dog*** *in the manger* ('someone who stops others enjoying something he cannot use or doesn't want'), *every* ***dog*** *has its day* ('everyone can be successful at something at some time in their life'), *give a* ***dog*** *a bad name* ('be blamed or suspected after losing your reputation'), *help a lame* ***dog*** *over stile* ('help someone who is in difficulty or trouble'), *like a* ***dog*** *with two tails* ('extremely happy'), *a* ***dog's*** *breakfast* ('a complete mess'), *go to the* ***dogs*** ('becoming less successful or efficient than before'), *let sleeping* ***dogs*** *lie* ('not interfering with a situation because one could cause problems'), *the tail wagging the* ***dog*** ('reversal of roles, with a small or minor element of something having a controlling influence on the most important element'), *why keep a* ***dog*** *and bark yourself?* ('if someone can do a task for you, there's no reason to do it yourself'), ***dog*** *and* ***pony*** *show* ('marketing event or presentation which has plenty of style but not much content'), ***donkey*** *work* ('unpleasant, boring parts of a job'), *talk the hind leg off a* ***donkey*** ('describing a very talkative person'), *take to something like a* ***duck*** *to water* ('doing something naturally and easily, without fear or hesitation'), *a sitting* ***duck*** ('an easy target, a person who is easy to deceive'), *like water off a* ***duck's*** *back* ('having no effect at all'), ***fish*** *in troubled waters* ('trying to gain advantages for yourself from a disturbed state of affairs'), ***fish*** *out of water* ('feeling uncomfortable because of an unfamiliar situation or unfamiliar surroundings'), *a different kettle of* ***fish*** ('something completely different from what was previously mentioned'), *drink like a* ***fish*** ('drink heavily'), *have other* ***fish*** *to fry* ('having more important things to do'), *there are (plenty of) other* ***fish*** *in the sea* ('that there are many other people just as good as the one somebody failed to get'), *neither* ***fish*** *nor* ***fowl*** ('describing people or things that are difficult to classify, that are neither one thing nor another'), *have a* ***flea*** *in one's ear* ('be angrily reprimanded or humiliated'), *drop like* ***flies*** ('fall ill or die in large numbers'), ***fly*** *in the ointment* ('someone or something that prevents a situation from being completely satisfactory'), ***fly*** *on the wall* ('person who watches a situation without being noticed'), *have a* ***frog*** *in one's throat* ('have difficulty in speaking clearly, because of a cough or a sore throat'), *all his* ***geese*** *are* ***swans*** ('referring to someone who constantly exaggerates the importance of somebody or something'), *cook somebody's*

goose ('spoil a person's chance of success'), *have* ***goose*** *pimples* ('have the skin temporarily raised into little lumps because of cold or fear'), *a* ***wild goose*** *chase* ('wasting a lot of time looking for something you might not find'), ***guinea pig*** ('people on whom new methods, treatment or ideas are tested'), *back or bet on the wrong* ***horse*** ('support the wrong person'), *get on your high* ***horse*** ('start behaving in a haughty manner'), *a dark* ***horse*** ('a secretive person'), *I could eat a* ***horse*** ('I am very hungry'), *one-****horse*** *town* ('a small, boring town'), *walk into the* ***lion****'s den* ('get involved in a difficult situation'), *like a* ***moth*** *to a flame* ('irresistibly attracted to someone or something'), ***mouse*** *potato* ('someone spending a lot of time in front of the computer'), *as stubborn as a* ***mule*** ('very obstinate and not willing to listen to anyone or anything'), *the world is your* ***oyster*** ('you are free and able to enjoy the pleasures and opportunities that life has to offer'), *proud as a* ***peacock*** ('very proud'), ***pigs*** *might fly* ('disbelief in something'), *make a* ***pig*** *of yourself* ('eat and drink too much'), *Shank's* ***pony*** ('walking, by foot'), *smell a* ***rat*** ('suspect that something is wrong'), *packed like* ***sardines*** ('very crowded'), *a* ***snake*** *in the grass* ('someone who pretends to be your friend while actually betraying you'), *black* ***sheep*** ('someone very different from the others and thus the least respected one'), *like* ***turkeys*** *voting for Christmas* ('a particular option is unlikely to be chosen because it would not be in the interest of the people concerned'), *have a* ***whale*** *of a time* ('have a great time, enjoy oneself'), and ***worm****'s-eye view* ('a narrow view from the closest point').

While trying to find Romanian equivalents for these **animal idioms**, the research was carried into effect to the extent that most of them entail an equivalent. However, there are some instances in which they do not possess one. We refer hereby to the cultural differences between the English and Romanian societies, instances where one cannot find equivalents because of the lack of similarities: ***a cat can look at a king*** (in Romanian there is an idiom that means precisely the opposite: *la soare te poţi uita, dar la ea ba* 'one is too perfect or too beautiful to be looked at'), ***a cat in gloves catches no mice***, ***a dark horse***, ***a dog in the manger***, ***as the crow flies***, ***give a dog a bad name***, ***have a frog in one's throat***, ***mouse potato***, ***one-horse town***, ***Shank's pony***, etc.).

Further on we focus on those **idioms** which entail a Romanian equivalent. We can infer that there are two categories of equivalents: expressions which, while having a similar meaning, do not entail any animal name (here we might speak of referential or denotative equivalence) and expressions which include an animal name (may it refer to the same animal as in the English idiom or to another) – when we imply connotative or even dynamic equivalence.

In the first category of **idiom equivalency** we may include: ***a sitting duck*** (R *o țintă ușoară* – literal meaning 'easy target'), ***birds of a feather*** (R *pe aceeași lungime de undă* – the term used here means 'wave length'), ***chicken out of something*** (R *a da bir cu fugiții* 'joining the runaways'), ***get on your high horse*** (R *a fi cu nasul pe sus* – the term used means 'nose'), ***have a bee in one's bonnet*** (R *are un cui în cap* – the term used here is *nail*), ***like a bat out of hell*** (R *ca fulgerul* – reference to the speed of lightning), ***like a bear with a sore throat*** (R *cu capsa pusă*), ***like a cat on hot bricks*** (R *a sta ca pe jar* – here the idea of 'hot' is maintained within the semantics of *jar* 'embers', but the animal name is dropped), ***like a dog with two tails*** (R *in al nouălea cer* – the term used is *sky / heaven*), ***like a headless chicken*** (R *a se arunca cu capul înainte* – the term *head* is maintained), ***like turkeys voting for Christmas*** (R *a-și face iluzii deșarte* – keywords meaning 'illusions' and 'vain'), ***pigs might fly*** (R *la Sfântul Așteaptă* – a literal translation would be 'Saint Waiting' – the idea of waiting a long time or R *când o face plopul pere și răchita micșunele* – meaning *never*), ***raining cats and dogs*** (R *plouă cu găleata* – the term used here means 'bucket'), ***talk the hind leg off a donkey*** (R *a-i merge gura ca o moară stricată* – the term *mouth* is specified), ***the bee's knee*** (R *buricul pământului* 'the centre of the universe'), etc.

If we examine the Romanian **animal idioms**, there are **idioms** in which the same animal as in the English one occurs, and **idioms** that contain the names of different animals (instance based on the cultural differences already mentioned). The following pertain to **idioms** exhibiting (almost) the same animal: ***a different kettle of fish*** (R *o altă mâncare de pește*), ***black sheep*** (R *oaia neagră*), ***crocodile tears*** (R *lacrimi de crocodil*), ***dog eat dog*** (R *se mănâncă între ei ca și câinii*), ***dog's life*** (R *viață de câine*), ***fish in troubled water*** (R *a pescui în ape tulburi*), ***fish out of water*** (R *ca peștele pe uscat*), ***fly in the ointment*** (R *ca musca-n lapte*), ***packed like sardines*** (R *ca sardelele-n cutie*), ***play cat and mouse*** (R *a se juca cu cineva ca pisica cu șoarecele*), ***take the bull by the horns*** (R a *lua taurul de coarne*), ***there are (plenty of) other fish in the water*** (R *e plină balta de pește*), etc.

As for the **idioms** containing names of different animals, there are: ***a dog's breakfast*** (R *a o face de oaie* – the term used here is *oaie* 'sheep), ***flea in one's ear*** (R *cu coada între picioare* – if we were to give a literal translation we would say *with the tail between one's legs*, alluding to dog behaviour), ***have goose pimples*** (R *a avea piele de găină* – just the name is changed, *goose* being replaced by *găină* 'hen'), ***kill two birds with one***

stone (R *a împuşca doi iepuri dintr-un foc* – here *bird* is replaced by *iepure* 'rabbit', both terms referring to game), ***let the cat out of the bag*** (R *a scăpa porumbelul din gură* – instead of *cat* there is *porumbel* 'pigeon', a bird – the exact opposite), ***like a red flag to a bull*** (R *a întărâta câinii printre garduri* – *bull* is replaced by *câine* 'dog'), ***proud as a peacock*** (R *a se umfla în pene ca curcanul* – Romanian farmsteads included *curcani* 'turkeys', but rarely peacocks), ***take to something like duck to water*** (R *a se simţi ca peştele în apă* – the term *duck* is replaced by *peşte* 'fish', both denominating aquatic animals), ***the tail wagging the dog*** (R *lupul mâncat de oaie* – the reversed situation here is illustrated by *oaie* 'sheep' and *lup* 'wolf' in inversed roles), ***worm's eye view*** (R *vedere de cal* – here the difference is very acute, the English *worm* being replaced in the Romanian idiom by *cal* 'horse'), etc.

Discussion

The previous section brought forth an account of English **animal idioms** and their Romanian counterparts. We have seen that most English **animal idioms** entail an equivalent, but a number of **animal idioms** still remain which bear no representation in Romanian, a fact induced by cultural differences. These cultural differences also determine a shift in the use of animal names within the two languages, resulting in dynamic equivalents.

Of the 82 English **animal idioms** studied, only 16 have no Romanian equivalents, i.e. 20%. Of the 66 Romanian counterparts, 28 (i.e. 34%) do not contain animal names, though issuing the same meaning. Interesting enough, of the 38 remaining Romanian **animal idioms**, 19 (i.e. 23%) share the English choice of the animal, while the other 19 (23%) take a different animal name (Figure 7-4).

Our research has clearly shown that, though most English **animal idioms** entail a Romanian equivalent from the point of view of semantics, these counterparts do not always take an animal name. Animal names may be replaced by concrete or by abstract nouns, and though they infer the same meaning as the English one, the mirroring of this meaning is specific to the respective language.

Animal idioms are also characteristic to the country they pertain to. Thus, although most English **animal idioms** built with the help of the words "pig" and "fish" find an almost perfect literal translation in Romanian, those sharing the words "dog" and "horse", as well as **insect** and **bird idioms**, mostly take another animal name in Romanian. This is possible because the same animal can have different connotations in the minds of the people pertaining to different cultures, people not sharing the same mentality and life concepts.

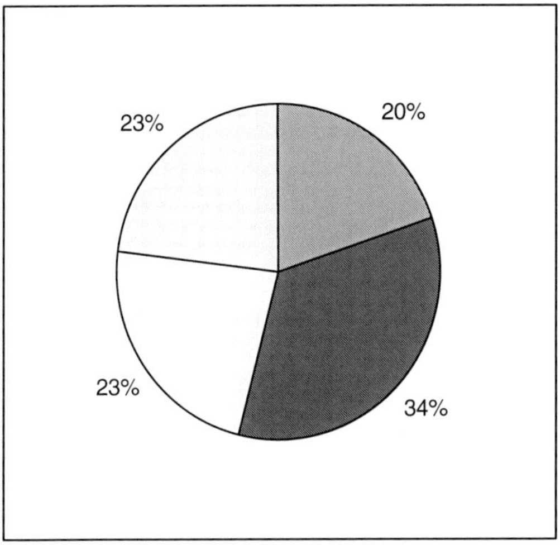

Figure 7-4. English Animal Idioms in Romanian: 20% - no Romanian counterparts; 34% - Romanian counterparts containing the same animal names; 23% - Romanian counterparts not containing animal names; 23% - Romanian counterparts containing different animal names

We also found that while some of the collocations in Romanian take an animal name, it is the exact opposite of the English one (***eat like a horse*** and "foame de *lup*" – *wolf*), there are also English **idioms** which do not show a Romanian equivalent, but a Romanian opponent (***a cat can look at a king***, the Romanian collocation inferring the exact opposite meaning: *la soare te puteai uita, dar la dânsa ba* – perfection which one does not dare to contemplate).

Conclusions

There is no perfect equivalency between English and Romanian **animal idioms**. We have proven this above. We could infer this from the analysis of the 82 English **animal idioms** presented, of which only 66 displayed Romanian counterparts. We could also see that only a small percentage (i.e. 23%) of the **animal idioms** actually respects the choice of the English animal name and preserves it within the Romanian language.

Another conclusion that can be drawn is that one of the determiners of this "imperfect" equivalency is the cultural difference between the English and the Romanian society, conferring different roles to the same animal. What we could clearly see was that **idioms** with "pig" and "fish" had an almost perfect word-for-word translation in Romanian.

We do believe that this kind of research is needed and can be very helpful in order to be as accurate as possible when translating English specialised or non-specialised texts into Romanian and vice versa. Our intention is to continue our research enlarging our scope beyond the animal world.

References

Baker, M. (Ed.). (2001). *Routledge Encyclopaedia of Translation Studies*. New York: Routledge.

Groszler, Astrid-Simone. (2008). On the Romanian Equivalency of Animal Idioms in English. *Journal of Linguistics Studies 1 (1)*: 65-70.

Idioms. Online: http://idioms.thefreedictionary.com.

Lungu Badea, G. (2003). *Dictionary of Terms Employed in the Theory, Practice and Didactics of Translation*. Timişoara: Ed. Orizonturi.

Perković, Anica, Raţă, Georgeta, Perković, Marina & Turalija, Alka. (2009). Animal Idioms in Different Languages. *Proceedings of the Second International Conference "Research People and Actual Tasks on Multidisciplinary Sciences" Lozenec, Bulgaria 3*: 120-124.

Varga, Andreea & Groszler, Astrid-Simone. (2008). Animal Idioms and Their Romanian Equivalents. *Lucrări ştiinţifice, Facultatea de Agricultură 40 (3)*: 369-373.

ENGLISH LOANWORDS IN THE ROMANIAN OF AGRI-TOURISM INTERNET SITES

GEORGETA RAȚĂ, CORNELIA PETROMAN AND IOAN PETROMAN

Introduction

Romanian Internet sites advertising agri-tourism should be quite different from other tourism advertising sites, i.e. there should not be as many English **loanwords** as in sites advertising travel agencies, tourist destinations or accommodation, for instance. Though less numerous, these **loanwords** seem to have had a life of their own, since they have either preserved or changed their original form, or even developed inflected or **derivative forms** in Romanian, thus showing that they have well adapted to the Romanian linguistic pattern.

Results and Discussion

Lexical Borrowings

Analysing the www.agroturism.com site, we have come across the following **loanwords** of English origin: ***agroturism*** (and its inflected form ***agroturismul*** and its **derivatives** ***agroturistic*** and ***agroturistice***), ***baruri***, ***bowling***, ***camping***, ***cicloturism***, ***cluburi***, ***duglas***, ***fast-fooduri***, ***folclor*** (and its inflective form ***folclorul*** and its **derivative form** ***folclorică***), ***meniul***, ***site-ului***, ***mouse-ul***, ***rafting***, ***ski***, ***ski extreme***, ***snowboard***, ***telecabină***, ***teleferice***, ***teleski***, ***turism*** (alone or in noun phrases such as ***turism agroturistic*** (*sic!*), ***turism balnear***, ***turism cultural***, ***turism de litoral*** and its inflective form ***turismul de litoral***, ***turism extrem***, ***turism local***, ***turism montan*** and its inflective form ***turismul montan***, ***turism neorganizat***, ***turism pentru sănătate***, ***turism rural***, ***turism rural cultural***, ***turism rural ecologic***, ***turism rural şi ecologic***, ***vitituri****smul*, and ***week-endul***.

Some of these **loanwords** have been borrowed as such, with no change whatsoever in spelling: ***bar*** (< E *bar*), ***bowling*** (< E *bowling (alley)*), ***camping*** (< E *camping (site)*), ***club*** (< E *club*), ***fast-food*** (E < *fast-food*),

mouse (< E *mouse*), ***rafting*** (< E *rafting*), ***site*** (< E *site*), and ***snowboard*** (< E *snowboard*), but with slight changes in pronunciation in almost all the cases (except, perhaps, for ***mouse***).

Other English **loanwords** have changed their form to better adapt to the Romanian spelling: ***agrotourism*** (< E *agritourism*), ***cicloturism*** (< Gk *kuklo* 'circle' + E *tourism*), ***duglas*** (< E *Douglas (fir)* (*Pseudotsuga menziesii*), ***folclor*** (< E *folklore*), ***meniu*** (< E *menu*), ***telecabină*** (< Gk *tele* 'far' + E *cabin* 'small room'), and ***teleferic*** (< E *telpherage* 'distance transport').

The omnipresence of the English language and the fashion of English **loanwords** have resulted in a strange case of over-correctness: the English spelling ***ski***, as in Rum ***ski*** and ***ski extrem***, and ***teleski***, though the Romanian language dictionary recommends the spelling ***schi*** (this is also the Romanian pronunciation for the English *ski*).

Some other English **loanwords** are mentioned by Romanian language dictionaries, indicating a double origin: English and French – ***camping*** (< E, F *camping*), or French and English – ***club*** (< F, E *club*) and ***folclor*** (< F, E *folklore*).

Some of these **loanwords** have come into Romanian vocabulary indirectly, through French, either entirely – ***bar*** (< F *bar* < E *bar*), ***folclor*** (< F *folklore* < E *folklore*), and ***teleferic*** (< F *téléférique* / *téléphérique* < E *telpherage* 'distance transport') – or partially – ***cicloturism*** (< F *cyclotourisme* < Gk *kuklo* 'circle' + E *tourism*) and ***telecabină*** (< F *télécabine* < Gk *tele* 'far' + E *cabin* 'small room').

It is interesting to note that most of these English **loanwords** are not mentioned by Romanian language dictionaries, though they have had a pretty long history with us: ***agroturism***, ***fast-food***, ***meniu***, ***mouse***, ***rafting***, ***site***, ***ski***, ***ski extrem***, ***snowboard***, and ***teleski***.

The following English **loanwords** have definite articles in Romanian: *agrotourism* – *agroturism**ul***, *meniu* – *meniu**l***, *mouse* – *mouse-**ul***, *site* – *site-**ul***, *turism* – *turism**ul***, *vititurism* – *vititurism**ul***, and *week-end* – *week-end**ul***, while the following English **loanwords** have plural forms in Romanian: *bar* – *bar**uri***, *club* – *club**uri***, *fast-food* – *fast-food**uri***, and *teleferic* – *teleferic**e***. These inflective forms show that the English **loanwords** have been adopted and made to sound as any other Romanian words of the kind, which was not very difficult since the definite article in Romanian has a postposition. The use of the hyphen is some of these **loanwords** proves that this is an ongoing process.

The following English **loanwords** have **derivative forms** in Romanian: *agrotourism* – *agroturis**tic*** (adjective in the masculine singular) and *agroturis**tice*** (adjective in the feminine plural), as in *distracţii* ***agroturistice*** *specific sezoniere* ('seasonal specific agri-tourism attractions'), *pensiunile* ***agroturistice*** ('agri-tourism guesthouses'), *puncte de atracţie* ***agroturistice*** ('agri-tourism attraction sites') – and *folclor* – *folklor**ică*** (adjective in the feminine singular) as in *zonă* ***folclorică*** ('folklore area'). This must also have been a natural process since gender and number marks are also post-positioned in Romanian.

Other **loanwords** are mentioned by Romanian language dictionaries together with their pronunciation: ***bowling*** [báuling], ***camping*** [chémping], and ***week-end*** [ŭíchend].

Syntactical Borrowings

If lexical borrowings are more or less part of our everyday life, together with such names of aquatic parks as ***AquaMagic*** or ***AquaPark Balada***, syntactical patterns have even worse consequences as they may end in serious distortions of the grammar structure of the Romanian language. Thus, under the influence of English: though we say, in Romanian, *Băile Băiţa* ('Băiţa Spa'), *Băile Someşeni* ('Someşeni Spa'), etc., we have started to also say ***Turda Băi*** ('Turda Spa') instead of *Băile Turda*, following the English pattern; we have also started to drop the preposition in noun phrases such as *pârtie ski* ('ski slope') instead of *pârtie* ***de*** *ski*, *tip turism* ('tourism type') instead of *pârtie* ***de*** *ski*, and *tip zonă* ('area type') instead of *tip* ***de*** *zonă*; we shorten the cardinal points as in *N. jud.* ('in the northern part of the county') or *E de* ('east from').

Conclusions

A few remarks on the impact of English **loanwords** on the Romanian of agri-tourism such as it is on Romanian Internet sites advertising this modern form of tourism:

- though not numerous, English **loanwords** appear on sites advertising agri-tourism in Romania;
- there are English **loanwords** that have been borrowed as such, with no changes whatsoever in spelling, except for some slight changes in pronunciation in almost all the cases, which speaks of the Romanians' receptiveness as far as foreign languages are concerned;
- there are English **loanwords** that have changed their form to better adapt to the Romanian spelling, which speaks of the Romanians' capacity of adapting almost anything to their language norms;
- there are few cases of over-correctness;

- some English **loanwords** are mentioned by Romanian language dictionaries as having two possible origin (English and French);
- some of the English **loanwords** have come into the Romanian vocabulary indirectly, through French, entirely or partially;
- most of the English **loanwords** are not mentioned by Romanian language dictionaries, though they have produced both inflective and **derivatives forms** in Romanian;
- some English **loanwords** are mentioned by Romanian language dictionaries together with their pronunciation, which is a good start in acknowledging this type of lexical items;
- borrowed syntactical patterns may end in serious distortions of the grammar structure of the Romanian language;
- phrases such as *turism rural (agrotourism)* ('rural tourism (agri-tourism)') can be misleading since *agri-tourism* is a form of *rural tourism*, not its synonym.

References

Agroturism. Online: www.agroturism.com.

Chalker, Sylvia & Weiner, E. (1994). *The Oxford Dictionary of English Grammar*. London - New York - Sydney - Toronto: BCA.

Dicţionarul explicativ al limbii române. [The Explanatory Dictionary of the Romanian Language]. (1998). Bucureşti: Editura Univers Enciclopedic.

Leviţchi, L. (Ed.). (2004). *Dicţionar englez-român.* [English-Romanian Dictionary].Bucureşti: Ed. Univers Enciclopedic.

Raţă, Georgeta & Petroman, I. (2006). On English Loanwords in the Romanian of Tourism (Names of Romanian Travel Agencies). *Lucrări ştiinţifice. Facultatea de Agricultură XXXVIII*: 507-510.

Raţă, Georgeta & Petroman, I. (2006). On English Loanwords in the Romanian of Tourism (Sites of Travel Agencies in the District of Timiş). *Lucrări ştiinţifice. Facultatea de Agricultură XXXVIII*: 503-506.

Raţă, Georgeta, Petroman, Cornelia & Petroman, I. (2007). On the English Loanwords in the Romanian of Agri-Tourism Internet Sites. *Proceedings of Lingua Summit 2007, Trencin, Slovakia*: 4 p.

CONTRIBUTORS

Diana-Andreea BOC-SÎNMĂRGHIȚAN has a PhD in Linguistics from the University of the West in Timişoara (Romania). She is a qualified FFL teacher and has been teaching French for 9 years. Her experience includes both French practical courses and several areas of FSP, and she has already published in these areas. She is a researcher in the field of cross-linguistics, having worked in French and Romanian, and understands the practical issues involved in teaching and learning different languages and transferring research ideas from one language to another.

Oana BOLDEA has been a qualified teacher of English and Romanian for 12 years. Her interest lies both in general English and in ESP, and her research in these fields has materialized in several research papers, published in Romania and abroad. She is an assistant lecturer at BUASVM Timisoara and helps organize the annual international conference on language education at this university. She has co-authored four university textbooks and authored *English for Vets*.

Alina-Andreea DRAGOESCU has a PhD in Philology from the University of the West in Timişoara (Romania). She is a qualified EFL teacher and has been teaching languages for 6 years. Her expertise includes general English and several areas of ESP. The latter have also provided topics for two university textbooks and a number of published papers. Her interests are in the field of conceptual metaphor and other applied linguistics areas. She contributes papers to the *Journal of Linguistic Studies* of her Department and helps organise the annual international conference on language education at the B.U.A.S.V.M. in Timişoara.

Petru DRAGOESCU has a BA in agriculture from the Agronomical Institute in Timişoara (Romania) and an MA in Farm management sustainable development from the BUASVM in Timişoara. He attended post-university studies in Forage production and conservation, as well as several other courses and Phare-vet programmes. He has worked as a qualified teacher at the Iulian Drăcea High School in Timişoara for 31 years. His areas of interest include agriculture, horticulture, environmental engineering, ecology, management, and marketing.

Camelia GIUCHICI has a BSc in Agriculture and a PhD in Agriculture from the Banat University of Agricultural Science and Veterinary Medicine in Timişoara (Romania). She has taught for over 18 years in the field of plant protection. She is qualified in seed certification and in field and seed evaluation. She has published widely in these areas. She also has expertise in coordinating and managing people and budgets. She is executive manager of the Territorial Inspectorate for Seed Certification and Propagating Materials. She is a member of international projects in the field of education and of EU regulation concerning food safety and phytosanitary services. She has published textbooks in Romania.

Astrid-Simone GROSZLER has an MA in Linguistics, an MA in Political Sciences, and an MA in Marketing Management granted by the University of the West in Timişoara (Romania). A PhD student in philology at the moment, she is a qualified EFL teacher and has been teaching languages for over 4 years. Her experience includes both English and German practical courses and ESP courses, having already published in these areas. She understands the practical issues involved in teaching and learning different languages and transferring research ideas from one language to another. She has also worked as an interpreter for German, English, and Dutch.

Scott HOLLIFIELD, a doctoral candidate at the University of Nevada in Las Vegas (U.S.A.), earned a BA in Film Studies and an MA in English Literature from the Wayne State University in Detroit, MI (U.S.A.). Recent UNLV teaching assignments have included Shakespeare's Comedies and Histories, British Literature c. 1000-1700, and World Literature. An intense interest in the process of adapting Shakespeare to film led to his doctoral dissertation-in-progress: *'Myn auctour shal I folwen, if I konne': Shakespeare Adapting Chaucer*, under the supervision of Dr. Evelyn Gajowski. He has participated in recent linguistic inquiries while producing solo essays on Kenneth Branagh's *As You Like It* (2006) and *Ye yan* ("The Banquet" 2006) for the *Borrowers and Lenders: The Journal of Shakespeare and Appropriation*. To keep these several interests in order, he maintains active membership in the Shakespeare Association of America, The New Chaucer Society, and The Modern Language Association.

Iasmina IOSIM has a BA in Communication and Languages and an MA in Applied Linguistics from the Tibiscus University in Timisoara, and a BSc in Management, an MSc in Agronomy, and a PhD in Agronomy from the

Banat University of Agricultural Sciences and Veterinary Medicine in Timişoara (Romania) with a thesis titled "Contributions to the communication policy in agri-food product marketing". She has been in the Romanian educational system since 2000. Her experience includes English, Organizational communication, Public relations, and Negotiation techniques, domains in which she has published scientific papers and books. She co-authored the books *Communication and Marketing*, *How do we communicate?*, *Communication and management*, and *English for Environmental Engineers*, and authored *Communication Policy in Agri-Food Product Marketing*.

Biljana IVANOVSKA has a BA in German and English language and literature from the University in Skopje (F.Y.R. of Macedonia) and got a PhD in contrastive analysis of contemporary German and Macedonian. She is currently working as Assistant Professor at the Goce Delcev University in Stip (F.Y.R. of Macedonia). She has been teaching contemporary German grammar (morphology and phonetics), and contrastive analysis (German and Macedonian), and she shows interest in teaching foreign languages for specific purposes. Her research and teaching interest focus on corpus linguistics, contrastive linguistics and translation studies, especially on theoretical, descriptive and applied perspectives. She is a member of the English Language Teachers' Association of the F.Y.R. of Macedonia, a member of the "Verein des Instituts für Deutsche Sprache" [Union of Institutes of German Language], and a member of the Editorial Board of the *Annual Proceedings* published by the Faculty of Philology in Stip.

Laura-Constantina MICU has a BA, a MA and a PhD in Philology from the University of the West, Timişoara, Faculty of Letters, History and Theology (Romania). She used to teach foreign languages (English and German) at the Banat University of Agricultural Sciences and Veterinary Medicine in Timişoara. She has presented at international conferences and has published articles and scientific papers in the field of agricultural languages.

Anica PERKOVIĆ, a PhD student in Linguistics, graduated from the Teacher Training College in Osijek (Croatia) and was appointed senior lecturer of English in 2004. As a qualified EFL teacher, she has been lecturing both English and Croatian for 34 years. During this period, her papers dealing with general English, ESP, and Croatian have been widely published. She has also been an active participant in many national and international conferences as well as a CEEPUS project co-ordinator. At the moment, she is Head of the Department of Informatics and Languages at the Faculty of Agriculture in Osijek where she gives English language courses.

Martina PERKOVIĆ has a BSc and an MA in Animal Husbandry from the Josip Juraj Strossmayer University in Osijek (Croatia). Her special interests are animals, ecology, languages and IT technologies. She used to be a teacher at the Trade and Commercial School in Osijek. At the moment she is in charge of an Eco-group.

Cornelia PETROMAN is a Professor at the Banat University of Agricultural Sciences and Veterinary Medicine of Timişoara (Romania). She has a BSc in Animal Husbandry and a BSc in Hospitality. She has a PhD from the University of Agricultural Sciences and Veterinary Medicine of Bucharest (Romania). She is specialised in food service, food processing, and hospitality. She has published 6 books and over 100 papers in the fields of animal husbandry, food processing, hospitality, and food service.

Ioan PETROMAN is a PhD Professor at the Banat University of Agricultural and Veterinary Medicine of Timişoara (Romania). He has a BSc in Animal Husbandry from the BUASVM in Timisoara and a PhD in Animal Husbandry from the UASVM in Bucharest (Romania). He also has a BSc in Management and a PhD in Management from the University of the West in Timişoara. He has published widely in the field of animal husbandry, food processing, and tourism.

Maria-Adriana PROCA has a BSc in Biology and a PhD in Agriculture from the Banat University of Agricultural Sciences and Veterinary Medicine of Timişoara (Romania). She has published in the fields of botany and agriculture (cultivation practices).

Georgeta RAŢĂ has a PhD in Linguistics from the University in Bucharest (Romania). She is a qualified EFL teacher and has been teaching languages for 35 years. Her extensive experience includes both general English and several areas of ESP, and she has published widely in these areas. She also has cross-linguistic expertise, having worked in French and Romanian as well as English, and understands the practical issues involved in teaching and learning different languages and transferring research ideas from one language to another. She used to be the Scientific Secretary of her Department, Editor-in-chief of the *Journal of Linguistic Studies*, a contributor to the *English Profile* research project, and regularly organises international events in the field of language education. She participates in the project *Languages and Cultures in Time and Space* carried out in Serbia. She has published textbooks and dictionaries in Croatia, Romania, and Serbia. She has co-authored *Compendium of Forage Technical Terms in English, French and Romanian* (Cambridge Scholars Publishing, 2012), she has edited *Language Education Today: Between Theory and Practice* (CSP, 2010), *Teaching Foreign Languages: Languages for Special Purposes* (CSP, 2010), and *Academic Days of Timişoara: Language Education Today* (CSP, 2011), and co-edited *Social Sciences Today: Between Theory and Practice* (CSP, 2010) and *Academic Days of Timişoara: Social Sciences Today* (CSP, 2011).

Florin SALA has a PhD in Agronomy (Agro chemistry) from the Banat University of Agricultural Sciences and Veterinary Medicine (B.U.A.S.V.M.) in Timişoara (Romania). He is a professor and he has taught and researched for 21 years. His extensive experience includes Soil Science, Agrochemistry and Plant Nutrition, mathematical models of fertilization, nanomaterials and nanotechnology in plant nutrition and environmental protection. He has been a director and member 25 national and international research projects in human resource development and training. He is the scientific secretary of the B.U.A.S.V.M. Since 2008, he has been a PhD adviser. He has published 6 ISI-indexed papers, 37 IDB-indexed papers, 25 papers in books of proceedings of international conferences and 39 papers in national journals. He also published 12 books, including: *Agro chemistry: methods of analysis* (2011), *Agro chemistry: practical applications* (2010), *Agro chemistry* (2007, 2008) *Systems of Agriculture* (2002), *Compendium Horticulture* (2000), *Pesticides* (1998), *Fertilization and protection of horticultural plants* (1997). He is a member of the editorial board of *Research Journal of Agricultural Sciences of the B.U.A.S.V.M.*. he it a scientific reviewer of the *African Journal of Biotechnology* (AJB – ISI) and of the *International Research Journal of Agricultural Science* (IRJAS). He is a member of the Romanian Association for the Promotion of Magnetic Liquids, of the Romanian National Society of Soil Science, of the Romanian Association for Sustainable Agriculture, and of the International Society for Horticultural Science (ISHS).

Elena-Mirela SAMFIRA has a BA in Psychology from the Lucian Blaga University in Sibiu (Romania) after working as a nurse at the County Hospital in Timisoara (Romania). She currently teaches Educational psychology at the Banat University of Agricultural Science and Veterinary Medicine in Timişoara. She attended courses in systemic family and couple psycho-therapy in 2008, she got her licence as autonomous psychotherapist in 2009, and she is a member of the Romanian Psychologists College and of the CRIS-DU Areopagus. She is a PhD Student in Consumer behaviour. She also works as a psychotherapist in family and couple systemic psychotherapy. She co-authored a Student's book of educational psychology and she authored and co-authored 15 scientific papers.

Ionel SAMFIRA has a PhD in Agriculture from the Banat University of Agricultural Sciences and Veterinary Medicine in Timişoara (Romania). He is a professor at the same university. His teaching and research competences resulted in studies on natural pastures, on the interaction between soil and vegetal sward, on the management and biodiversity of permanent pastures, and on the breeding of forage plants. He has been a member of several research teams financed by national and international grants. In the last years he has been involved in some research and academic development project such as the Postdoctoral School in Agriculture and Veterinary Medicine POSDRU/89/1.5/S/62371, Romania 2010-2012; Academic online networks for the development of competences and competitiveness on the job market POSDRU 86/1.2./S/6273 Romania, 2010-2012; Sustainable preservation of indigenous South East European legumes and traditional feed and food products, Regional Programme for Cooperation with South-East Europe, ERA168/01, Germany, 2010-2012, Sub-project Leader; Monitoring and control of the invasive biological pollutant *Ambrosia artemisiifolia* l. Acronym Stop Ambrosia, Romania-Serbia, Trans-border programme nr. 39, COD 408/ 31.12.2010, 2010-2011. He has also been a reviewer for the journal issued by his college, and a member of the scientific board of the *Agro Buletin AGIR*. He organises the symposium for young researchers in agriculture. He has been a vice-dean of his college since 2004. He has authored and co-authored 10 textbooks and research books, as well as over 100 research papers, among which *Compendium of Forage Technical Terms in English, French and Romanian* (Cambridge Scholars Publishing, 2012).

Snežana TOLIĆ has a PhD in economics from the Josip Juraj Strossmayer University in Osijek (Croatia). As an assistant professor, she lectures and coordinates modules at university undergraduate, graduate, postgraduate, and lifelong interdisciplinary studies. Her interests are scientific work in agro economics, rural and local development as well as in the implementation of LEADER programmes. Since 2004, she has been the manager of *SLAGALICA production and service family cooperatives* dealing with products and services in agriculture, tourism and trade.